Extension for Rural Entrepreneurship Development

NIPA® GENX ELECTRONIC RESOURCES & SOLUTIONS P. LTD.
New Delhi-110 034

About the Editor

Prof. Kalyan Ghadei Ph.D (BHU), B.Mus (ABGMM), C.C (PI) has a good academic career throughout. Prof. Ghadei has successful record activities at All India Radio, Cuttack, CIFA(ICAR), Bhubaneswar and in OUAT Odisha for transfer of Agricultural, aquaculture and pearl culture technology in Odisha and India. He is serving as a Professor & Head in the Department of Extension Education in Banaras Hindu University, BHU, Varanasi. He has published more than 60 research papers , authored 09 books and about 25 book chapters. He is a member of many national and international professional societies. Prof. Ghadei has received many awards of national and international repute. The most significant award to his credit is Raman fellowship as a visiting scholar to USA and Post Doctoral fellow at Virginia Tech, USA to study climate change issues and its mitigation strategies.

Extension for Rural Entrepreneurship Development

Editor
Kalyan Ghadei
Associate editors
Sushree Purabi Panigrahi
Gyan Shukla
Musani Chennamadhava
Nikhil J
Priyanka Roy
Suman Roy
Waris Ali
Mani Prakash Shukla

NIPA® GENX ELECTRONIC RESOURCES & SOLUTIONS P. LTD.
New Delhi-110 034

NIPA® GENX ELECTRONIC RESOURCES & SOLUTIONS P. LTD.

101,103, Vikas Surya Plaza, CU Block
L.S.C.Market, Pitam Pura, New Delhi-110 034
Ph : +91 11 4386 0225, 9717133558, 9540816132
E-mail: newindiapublishingagency@gmail.com
Website: www. niparesources.com

Print ISBN: 978-93-58877-98-4

ebook ISBN: 978-93-58874-02-0

NIPA® also publishes books in a variety of electronic formats. Some content that appears in print may not be available in electronic books, and vice versa

Composed and Designed by NIPA®.

Dedicated to
All Extension Professionals

Preface

Entrepreneurship, Learning, and Extension are three pivotal pillars that shape the landscape of sustainable development, particularly in the realm of agriculture, across the Indian subcontinent. It is with great pleasure and a profound sense of purpose that we present this compilation of significant research and review papers contributed by distinguished scholars deeply engaged in extension activities throughout India.

This book is a testament to the collective dedication and relentless efforts of numerous individuals who have tirelessly worked to unravel the complexities of entrepreneurship in the context of Indian agricultural extension. It is the fruit of their commitment to advancing knowledge and fostering innovation, which have the potential to catalyze positive change in both rural and urban settings.

The pages within contain a treasure trove of research findings and theoretical foundations that are not only instructive but also highly relevant for those involved in the fields of teaching, research, and extension activities in India and beyond. As the authors, we earnestly hope that this volume will serve as an invaluable resource for extension scientists, educators, students, and all those who seek to promote entrepreneurship and agricultural extension for the betterment of society.

We would like to extend our heartfelt gratitude to the esteemed contributors who have shared their insights and expertise, making this publication possible. Their unwavering commitment to pushing the boundaries of knowledge has enriched this book immensely.

We are also deeply appreciative of the countless volunteers and supporters, both directly and indirectly involved in this endeavor. Your assistance, whether through guidance, encouragement, or practical support, has been instrumental in bringing this manuscript to fruition.

A special thanks is owed to the associate editors and research scholars of the Department of Agricultural Extension Education at Banaras Hindu University. Your relentless dedication to the cause of agricultural extension and your unwavering support in the development of this book are sincerely appreciated.

As we embark on this journey through the pages of Entrepreneurship, Learning, and Extension, we encourage readers to explore the depths of knowledge contained herein. It is our hope that this book will not only inform but also inspire new generations of entrepreneurs and extension professionals to shape a more vibrant and prosperous future for agriculture and rural development.

Kalyan Ghadei

Contents

1

Knowledge of the Gram Panchayat Members About MGNREGA Activities

[1]Anurag Shankar Singh, [2]N.R. Meena, [3]Sanskriti Sindhu and [4]Ritesh Singh

[1]Extension Education, ANDUAT, Kumarganj, Ayodhya U.P.
[2]Department of Extension Education, ANDUAT, Kumarganj, Ayodhya U.P.
[3]Krishi Vigyan Kerndra, Parsauni, Bihar
[4]Extension Education, ANDUAT, Kumarganj,Ayodhya U.P.

Abstract

Wage employment is monetary compensation paid by an employer to an employee in exchange for work done. Payment may be calculated as a fixed amount for each task completed (a task wage or piece rate), or at an hourly or daily rate, or based on an easily measured quantity of work done. The present study was aimed to assess the knowledge level of Gram Panchayat Members about MGNREGA activitries. Descriptive reseach design has been used for present study. This study is based on both primary and secondary data. 21 villages from Hanumanganj block were selected by Random sampling. Hanumanganj block and 126 respondents (6 respondents from each village) were selected for study. It is also reported that majority of Gram Panchayat Members were having medium level of knowledge about MGNREGA activities.

Introduction

The Government of India created a historic act, by enacting the Mahatma Gandhi National Rural Employment Guarantee Act (MGNREGA), the largest employment generating programme in the world, ensuring the right to work in a country with a population of over one billion. The Government of India passed the NREGA 2005, (Central Act No. 42 of 2005). NREGA was renamed as Mahatma Gandhi National Rural Employment Guarantee Act (MGNREGA) on October 2nd, 2009. MGNREGA is the first ever law, in the world that guarantees wage employment at an unprecedented scale. This Act gives legal guarantee of at least one hundred days of wage employment in a financial year to a rural household, whose adult members volunteer to do unskilled and manual work. The Act is applicable in the Districts notified by the Government of India, the implementation of the Act calls for the formulation of Rural Employment Guarantee Scheme by the State Governments.

The key processes in the implementation of MGNREGA are the adult members of rural households submit their name, age and address with photo to the Gram Panchayat, which in turn registers households after making enquiry and issues a job card which contains the details of adult member enrolled and his/her photo. Registered person can submit an application for work in writing (for at least fourteen days of continuous work) either to Panchayat or to Programme Officer, who will accept the valid application and issue dated receipt of application, letter-providing work will be sent to the applicant and displayed at Panchayat office. The employment will be provided within a radius of 5 kilometers and if it is above 5 kilometers extra wage will be paid. If employment under the scheme is not provided within fifteen days of receipt of the application, daily unemployment allowance will be paid to the applicant.

Mandate of Mgnrega

The mandate of Mahatma Gandhi National Rural Employment Guarantee (MGNREGA) is to provide at least 100 days of guaranteed wage employment in a financial year to every rural household whose adult members volunteer to do unskilled manual work. MGNREGA is implemented in all rural districts of the country. MGNREGA has an objective to enhance livelihood security to the rural poor by providing at least one hundred days of wage employment to every household whose adult members are ready to do unskilled manual work. MGNREGA not only provide wage employment as an alternative source of livelihood but also create durable assets such as road construction, land development, water conservation and irrigation facility, which has tremendous influence on different sectors of village economy.

Research Methodology

The present study was conducted in Hanumanganj block of Ballia District of Uttar Pradesh. 21 villages from the Hanumanganj block were selected through random sampling method. The study was based onboth primary and secondary data. Descriptive research design was used for the present study. Data analysis is done through various statiticasl tools listed below. On the basis of minimum and maximum possible score, the categories were developed as:

Category	Score
Not correct	1
Partially correct	2
Fully correct	3

Category	Score
Low	(<Mean-SD)
Medium	(in between mean ±SD)
High	(>Mean + SD)

For calculating percentage, frequency was multiplied by 100 and divided total Respondents.

P=X/N×100

Where,

P = Percentage

X = Frequency of the Respondents

N = Total number of Respondents

Mean

Mean was obtained by dividing the sum of the scores by the total number of cases involved. The formula for determining mean is

$$\text{Mean } (\bar{X}) = \frac{\sum_{i=0}^{n} Xi}{n} \quad [i=1, 2, 3..........n]$$

Where,

$\bar{X}$ = mean

$\sum Xi$ = sum of scores

n = no. of cases

Standard Deviation

The standard deviation is the square root of the arithmetic average of the square of various from their arithmetic mean.

$$\text{S.D.} = \sqrt{\frac{1}{n}[\sum x^2 - \frac{(\sum x)^2}{n}]}$$

Where,

$\sum X$ = Deviation of the score from mean

n = Number of observation

Results and Discussion

Table 1: Distribution of respondents based on knowledge regarding MGNREGA activities

S. No.	Activities	Evaluation		
		FC	PC	NC
1.	Job cards			

	Registration	121 (96.03)	05 (3.96)	00 (00.00)
	Application	122 (96.82)	04 (3.17)	00 (00.00)
	Verification	118 (93.65)	08 (6.34)	00 (00.00)
	Issuing of job cards	123 (97.61)	03 (2.38)	00 (00.00)
	Evaluation	115 (91.26)	11 (8.73)	00 (00.00)
	Monitoring	113 (89.68)	11 (8.73)	02 (1.58)
2.	**Muster rolls**			
	Programming	107 (84.92)	13 (10.31)	06 (4.76)
	Implementation	101 (80.15)	18 (14.28)	07 (5.55)
	Participation	106 (84.12)	16 (12.69)	04 (3.17)
3.	**Watershed related works**			
	Contour trenches	14 (11.11)	38 (30.15)	74 (58.73)
	Contour bunds	86 (68.25)	31 (24.60)	09 (7.14)
	Boulder checks	63 (50.00)	34 (26.98)	29 (23.01)
	Farm buildings	57 (45.23)	42 (33.33)	27 (21.42)
	Gabian structure	13 (10.31)	19 (15.07)	94 (74.60)
	Underground dykes	37 (29.36)	12 (9.52)	77 (61.11)
	Earthen dams	39 (30.95)	23 (18.25)	64 (50.79)
	Dugout farm ponds	89 (70.63)	28 (22.22)	09 (7.14)
	Stop dams	39 (30.95)	26 (20.63)	61 (48.41)
4.	**Watershed related works in mountain regions**			
	Spring shed development	07 (5.55)	06 (4.76)	113 (89.68)
5.	**Agriculture related works**			
	NADEP composting	17 (13.49)	12 (9.52)	97 (76.98)
	Vermi- composting	31 (24.60)	21 (16.66)	74 (58.73)
	Liquid bio- manures	14 (11.11)	16 (12.69)	96 (76.19)
6.	**Livestock- related works**			
	Poultry	67 (53.17)	34 (26.98)	25 (19.84)
	Goat shelter	29 (23.05)	13 (10.31)	84 (66.66)
	Azolla as cattle- feed supplement	03 (2.38)	02 (1.58)	121 (96.03)
	Construction of pucca floor tank & fodder trough for cattle	89 (70.63)	34 (26.98)	03 (2.38)
7.	**Fisheries- related works**			
	Fisheries in seasonal water bodies on public demand	77 (61.11)	38 (30.15)	11 (8.73)
8.	**Works in coastal area**			
	Fish drying yards	01 (0.79)	03 (2.38)	122 (96.82)
	Belt vegetation	00 (00.00)	02 (1.58)	124 (98.41)
	Construction of storm water drains for coastal protection	01 (0.79)	01 (0.79)	124 (98.41)
9.	**Rural drinking water related works**			

	Soak pits	73 (57.93)	39 (30.95)	14 (11.11)
	Recharge pits	56 (44.44)	28 (22.22)	42 (33.33)
	Rural sanitation related works	98 (77.77)	14 (11.11)	14 (11.11)
	Individual household	96 (76.19)	19 (15.07)	11 (8.73)
	School toilets	76 (60.31)	34 (26.98)	16 (12.69)
	Solid and liquid waste management	29 (23.01)	37 (29.36)	60 (47.61)
10.	**Flood management related works**			
	Deepening & repair of flood channels	114 (90.47)	09 (7.14)	03 (2.38)
	Chaur renovation	86 (68.25)	19 (15.07)	21 (16.66)
11.	**Irrigation command related works**			
	Rehabilitation of minors, sub minors and field channels	106 (84.12)	13 (10.31)	07 (5.55)

FC- Fully correct, PC- Partially correct, NC- Not correct

Frequency (Percentage)

The above table, shows that a majority of 96.03% respondents were fully correctabout the registration of job cards in MGNREGA activities. whereas 3.96% were partially correct about the registration of job cards in MGNREGA activities. About 96.82% of the respondents fully know about the application of job cards in MGNREGA activities while 3.17% of respondents partially know about the application of job cards in MGNREGA activities. About 93.65% of the respondents fully know about the verification of job cards in MGNREGA activities while 6.34% of respondents partially know about the verification of job cards in MGNREGA activities. About 97.61% of the respondents fully know about the issuing of job cards in MGNREGA activities while 2.38% of respondents partially know about the issuing of job cards in MGNREGA activities. About 91.26% of the respondents fully know about the evaluation of job cards in MGNREGA activities while 8.73% of respondents partially know about the evaluation of job cards in MGNREGA activities and so on.

Table 1.1: Distribution of the respondents based on the level of knowledge regarding MGNREGA activities

S. No	Category	Frequency	Percentage
1	Low (16-21)	32	25.39
2	Medium (22-27)	53	42.06
3	High (28-32)	41	32.53
	Total	126	100.00

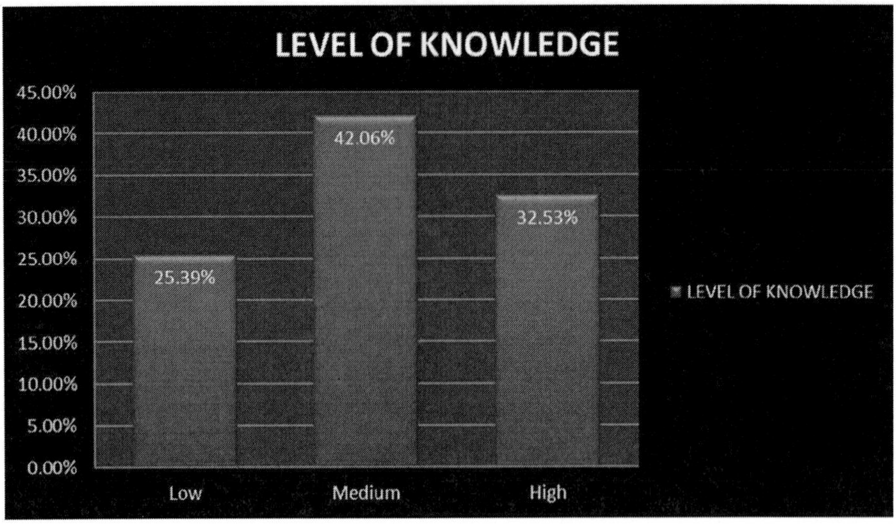

Fig. 1: Distribution of the respondents based on the level of knowledge regarding MGNREGA activities

From the above table and figure, it is found that the level of knowledge of respondents regarding MGNREGA activities is medium 42.06% followed by high 32.53% and low 25.39% respectively. Similar findings were also reported by **Shukla & Sidhu in 2018.**

Conclusion

PanchayatRaj system facilitates easy and effective implementation of developmental programs in remote village areas. MGNREGA gives legal guarantee of at least one hundred days of wage employment in a financial year to a rural household, whose adult members volunteer to do unskilled and manual work. The Act is applicable in the Districts notified by the Government of India, the implementation of the Act calls for the formulation of Rural Employment Guarantee Scheme by the State Governments. It is also reported that majority of Gram Panchayat Members were having medium level of knowledge about MGNREGA activities.

References

Bhosale, S.G. 2007 A study of role perception and role performance of women members in Gram Panchayat from Rahuri Taluka of Ahmednagar District. M.Sc. (Agri.) Thesis,MPKV,Rahuri.

Dedun, V.S. Naik, R.M. and Gamit, P.S. 2016 Constraints Faced by Gram Panchayat Members in Performing their Role for Village Development. International J.Agric. Sci., ISSN:0975-3710&EISSN:0975-9107,(8),(21).

Deshpande, A.R.; Soni, M.C. and Shekhawat, S.S. 2013 Role Performance of Gram Panchayat Members in Agricultural Development Programmes, Ind. Res. J. Ext. Edu. 13(2).

Mankar, D. M. Hireven Kangouda, L. V. and Manjunath, L. 2011 Awareness and performance of the roles in agricultural development by panchayat members, Karnataka J. Agric. Sci., 24 (2).

Sharma R. & Didwania M., 2013 Performance analysis of MGNREGA: A case study of district Jind, International Journal of business economics & management Research Vol. 3(8) ISSN: 2249- 8826.

Shukla N. & Sidhu K. 2018, Knowledge level of PRI members regarding selected schemes, Asian Journal of Agricultural Extension, Economics & Sociology. Vol. 2510. Pp. 1-7

Singh, Rahul, Singh, RP, Singh, Avanish, Upadhyay, Sandeep, Singh, Arvind, Singh, Chandan, 2019. Role performance of gram panchayat members about agriculture development programmes in Faizabad district of Uttar Pradesh

2

Gender & Youth Issues in Agriculture

***Asha Rani*[1], *Prasanlata Arya*[1], *Dipankar Saikia*[2]**

[1]*Department of Extension Education and Communication Management Swami Keshwanand Rajasthan Agricultural University, Bikaner, Rajasthan*
[2]*ICAR-DRMR-APART, Jorhat, Assam*

Abstract

Gender and youth both have issues in terms of rural and agricultural development because both have high rate of rural- urban migration and youth also have negative attitude towards agriculture. Most common excuses given for denying women and youth over productive resources is that they considered they are unable to handle the challenges posed by the increased technicality of equipment. In most cases, women and youth have uncertain access to land; women's rights to land are often dependent upon the relationship to a husband. Lack of interest of youth there is degradation in agriculture. Lack of labour is common problem in rural area, and more critical for women and youth they are unable to determine the use of even their own labour. Lack of capital or credit is a major limitation for both male, female and whether young or adult. Credit programmes are good for women but women are usually afraid to take a formal loan so women and youth are at disadvantage. Lack of control over decision-making unfortunately, even when women or youth may "own" the other resources, they might not have the power to make independent decisions concerning their use. Increasing responsibilities for women and youth this has greatly increased the work load, particularly upon women and youth still remaining in the community. Due to lack of poor health and overworking and taking too many tasks for women and create constraints that they become ill. Social restrictions have limited youth and female participation. Women and youth are sometimes not able to be members of groups.

Keywords: *Gender, Youth, Agriculture, Issues, Challenges, Constraints*

Introduction

Gender is a topic that cuts across all subjects and should be considered. The generational dimensions of rural and agricultural growth, whether we're talking about kids, teenagers, or the elderly, have also grown in importance in recent years. Therefore, it is not surprising that a specific focus should be given on the issues of gender and youth and their relevance to the growth of agriculture. However, it is our custom to start with a few points to lay a foundation of thoughts. Nations Like India where majority of farmers practise subsistence agriculture, rural development is the cornerstone of agricultural development. According to the type of the problem in each country, an assessment of the present global response to the food crisis has indicated that various strategies are being implemented with varying degrees of success and failure. Different tactics are being used to shield the underprivileged from the rise in price shocks. Investment in the agricultural sector is consequently important for rural development and a rise in agricultural output. The contribution of women is also essential to the overall success of initiatives aimed at rural development to boost agricultural production in India. In sub-Saharan Africa, Women engage in more agricultural work than men do and supply the bulk of the labour for a variety of agricultural tasks. Women make up around 90% of the workers directly involved in producing arable crops (Ogunlela and Aisha, 2009). Therefore, rural development will occur when different pertinent institutions are given attention, including cooperative societies, the system of land tenure, banks and credit, the creation of sustainable local governments, educational institutions, medical institutions, and agricultural marketing outlets.

Gender Related Terms

Gender: This is the group of culturally particular traits that characterises how men and women behave in social situations and how they relate to one another. Specifically, it is the socially constructed male and female relationships.

Sex: The biological variations between men and women are discussed here. Regardless of ethnicity, class, or age, a person is either male or female. But that person's sex may have a different social connotation.

Gender Equity: This is the process of treating both men and women fairly. Equity typically results in equality. Therefore, gender equality refers to the equalisation of life outcomes for men and women while taking into account their disparate needs and interests (University of Ibadan Draft Gender policy,2011)

Gender Equality: This refers to the distribution of resources, opportunity, support, and encouragement between men and women without any distinction on the basis of genetics. Specifically, it means that both women and men have an equal opportunity to realise their human rights and potential, as well as to contribute to and benefit from, national, political, economic, social, and cultural development. Therefore, gender equality refers to giving women the same opportunities in life as men, including the freedom to engage in public life. For instance, equality will be largely realised if men and women are given equal opportunities, i. e. when society values both the similarities and differences between men and women as well as the various roles they play equally.

Gender roles: These are collections of learnt, socially or culturally determined expectations about how people should act in particular circumstances.

Gender mainstreaming: Through the development of gender capacity and responsibility, this organisational strategy aims to integrate a gender perspective into all facets of an institution's policies and operations.

Gender Needs: Shared requirements that women have defined as priorities that result from their shared experiences.

Gender Planning: This involves the technical and political steps required to put gender-sensitive policy into practise.

Gender Relations: Hierarchical power relationships between men and women are involved, which tend to dissuade women.

Gender Violence: A male or male-dominated institution commits or threatens to commit an act that causes bodily, sexual, or psychological harm to a woman or a girl solely because of their gender. This happens when men and women are vying for the same resources, social status, and opportunities to exercise power. The uneven distribution of resources and systematic exclusion of women from exclusion are both present.

Gender Training: A procedure for increasing one's understanding of and ability to address gender issues in order to effect personal or organisational change for gender equality.

Gender Discrimination:, This is the systematic denial of resources, opportunities, or rights to individuals based on their gender.

Women Empowerment: By raising people's awareness of women's subordination and enhancing their ability to challenge it, individuals or organisations can change gender power dynamics from the bottom up. It is about empowering both men and women to take charge of their life, set their own goals, acquire new skills, boost their self-confidence, resolve issues on their own, and become more self-reliant.

Gender division of labour: This is how labour is organised under the premise that males fill particular jobs and that women perform domestic work as housewives.

Gender awareness: In this scenario, every participant in the company or institution is aware of the significance of gender and how it affects their goals, strategies, and plans.

Gender sensitivity: This is the conversion of awareness into actions, which affects how institutions and organisations perceive the world and make plans and carry out their daily operations.

Gender blindness: This is the intentional creation of goals, plans, and programmes within an organisation without making any attempt to acknowledge or take into account gender issues that may have an impact on how that organisation operates, the creation of plans, the implementation of programmes, and the results of those programmes. People who do not or will not consider gender as an issue in institutional contexts propagate or practise it.

Gender Gap: The difference between men and women's ratings on attitudes, interests, actions, knowledge, and viewpoints on certain problems, such as political preferences and voting preferences, is referred to as this. Women's perspectives, independence, and differences from men in their beliefs, perceptions, economic goals, and social and psychological autonomy may all contribute to the gender gap. It differs depending on things like class, race, age, marital status, and religion. (Gender Equity policy,2011)

Gender Audit:. This gender audit enables an organisation to better understand how women are faring in comparison to men in that organisation. This may be regarding enrolment, retention, and male and female academic accomplishment.

Gender Issues in Agriculture and Rural Development

Demonstrated capability of women and youth in productive activities: One of the most frequently used justifications for denying women and young people access to productive resources or benefits from development initiatives is that it is assumed that they are unable to handle the difficulties brought on by the increased complexity of procedures or tools. Men may argue that giving women or young people access to less desirable farmland at the family level is justified because they won't use the better fields as effectively. Gladwin (1996) draws the conclusion that because it is falsely believed that male farmers are more productive than female farmers, women have not received enough consideration when allocating resources and inputs. Multiple studies have discovered, however, that "female farmers are equally efficient as male farmers after individual traits and input levels are adjusted for" (Quisumbing 1994). Gladwin (1996) observed that most studies reveal no difference between male and female farmers in terms of their capacity to manage their farms when researchers estimate a production function that controls for other explanatory variables besides gender, such as input levels. Studies including both male and female youth in their creative activities could lead to similar outcomes.

Lack of control over productive resources: According to Tsikata *et al.*, it has been demonstrated in numerous studies that gender is a key factor in determining access to or control over factors such as the size and quality of agricultural land, paid and unpaid labour, production revenues, free time for leisure or productive pursuits, membership in producer organisations, information, subsidies, credit, and training. Similar to this, youth face obstacles when trying to gain control over resources that are typically held or managed by male adults. The issue of female producers' lack of control over productive resources, which also affects young people, is further explored in the parts that follow.

Lack of control over land: Women and young people typically lack secure access to land and have a history of losing their land rights. This has, naturally, made it less appealing to make long-term investments in the land (Chitepo,1991). Women and youth would not consider the additional effort necessary for sustainable land use to be worthwhile because they understand they would not have access to that plot in the next season. Additionally, a woman's access to land is frequently based on her relationship with her husband or another male relative. Even temporary access to land may be quite difficult to come by for young people or single women.

Lack of control over labour: Rural communities frequently struggle with a labour shortage, but this issue is particularly acute for women and young people who may not be able to select how their own or other family members' work will be used. It has also been discovered that the majority of the research-proposed farming improvements will result in a rise in the demand for young people's and women's labour. For instance, adding fertiliser necessitates more weeding, which is typically carried out by women or young people. However, traditionally masculine jobs like clearing and preparing land are now being automated. Therefore, innovative initiatives must take this into account before promoting new techniques that could result in an excessive workload.

Lack of control over capital: For the majority of rural producers—male and female, young and old—lack of finance or credit is a significant barrier. It is necessary to employ better inputs, hire personnel, use better equipment, and expand the production area in order to increase productivity. The majority of rural residents take part in native group-based savings and credit systems. Given that women typically fear debt more than men do, several credit programmes are discovering that women are good credit risks with considerably lower default rates than males. Women and young people, however, are at a disadvantage when applying for a formal loan.

Lack of control over decision-making: An essential resource in and of itself is the gathering and selection of the productive resources. Unfortunately, even while women or young people may "possess" the other resources, they might not have the authority to decide on their own how to use them.

Increasing responsibilities for women and youth without increasing returns: Greater home responsibilities for rural women have occasionally been a result of social development. For example, growing population density in southeast Nigeria has placed strain on arable lands that have also seen soil erosion. The available productive resources for many rural households are insufficient to meet their needs for a living. In order to augment what the wife and kids can generate on the farm, the male head may then be forced to leave rural areas in pursuit of salaried job. The labour burden has significantly grown as a result, especially for women and young people who still live in the neighbourhood.

Poor health and lack of time*:* Poor health and a lack of time are closely associated. Time limits result from overworking and taking on too many responsibilities, especially for women who physically push themselves to the point of illness. Women frequently find themselves unable to increase the scope of their current endeavours or take advantage of additional opportunities to generate revenue due to time constraints. Women just cannot afford the additional time commitment that such scaling up would require. Good health is essential for being active, effective, and having a respectable standard of living: Achieving good health is a feat in and of itself, and it also increases productivity and the capacity to turn earnings and resources into a comfortable lifestyle (Sen,A.,1989). In female-headed households, the combination of poverty, increased demands on women's time, and increased reliance on child labour has a detrimental effect on family food security, nutrition, and child welfare (World Bank, 1992)

Due to their involvement in several duties, rural women do not have enough time for other economic activities, which has hindered the expansion of their scope of activities. This restricts their capacity to engage in additional income-generating activities (Odejide, 1996). According to Jiggens (1989), the challenges faced by rural women in agriculture include seasonal variations in supply and demand, frequent product market ventures, limited options for diversification, and many roles that limit their ability to take use of commercial assets. Designing programmes that will increase rural women's income and productivity needs to receive more attention

Exploitation by marketers or other service providers: The likelihood of being taken advantage of by people who have such resources and influence increases when one lacks economic and social power. Women, and men for that matter, may be at the mercy of marketers, transporters, credit and loan providers, input suppliers, and other service providers who frequently use the situation for their own financial benefit in addition to the issue of low education. Given that the typical rural resident lacks access to such services elsewhere, service providers can choose the price that those in need of them must pay. Such exploitation is not always committed by strangers or by people of the opposing sex. The production of other women in rural villages is frequently purchased at a very low price by wealthy rural women or urban female traders, who then take the goods to the city where they are sold for a much greater price.

Social restrictions: Youth and female participation in several social and economic activities has been restricted by social constraints in some localities. Women and young people occasionally aren't allowed to participate in groups or receive benefits from development initiatives. Even in purdah, the most severe form of social segregation, women are still economically active. They occasionally have their own properties that are managed by a male relative, and almost always engage in processing, livestock rearing, crafts, and the preparation of snacks or meals for sale from within the walls of their compounds. Young and female farmers must have the chance to take part in cutting-edge strategies to advance agriculture and further their own interests.

Food security: Gender and food security are closely related because mothers are typically the primary food providers for families. Homestead farming offers a safety net for the family to smallholder farmers. Women frequently maintain cattle or home gardens, with a lot of help from household water supplies. The improvement of dependable water/ irrigation supply could have a significant positive impact on household food security. Therefore, planners must take this into account and offer water access that matches the variety of uses that women make of it.

Migration: In Asia, there are around 1 billion job seekers, and this has a historically significant impact on rural economies. Particularly men are leaving rural areas in quest of opportunities for better work. Women are left with a heavier labour and more obligations when males migrate, yet they do not have equal or direct access to financial, social, and technological resources. When left in control of houses, women are expected to carry out both their traditional duties and those of males. IWMI conducts research to better understand how migration affects rural development, water resources, and livelihoods.

Lack of Education and Training

Education is a crucial component of development. It increases engagement and aids in female confidence-building. Progress is hampered by inequality with men's counterparts. Women's productivity is influenced by their access to education and training (Adeyeye, 1991). The vicious cycle of underdevelopment, low productivity, and terrible conditions for women's health and welfare is exacerbated by a lack of basic education and training (Odejide, 1988). Due to a lack of education, rural women typically rely on their service providers. Transporters, marketers, input suppliers, and loan lenders are just a few of these service providers.

Challenges with Gender in Agriculture

In recent years, there has been a rising understanding of the significance of gender in several particular development areas. The development of agriculture, health care, and infrastructure are now all related issues. This is due to considerable changes in gender that have occurred. Women have a history of contributing significantly to rural agriculture during the previous few decades. They played a part in agricultural activities, but because their contributions went unappreciated, they were excluded from development plans. Consequently, agricultural productivity has decreased (Odebode, 2008).

In the majority of areas, women perform the majority of the work in the subsistence farming sector. Women are in charge of taking crops to the mill for processing, carrying harvested produce to the homestead and market, carrying harvested produce to the homestead and gathering firewood. Due to their involvement in several duties, rural women do not have enough time for other economic activities, which has hindered the expansion of their scope of activities. This restricts their capacity to engage in additional income-generating activities Odejide (1996). According to Jiggens (1989), the challenges faced by rural women in agriculture include seasonal variations in supply and demand, frequent product market ventures, limited options for diversification, and many roles that limit their ability to take use of commercial assets. Designing programmes that will increase rural women's income and productivity needs to receive more attention.

Social sciences employ the idea of gender to identify the jobs and pursuits of men and women. A given culture's customs and beliefs serve to socially define these positions. Therefore, gender is not the same as sexual differences that are only based on biological traits. Culture-specific gender constructs provide men and women separate identities and duties (Olawoye,2002)

Gender analysis is a strategy for studying and learning more about the roles played by men and women in society, as well as the possibilities and challenges they encounter. Gender is a social construct that describes the socially expected rights, responsibilities, privileges, and obligations of males and females, according to Olawoye*et al.* (1994). She claims that each culture or group determines which responsibilities and opportunities should be given to men and women, young and old. According to Olawoye*et al.* (1994), gender analysis includes examining the interrelationships between the duties and activities performed by men and women.

Understanding socio-cultural differences will help us better understand how gender division of labour operates. Consequently, contrary to what most people in society believe, gender analysis is not always about creating programmes for women. It focuses on how men and women interact with the social and economic system of the society.

Men and women in various communities have varied duties and obligations, and frequently, these disparate positions are explained by the biological differences between men and women. For instance, men have typically been involved in cloth weaving in Ghana. Women work as labourers in the building industry in India. As a result, biological differences do not always dictate what men and women can or should do (Olawoye, 2005). The customs and values of a specific culture frequently define social roles. As a result of the fact that women often shoulder more responsibility than males, there are numerous possible implications of gender inequality.

Importance of Involvement of all Stakeholders for Sustainable Development Efforts

In recent years, the emphasis on development has evolved to a participatory approach, where the development programmes' intended recipients can influence decisions made about them and participate in project implementation, giving them some control over the development process as it impacts them. Many have already passed because a sizable segment of the population, particularly.

Getting information on various groups is one of the key elements in involving all stakeholders in development activities. It seems doubtful that women or young people would be able to freely express their ideas in a mixed crowd unless opportunities are made to collect data directly from them. Because of this, it is best to collect information through separate focus groups for adult males, adult females, male youth, and female youth; stratified sampling that includes women and youth groups; later sex and age-group disaggregation of findings; and in-depth interviews with women and youth leaders as well as traditional or male leaders to determine the perspectives of all groups. Females and young people, did not participate beyond being counted as welfare beneficiaries.

Significant disparities in viewpoints, restrictions, actions, and ambitions have been discovered in research where results are interpreted in the context of gender and generation. Venn diagrams, for instance, indicate that different reference groups are significant to women and youth, demonstrating that different techniques are required to mobilise members and get legitimacy from key leaders. Similar to this, the results of the paired needs priority ranking show that the most important needs from the perspective of male adults, such as roads, are likely to differ from those of women, who may prioritise markets and a reliable supply of water, as well as from young people, who may prioritise jobs and training.

Many development initiatives aim to increase rural residents' capability for entrepreneurship and economic empowerment, which will improve their ability to generate revenue. In essence, this implies turning from recipients of others' charity to providers of their own welfare. There are challenges to be overcome; therefore achieving this goal is not simple. At the same time, a valuable resource that is sometimes ignored is the local potential for expanding the livelihood activities of rural residents. The following ideas centre on underutilised local resources and the barriers women and young people face to fully participating in cutting-edge agriculture initiatives.

Conclusion

In many regions of the world, women's place in farming still has been disregarded. Women produce more food than men do globally, especially areas like sub-Saharan Africa where food security is still a major issue. The previous couple of decades have seen improvements and increased importance in all facets of human efforts with regard to the study of gender considerations in the conduct of research, programmes, and government policy creation. Almost every industry now recognises gender as a universal problem. It is now crucial to take into account gender concerns in agriculture in order to fully comprehend social dynamics at all scales, from the global to the domestic. Therefore, planners must take this into account and offer all access that matches the variety of uses that women make of it.

References

Adeyeye,V.A. (1991). Group Farming in Oyo State, Nigeria. A comparison of male and female participants. Nigeria Institute of Social and Economic Research;5:75.

Chitepo,V. (1991).Population,women and environment. Development,1: 124-128.

Gender Policy (2011). (Draft) Glossary of Gender terms. Submitted by the Gender Mainstreaming Project Committee. Sponsored by Macarthur Grant project, University of Ibadan Nigeria.

Gladwin,C.H.(1996).Women and sustainable food production in Africa. Paper presented at the Sasakawa Global 2000 International Workshop on Women Agricultural Intensification and Household Food Security: Developing Gender Sensitizing.

Jiggens. (1989). "How poor women earn income in sub-Saharan Africa and what works against them". World Development, 17(7):953-963.

Odebode,S.O.(2008). Appropriate Technology for Cassava Processing in Nigeria: Users Point of View Journal of International Women Studies, 9,326: 9–283.

Odejide,A.(ed.)(1996).Women and the MediainNigeria.Women's Research and Documentation Centre (WORDOC) Ibadan,:10-16.

Odejide.A.(1998). "Income Generation Programmes for Women in Development. An Appraisal, Workshop on Women in Economic Development NCEMA, September7–11,:4-10.

Ogunlela, Y.I. and Mukhtar, A.A. (2009) Gender Issues in Agriculture and Rural Development in Nigeria: The Role of Women. Humanity & Social Sciences Journal;4(1):19-30.

Olawoye,J.E. (2005). "Entrepreneurship among Rural Women: Problems and Prospects" in Empowering Rural Women Issues, Opportunities and Approaches by R.K. Samantha(ed.) Dehli,The Women Press.: 175-194.

Olawoye, J.E., Ladele, A.A., Oladeji, J.O., Akinbile L.A., Odebode, S.O., Olujide, M.G., Oyesola, O.B and Fakoya, E.O. (2002). Fadama II preparation studies: Social assessment for the southern states of Nigeria. Report submitted to The World Bank.

Olawoye, J. E.; Agumagu, A. C.; Gbadegesin, A; Akande, S. O.; and Yahaya, M.(1994).Situation Analysis of Nigerian Women and Girls. Agricultural Production and Rural Productivity Sector for WORDOC coordinated research project for UNICEF.Ibadan.

Quisumbing,A.R.(1994).Gender differences in agricultural productivity: A survey of empirical evidence. ESP Discussion Paper Series No.36, The World Bank, Washington DC.

Sen, A.(1989).Human development in the 1990s and beyond. Journal of Development Planning(Nov.): 55,

Tsikata, D., Kerr J., Blacklock C., and Laforce, F. (eds) (n.d.)(2000).Demanding dignity: Women confronting economic reforms in Africa. The North-South Institute and Third World Network–Africa.

World Bank (1992). Designing and implementing agricultural extension for women farmers. World Bank, Washington DC.

3

Utilization Pattern of Indigenous Technical Knowledge (ITK) Practices by Farm Families in Patna District of Bihar

Ashish Anand[1], Sarbani Das[2], Sweta Sahoo[3] and Jeebanjyoti Behera[4]

[1,2,4]Department of Extension Education, CoA, OUAT, Bhubaneswar, Odisha
[3]Department of Agricultural Extension, IAS, SoA, Bhubaneswar, Odisha

Abstract

Indigenous technical knowledge has been traditionally owned wisdom from time to time, but as the years passed owing to the invasion of modern technologies, it is likely to be forgotten by the society. Due to ecological imbalance and overexploitation of land, water and other natural resources, their ecology is disturbed which ultimately brings down agricultural productivity.Also, 12th Sustainable Development Goal (SDG) of Vision 2030 of United Nations envisions for 'Responsible consumption and production'. Taking this into consideration, this study focused on the utilization pattern of ITK practices relating to the farming enterprises, undertaken in Patna district of Bihar state, India. For this present study, primary data has been collected from 120 respondents with the help of pre-structured interview schedule through personal interview.Fourteen ITK practices has been selected to quantify utilization practices by farm families. It has been found that majority of the respondents (67.50%) were having medium level of utilization pattern followed by high (18.33%) and low (14.17%) level. "Use of Bijamrut" was the most followed ITK practices among the respondents. These results will be helpful for policy makers to document these ITK practices for sustaining the food production.

Keywords: Document, ITK, Quantify, Traditional, Utilization

Introduction

A concept focused on crop and livestock production with minimal environmental impact has emerged as a result of the paradigm shift in Indian agriculture from traditional to sustainable agriculture. This concept also supports the farm economy and helps farmers improve their tactics and quality of life. This has sparked farmers' interest in learning traditional technology skills that draw on local knowledge that people have acquired from their ancestors to address issues in agriculture and activities associated with natural processes. Indigenous Technical Knowledge (ITK), which is a component of using natural products to address agricultural problems and integrate activities, has attracted interest since the concept of sustainable agriculture began to emerge in the Indian state of agriculture in the late 1980s. ITK is regional knowledge unique to a particular culture or community.

At the beginning of the 21st century, the world is facing a number of major global challenges. On the development side one of these challenges is the establishment of sustainability. In the context of agricultural development, a new agenda has been set to achieve the goals of producing and disseminating economically viable, socially acceptable and environmentally sound technologies that are geographically sound. , even local agricultural schools cannot teach such knowledge because of its size and diversity (Pereira, 1993).

Developing countries like India have a repository of traditional knowledge that may seem to be very important for all of us in redesigning our efforts to develop the country to our satisfaction.As an agricultural country, India, our development efforts are focused on transforming traditional agricultural technology into a modern technology but the potential of traditional farming has been increasingly recognized over the past few years. "Developing countries have an important but underutilized source of traditional agricultural knowledge." (Warren, 1993). Such knowledge or experience has been passed down from generation to generation, commonly known as "Traditional Agricultural Practices." The term "Indigenous" is often used interchangeably with "Traditional" or "Local" (Chithraichelvan, 1994).

A cultural idea deriving from the Latin "Tradere," which means "transfer," is what the term "Traditional" refers to. In this sense, culture can be defined as the transmission of values and social standards that have their origins in the past. Due to many internal and external influences, new traditions frequently replace old ones. In this sense, culture has a fluid nature and is not static. Locals produce indigenous agricultural knowledge using their knowledge and evaluation to satisfy their needs and ensure sustainability. To preserve their identity, this knowledge is strongly correlated with their practises, beliefs, and traditions. Indigenous agricultural knowledge provides a few advantages, including a low risk profile, an excessive dependence on genetic and physical diversity, the utilisation of local resources, natural health, and the ease with which it can be accessed and comprehended.

A variety of socioeconomic factors may also influence how well-informed rural farm families are about ITK. It also investigated the connection. Additionally, it requires more employees, participates in the neighbourhood farming system, and can provide a wider range of community demands based on local customs, as shown by evidence from trustworthy sources, and it had exceptional cooking abilities.

Farmers should be incredibly happy of being acknowledged as the guardians of their knowledge, as participants in development initiatives, and as owners of their own assets, infrastructure, and expertise.

Aim and Advantages of the Study

The study has been aimed to find out the utilization pattern of ITK among farm families which will then be helpful in increasing sustainability of agriculture through proper documentation and blending of traditional knowledge with modern knowledge. Apart from this, the findings of the present study will be helpful in providing scientific explanation to the experientially generated knowledge and age-old farming practices, of which adoption rate must be high.

Materials and Method

The present study was conducted in Patna district of the Bihar state. They prominently and actively participate in various farming activities and contribute in their household economy.There are 23 blocks functioning in the District. Sampatchak block was selected by purposive sampling for the research due to more number of ITK users as per pilot survey. There are 47 villages in Sampatchak Block. Out of these, 10 villages were selected purposively based on maximum number of ITK users for the present study.

From the 10 villages, 120 respondents were selected randomly. A structured interview schedule was used for data collection. Descriptive research design was used for this study. The statistical tools such as frequency, percentage and chi-square test were used to interpret the data and for drawing logical conclusion.

Results and Discussion

Particularly at the grassroots level, Indigenous Technical Knowledge (ITK) holds enormous potential for innovation. Numerous indigenous communities live in India, and the majority of them have their own distinct traditional knowledge and technological foundation. Many of these skills and technologies are on par with the systems of skills and knowledge used today, and they have helped the indigenous populations live comfortably and independently. The entire socio-economic growth of the communities has benefited greatly from these ancient knowledge and skills.

Results portrayed in Table 1 show that use of *"Bijamrut"*which is used for seed treatment before sowing was most prominent ITK practised by farmers among fourteen ITK practises which were identified. Salt added to mustard oil (*Brassicacampestris)* applied all over the body for control of ectoparasites and use of scarecrow both held second most position among ITK which were followed.

ITK which took third position was use of "*Panchgavya*" for seed and seedling treatment and also as a soil fertility enhancer by applying through irrigation water. Next came use of "Sanjivak" which is used as soil fertility enhancer followed by use of neem extract for controlling pest. Sixth position was held by use of rat traps followed by use of chapatti made from mixture of 250 g jaggery ,150 g ghee and wheat flour (*Triticumaestivum*) to feed animal once a day for inducing heat and planting of redgram on bunds both of which shared seventh position.The ITK which held eighth position was green manuring which helps in increasing soil fertility followed by use of enriched "*Amrut Ghol*" used as soil fertility enhancer and growth and flowering enhancer.Tenth position was held use of "*Amritpani*" used as soil fertility enhancer (@ 200 ltrs per acre along with irrigation water). Eleventh most preferred ITK among the fourteen ITKs was use of powder of half ripe bael (*Aeglemarmelos*) fruit + powder of sheesam (*Dalbergiasissu*) leaves which is given to animal for diarrhoeacontrol.Use of sheep manure for making heavy soil more crumbly was the least preferred ITK among all ITKs that were identified. Table 2 results show level of utilization of ITK.Regardingthis,it was observed that majority i.e. 67.50 per cent of the ITK practitioners had medium level of utilization followed by 18.33 per cent of respondents having high level of utilization and 14.17 per cent of respondents having low level of utilization of ITK.

Table 1: Utilization pattern of ITK

S.No.	Activities	Always	Sometimes	Never	Mean score
1.	Usage of neem extract for controlling pest	71(59.17)	38(31.66)	11(9.17)	2.50(V)
2.	Use of sheep manure	13(10.83)	36(30.00)	71(59.17)	1.52(XII)
3.	Use of rat traps	59(49.17)	59(49.16)	02(1.67)	2.48(VI)
4.	Use of amritpani	20(16.67)	66(55.00)	34(28.33)	1.88(X)
5.	Use of panchgavya	76(63.33)	36(30.00)	08(6.67)	2.57(III)
6.	Use of scarecrow	73(60.83)	45(37.50)	02(1.67)	2.59(II)
7.	Green manuring	57(47.50)	43(35.83)	20(16.67)	2.31(VIII)
8.	Planting of red gram on bunds	53(44.16)	59(49.17)	08(6.67)	2.38(VII)
9.	Use of chapati made from mixture of jaggery,ghee and wheat flour to feed animal	69(57.50)	28(23.33)	23(19.17)	2.38(VII)
10.	Powder of half ripe *bael*fruit mixed with powder of *sheesam* leaves for feeding animal	21(17.50)	47(39.17)	52(43.33)	1.74(XI)
11.	Salt added to mustard oil applied all over the body of animal	76(63.33)	39(32.50)	05(4.17)	2.59(II)
12.	Use of enriched *Amrut Ghol*	46(38.33)	60(50.00)	14(11.67)	2.27(IX)
13.	Use of *Bijamrut*	81(67.50)	33(27.50)	06(5.00)	2.63(I)
14.	Use of *Sanjivak*	75(62.50)	33(27.50)	12(10.00)	2.53(IV)

Table 2: Level of utilization of ITK

S.No.	Range	Utilization level	Frequency	Percentage
1.	<29.73	Low	17	14.17
2.	29.73-34.95	Medium	81	67.50
3.	>34.95	High	22	18.33
		Total	120	100

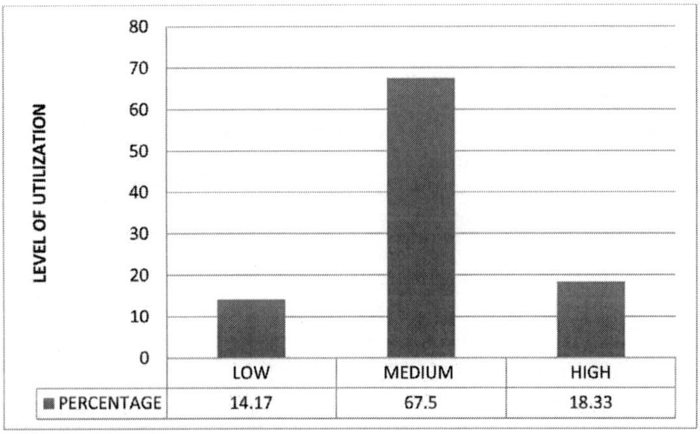

Fig. 1: Level of Utilization of ITK

Conclusion

Present study focused on assessing the association between socio economic profile and utilization level of ITK among farmers.It was found that use of Bijamrut was most prominent ITK practised among sampled ITK practitioners.It was also observed from the study that majority of respondents i.e. 70.83 per cent had medium level of utilization of ITK.

Future Scope of the Study

Studies can be designed to determine the association between the ITK respondents' profile traits and the extent of ITK use in agricultural operations.The perspectives of extensionists and researchers regarding some of the indigenous knowledge captured in this examination can be identified and compared through studies.

References

Chandola, M., Rathore, S. and Kumar B. (2011). Indigenous pest management practices prevalent among hill farmers of Uttarakhand. Indian Journal of Traditional Knowledge, 10(2): 311-315

Chithraichelvan R. (1994). Use of ITK in farming system research Proceedings of International workshop on Genetic resources, UPWARD, LOS Banos, Phillipines.

Christian, B.M. (2015). A study on extent of adoption of IPM strategy by cotton growers of Vadodara district of Gujrat State. Unpublished M.Sc. (Ag) Thesis, G.A.U., Anand 71 | Page Campus, Anand.

Deshmukh, P.S. (2015). A Case Study : Traditional Methods of Pest Control in Some Villages of Kolhapur District.International Interdisciplinary Research Journal, 5(3):87-92

Devi, R., Pandit, A. and Kashyap, D. (2014). Assessment of applicability of Indigenous Technical Knowledge (ITK) in aquaculture as perceived by fish farmers in Assam. Indian Journal of fisheries, 61(3):104-110

Firake, D. M. ,Lytan, D., Thubru, D. P ., Behere, G.T., . Firake, P.D, and Azad Thakur, N. S. (2013), Traditional Pest Management Practices and Beliefs of Different Ethnic Tribes of Meghalaya, North Eastern Himalaya, Indian Journal of Hill Farming 26(1):pp.58- 61.

Gahukar, R. T. (2014). Utilization of indigenous plant species in tobacco crop protection in India - a review. Agricultural Reviews, 35(12):142-147.

Hiwasaki L, Luna E, Syamsidik SR. (2014). Local and indigenous knowledge for community resilience: Hydro - meteorological disaster risk reduction and climate change adaptation in coastal and small island Communities Jakarta: UNESCO.

Khatri , Cogent Food & Agriculture. (2021). Assessment of applicability of Indigenous Technical Knowledge ITK in aquaculture as perceived by fish farmers in Assam, 7: 1914384.

Kumar Sujay, S. (2012). Participation and time utilization pattern of rural youth in organic sugarcane cultivation under cauvery command area of Karnataka. M.Sc. (Agri.) Thesis submitted to the University of Agricultural sciences, Bengaluru.

Naharki, K., & Jaishi, M. (2020). Documentation of Indigenous Technical Knowledge and Their Application in Pest Management in Western Mid Hill of Nepal. SAARC Journal of Agriculture, 18(1): 251-261.

Narayanasamy, P. (2001). Traditional Pest Control: A retrospection. Indian Journal of Traditional Knowledge, 1(1): 40-50

Panampitiya,G.(2018).A review of Indigenous knowledge related to the traditional agriculture in Srilanka. EPRA International Journal of Multidisciplinary Research,4(10)

Pereira , W. (1993). Tending the earth Earth care books, Bombay: 135-134.

Ponnusamy, K., Gupta, J. & Nagarajan, R. (2009). Indigenous Technical Knowledge (ITKs) in dairy enterprise in coastal Tamil Nadu, Indian Journal of Traditional Knowledge, 8 (2):206-11.

Ponnusamy, K., Kale, R.B., Ravi, K.N., Arulmozhi Devi.M.C. & Sharma, P. (2017) Cross-regional analysis on usage of Indigenous Technical Knowledge in dairy farming. Indian Journal of Animal Research, 51 (3): 549-556.

Pradhan, K., Radhan, K., Yolmo, Z., Saha A. and Prasad, C. (2017). Idenitification and documentation of indigenous technological knowledges regarding pest control methods in agriuclture. International Journal of Agriculture Sciences, 9(38): 4580-4584.

Rokonuzzaman, M. (2015). Training needs of tribal people regarding income generating activities.Indian Res. J. Extn. Edu., 13(2): 10- 16.

Sarkar, S., Padaria, R.N, Vijayragaran, K:, Pathak, H. and Jha J. K. (2015). Assessing the potential of ITK for adaption to climate change in the Himalaya and arid ecosystem, Indian journal of traditional knowledge,14(2):251-257.

Warren, D. M. (1993). Indigenous knowledge and development monitor. CIKARD, 1(1): 7.

4

Mobilization of Farm Women and Their Subsequent Mainstreaming Through Krishi Vigyan Kendras (KVKs) The Approaches and Executions

B.K. Mohanty[1*], Saumyesh Acharya[2] and S.K. Acharya[3]

[1]Department of Agricultural Extension & Communication
Faculty of Agricultural Sciences, IAS, S'O'A Deemed to be University
Bhubaneswar, Odisha
[2]Department of Agricultural Extension, Institute of Agriculture
Visva-Bharati University, Sriniketan, West Bengal, India
[3]Department of Agricultural Extension, BCKV, Mohanpur
Nadia, West Bengal

Abstract

Krishi Vigyan Kendra (KVK) provides multifaceted and beneficial services in enabling the farmers with recent innovation and methodologies. The present study deals with the gender issues of Krishi Vigyan Kendra and both the adopted and non-adopted villages have been selected for the study. Following multistage sampling procedure, 150 respondents have been selected. Correlation analysis revealed that the content of the training is well tuned with the Scientific aspiration of the respondents. The method of training is correlated with the education of the respondents while the subject area and the content of training have been closely tuned to scientific aspiration. Regression analysis elicited that all the independent variables jointly have contributed to 42.70 percent of the variance embedded with the consequent factor, training perception. Also, the canonical correlation analysis revealed that there have been conjoint interactions between x and y sets of variables based on cross-loading. It has been found that, in making a difference between adopted and non-adopted village, the role of occupation, farm power, media exposure, training content, training venue, subject area of training for the farm women have been dominant. While the factors, occupation, media exposure, training content, training venue have been unique with adopted villages, the other factors, farm power and training area have gone precisely attuned to the respondents of non-adopted village. The gender related studies, as has been conducted for these two adopted and non-adopted villages, has offered huge policy implications to make the lighthouse of farmers, the KVKs more active and effective towards mobilizing and attaining the goals of mainstreaming and empowerment through KVK participation and executions.

Keywords: *Empowerment, Gender, Krishi Vigyan Kendra, Mechanization, Mobilization, Socialization, Training*

Introduction

Krishi Vigyan Kendras (KVKs) have marked their presence across India by executing various scientific as well as institutional reforms for the agrarian mass of our country. Right from its beginning in 1970s, farmers were actively trained for adopting different innovative tools and methods and were also provided with proper assistance in implementing them. The farmers were adapted to technical know-how of an introduced technology and produced significant results in the real field conditions.

KVKs provided proper training programmes on need-based curriculum in order to cater to the local and regional needs. This has also led to need assessment of various training needs and efficacies to make improvement in policies and future interventions. KVK can be instrumental in nurturing the process of gender empowerment by inviting and involving the farm women in the process of technological adaptation, adoption and socialization. There are methodological innovations as to how we can involve the farm women by employing and applying gender-sensitive KVK technologies as suitable for their job environment and climate. Especially in areas of value addition, rural crafts, women crafts, fishery, horticulture and micro-finance, their roles are becoming more prominent and decisive. This is positive achievement in making them free from gender discrimination and impoverishment.

The knowledge score gained by respondents after training was more satisfactory in all aspects (58.33% had medium and 48.67% had high level of knowledge). The socio-economic factors of the participants viz. Education level played a significant positive role on the knowledge gain of the participants. (Upamanya et.al., 2020) Most of the farmers, i.e., 66.67 per cent with theoretical knowledge suggested that more field visits can be arranged, 63.33 per cent of the respondents suggested the training should be conducted in their own villages because the farm women cannot able to come to KVK by leaving their children and their regular household work and around sixty per cent (60.83 per cent) of the respondents suggested that organizing continuous and frequent training programmes for effective participation and retention of knowledge. (Chakkaravarthy and Balakrishnan, 2020)

According to Acharya et.al.(2020), multivariate analysis regarding KVK functioning found that the variable education, no. of training in KVK and mechanization in family has recorded a positive and prevalent impact on Attitude towards KVK activities (y) and exogenous variable age has recorded negative impact on attitude towards KVK activities(y). Education is an important factor for improvement of KVK activities. Training and mechanization in family helps people to adopt new technologies. Adoption of improved and newer technology requires decision by farmers, and scientific orientation is a degree to which respondents can oriented to the use of scientific methods in relation to adoption behavior. Training plays important role for improvement of KVK and makes people more optimistic and provides maximum information about new technologies.

It has been observed that 70% farm women were in the age group of 31-45 years, 50% illiterate, 80% having joint family system, 46% had farm area more than 2 ha and had agriculture as the main occupation with subsidiary occupations like pisciculture, horticulture and dairying. But their social, extension and media participation were far from satisfactory level. (Mishra et al., 2005)

Farm women should be empowered through effective training programmes. The challenges are mutually supportive manner, so that men and women can play active role in productivity. (Sailaja and Reddy, 2003)

Adoption of improved and newer technology requires decision by farmers, and scientific orientation is a degree to which respondents can oriented to the use of scientific methods in relation to adoption behavior. It is an important psychological factor in decision making process. It is concluded that people with progressive attitude will always try to involve themselves in all activities through which more annual income can be achieved and education plays a vital role for adoption process. Training plays important role for improvement of KVK. Training makes people more optimistic and provides maximum in formation about new technologies. (Biswas et al., 2020)

Table1, Presents the Coefficient of correlation between the Attitude towards KVK activities (y) and 10 independent variables in case of all the KVK adopted and non adopted respondents. It has been found that the variable education (x_2), No. of training in KVK(x_6), mechanization in family (x_7) has recorded a positive and prevalent impact on Attitude towards KVK activities (Y_1) and variable age (x_1) has recorded negative impact on attitude towards KVK activities(y).

Education is an important factor for improvement of KVK activities. Higher education level helps to improve KVK activities through knowledge. Training and mechanization in family helps people to adopt new technologies.

Multiple Regression Analysis

Attitude towards KVK Activities (y) Vs. 10 Causal Variables (x1-x10)

Table 2: Does present the Regression analysis; Attitude towards KVK activities (y) vs. 10 Causal variable (x1-x10). The full model on regression analysis depicts that, with the communication of 10 causal variables together; only 94.60% of the variance in nutritional availability has been explained. This suggests that more number of relevant variables caused have been included in this study.

Stepwise Regression Analysis

Attitude towards KVK Activities (y) Vs. 10 Causal Variables

Table 3 does present the Stepwise regression analysis suggest that age (x1), no of training in KVK (X6) and education (x2) these three variables retained in the last step and has contributed 93.50% of the variable explain With these background the general objective of the study was to assess the Impact of Krishi Vigyan Kendra on farm women.

The specific objectives of the study were to study the Socio-personal profile of woman trained at KVK during last 5 years. And to determine the perception of women trainees and non-trainees about the programme of KVK for women in relation to training programme, content, venue, duration, use of training method and motivation to put learning into action.

Materials and Methods

The study was conducted in 20 (twenty) different village out of which 10 from adopted and 10 from non-adopted villages located in 10 block at 10 districts of Odisha. Both purposive and random sampling were followed to select the respondents. First the KVK was selected purposively. Then random sampling technique was adopted to select the blocks, villages and respondents. from the ten KVKs 150 trained and 150 untrained farm women were selected as sample respondent of the study. The focus of the study was to assess the change in the socio-economic status of the beneficiaries due to their involvement in KVK training programme and to estimate their knowledge about KVK and constraints associated with the institutions and to illicit suggestion if any from the trainees for conducting better and efficient training programme. Appropriate statistical tools have been used to carry out the study viz, Rank order correlation, Mann-Whitney test and Step wise regression analysis. Appropriate operationalization and measurement of the variables have helped the researcher to land upon the accurate conclusions. Therefore, the selected variables for this study had been operationalized and measured in the following manner:

1) Independent variables 2) Dependent variables. Independent variables selected for the study were age (x_1), education (x_2), occupation (x_3), family size (x_4), holding size (x_5), house status (x_6), material possession (x_7), income (x_8), farm power (x_9), social participation (x_{10}), scientific aspiration (x_{11}) and media exposure (x_{12}).

Dependent variable selected for the study was training perception(y)

Results and Discussion

The subjective information is measured utilizing explicit numerical methodology. Then data analysis through appropriate statistical tools i.e. Mann-whitney test,

Rank order correlation, Step wise regression analysis and Path analysis has been done to evaluate the information .

Table 1: Mann-Whitney Test: Training perception (y) and Impact of KVK on both adopted and non-adopted villages

Variable	Test Statistics							
	NA	A	Mann-whitney U	Wilcoxon W	Z	Sig	Kolmogorov-Smirnow	Sig
Management	2.12	2.27	1018.00	2293.00	-1.61	0.11	1.00	0.27
Content	2.05	2.22	968.00	2243.00	-1.96	0.05	1.10	0.18
Venue	2.19	2.44	765.50	2040.50	-3.37	0.001	2.20	0.00
Duration	2.15	2.10	1152.00	2427.00	-0.68	0.50	0.60	0.86
Method	2.16	2.30	1040.00	2315.00	-1.45	0.15	0.80	0.54
Area	2.23	2.21	1212.50	2487.50	-0.26	0.80	0.80	0.54
Perception score	2.15	2.26	926.00	2201.00	-2.24	0.03	1.50	0.02

Table-1 presents the Mann-Whitney test to delineate the comparable differences of different predictor variables between adopted and non-adopted villages in terms of training perception.

It has been found that the two category of villages, adopted and non-adopted villages, bestowed a significant differences in terms of content, venue and in training perception score, while a KVK in envisages to make a rapid progress in the adopted villages, the villages strategically selected and the non-adopted villages, the recipient of spill- over effect of KVK vis-à-vis adopted villages, These indicator factors /predictors need to be considered.

Table 2: Rank Order Correlation: KVK adopted and non-adopted villages: Components of training perception (y) and 12 predictor characters

Variable	Correlation				Spearman's Rho		
	Management	Content	Venue	Duration	Method	Area	Percp Score
Age (X_1)	0.326	0.103	0.037	-0.004	-0.057	0.114	0.156
Education (X_2)	0.101	0.064	0.062	0.034	0.233	-0.002	0.123
Occupation (X_3)	-0.114	0.107	0.068	0.228	-0.056	0.126	0.088
Family size (X_4)	-0.001	-0.062	0.044	-0.179	-0.141	-0.044	-0.089
Holding size (X_5)	-0.007	-0.001	-0.024	0.018	-0.005	0.017	0.031
House status (X_6)	0.265	0.124	0.007	-0.059	0.058	0.017	0.13
Material possession (X_7)	-0.02	-0.187	0.052	0.048	-0.024	-0.003	-0.026
Income (X_8)	0.047	0.02	0.015	0.06	0.076	0.084	0.104
Farm power (X_9)	-0.159	-0.041	-0.017	-0.034	-0.048	-0.15	-0.123

Social participation (X_{10})	-0.132	0	0.034	0.037	-0.076	-0.077	-0.058
Scientific Aspiration (x_{11})	0.082	0.255	-0.041	-0.003	0.186	0.205	0.173
Media Exposure (X_{12})	0.13	0.047	0.125	-0.139	0.136	0.052	0.114

$P < 0.01$ $P < 0.05$

Table 2 depicted that in both the category of village, the component management is correlated with the age and House status of the respondents.

The content of the training is well tuned with the scientific aspiration of the Farmwomen, the respondents. The training component, venue is not that important a perception. The method of training is correlation with the education of the respondents while the subject area and the content of training has been closely tuned to scientific aspiration. However, media exposure could not make any take with any of the component of training perception. In other way, the role of media has been universal for any for training component per se.

Table 3: Regression Analysis: Isolating critical variables : Training perception (y) and 12 predictor characters

Variable	**Un standardized Coefficients**		**standardized Coefficients**		**Significant**
	B	Std Error	Beta	+	
(constant)	1.715	0.238		7.200	0.000
Age (X_1)	0.061	0.043	-0.147	1.405	0.163
Education (X_2)	0.038	0.021	0.196	1.818	0.072
Occupation(X_3)	0.010	0.008	0.123	1.243	0.217
Family size(X_4)	-0.053	0.055	-0.095	-0.971	0.334
Farm power(X_9)	-0.033	0.016	-0.205	-2.010	**0.047**
Scientific aspiration(X_{11})	0.134	0.078	0.171	1.713	0.090

R square value = 0.427

Table-3 presents the regression analysis for selecting some few causal factors out of the plethora having conspicuous efficacy in characterizing, the training perception and these are: age, education, occupation, family size, farm power and scientific aspiration.

The value stand at 0.427, it is to conclude that 42.70 percent of the variance embedded with the consequent factor training perception has been interpreted so far.

Table 4: Path Analysis: Total effect of exogenous variables on training perception (y): KVK adopted villages

Exogenous variables	Total effect (r)	Total direct effect (TDE)	Total indirect effect (TIE)	Substantial indirect effect		
				I	II	III
Age (X_1)	0.273	0.098	0.175	0.123 (x_9)	0.049 (x_2)	0.048 (x_4)
Education (X_2)	0.049	0.195	-0.146	0.152 (x_9)	0.025 (x_4)	0.021(x_6)
Occupation (X_3)	0.071	0.029	0.042	0.093 (x_{12})	0.064 (x_4)	0.053(x_6)
Family size (X_4)	0.183	0.241	-0.058	0.084 (x_9)	0.025 (x_{12})	0.019(x_1)
Holding size (X_5)	0.053	-0.016	0.069	0.058 (x_{10})	0.048 (x_9)	0.043 (x_2)
House status (X_6)	0.155	0.134	0.021	0.057 (x_{12})	0.042 (x_{10})	0.030(x_2)
Material possession (X_7)	-0.106	-0.029	-0.077	0.061 (x_9)	0.021 (x_6)	0.021(x_1)
Income (X_8)	-0.034	0.017	-0.051	0.075 (x_9)	0.041(x_4)	0.015(x_{12})
Farm power (X_9)	-0.236	-0.361	0.125	0.082 (x_2)	0.056 (x_8)	0.033(x_1)
Social participation (X_{10})	-0.099	-0.125	0.026	0.045(x_6)	0.043(x_{12})	0.036(x_4)
Scientific Aspiration (x_{11})	0.116	0.045	0.061	0.076 (x_9)	0.075(x_2)	0.059 (x_4)
Media Exposure (X_{12})	-0.101	-0.168	0.067	0.046(x_6)	0.036(x_4)	0.032 (x_{10})

Residual effect : 0.6805; Highest indirect effect (x_9) (6times)

Table 4 presents the path analysis towards delineating the effect of twelve exogenous variables of the sub-consequent variable, training perception (y_3) in adopted village.

The following variables vizAge (x_1), Farm power (x_9) have wielded a comprehensive associational impact on training perception. The training perception has gone stronger here for the farm family having lower level of farm power. This might be due to the fact poor level of farm power begets higher desire for receiving training and, there by, a positive perception on training. The variables, Education (x_2), Family size (x_4) and Farm power (x_9), have recorded comprehensive and direct impact on training perception. Family size generates a better urge for undergoing training programme. The education here has generated an inclination for undergoing a training experience for the up- gradation of knowledge and scaling up of skills.

So far as the indirect effect is in concern, Age (x_1) has generated the highest associational property to characterise the behaviour of the training perception. The variable farm power (x_9) has routed the highest indirect effect of as many as six variables, to characterise the performance of training perception. That value of residual effect being 0.6805, it is to conclude that 68.05 per cent of the variance embedded with the consequent variable, training perception could not be explained only.

Table 5: Path Analysis: Total effect of exogenous variables on training perception (y) : KVK non-adopted villages

Exogenous variables	Total effect (r)	Total direct effect (TDE)	Total indirect effect (TIE)	Substantial indirect effect		
				I	II	III
Age (X_1)	-0.049	0.054	-0.103	0.104 (x_2)	0.064(x_{12})	0.052 (x_4)
Education (X_2)	0.203	0.249	-0.046	0.065 (x_{12})	0.049 (x_9)	0.040 (x_{10})
Occupation (X_3)	0.132	0.144	-0.012	0.043 (x_2)	0.039 (x_7)	0.033 (x_9)
Family size (X_4)	-0.305	-0.249	-0.056	0.100(x_{11})	0.052 (x_{12})	0.036 (x_2)
Holding size (X_5)	-0.124	-0.229	0.105	0.107 (x_{11})	0.047 (x_9)	0.034 (x_{10})
House status (X_6)	-0.006	-0.152	0.146	0.106 (x_{11})	0.070 (x_7)	0.048 (x_2)
Material possession (X_7)	0.056	0.232	-0.176	0.098 (x_9)	0.046 (x_6)	0.033 (x_5)
Income (X_8)	0.128	0.085	0.043	0.119 (x_9)	0.068 (x_2)	0.056 (x_{11})
Farm power (X_9)	-0.072	-0.279	0.207	0.081 (x_7)	0.065 (x_{12})	0.044 (x_2)
Social participation (X_{10})	-0.039	-0.169	0.130	0.065 (x_{11})	0.060 (x_2)	0.046 (x_5)
Scientific Aspiration (x_{11})	0.292	0.433	-0.141	0.125 (x_{12})	0.058 (x_4)	0.056(x_5)
Media Exposure (X_{12})	0.122	0.352	-0.230	0.154 (x_{11})	0.052 (x_9)	0.046(x_2)

Residual effect: 0.5795; Highest indirect effect (X_{11}) (5 times)

Table - 5 presents the path analysis for depicting the total direct, indirect and residual effect of the 12 exogenous variables on the consequent variable training perception (y). In non-adopted village

It has been found that the variables Education, Family size, Scientific Aspiration have wielded comprehensive associational impact on the perception of training.

Education begets a need for training and so also may be done by Scientific Aspiration, Here, farm women for smaller family size bestows a demand for training since they can spend some amount time for a training exposure.

The variables, scientific aspiration, Media exposure, Education, Family size, Holding size, Material possession, Farm power, Scientific Aspiration and Media exposure have wielded the substantial direct effect on the variable Scientific Aspiration has routed the highest indirect effect of as many as 5 variables to characterise the behaviour of the consequent variable, Training Perception.

The value of residual effect being 0.5795, it is to infer that 57.95 per cent of the variance embedded with the consequent variable could not be interpreted even with the combination of the 12 exogenous variables.

Conclusion

Krishi Vigyan Kendra (KVK) is the lighthouse that shows and directs the pathway for achieving income, livelihood and empowerment. It is precisely true for the women folk of the India. India is having close to 800 KVK across the terrains of agroclimatic zones and regions, cutting across different community, income and culture layers and its constantly serving the rural people towards achieving a sustainable development. Training or capacity building is the prime driver for KVK functioning and in the present study it is focusing on women empowerment and livelihood generation through KVK by imparting well targeted training and capacity building. For the farm women folk, it is not that easy to access all these training facilities, nevertheless they are joining in increasing number to access the training facilities as deflected through their seamless capacity building and perceptual change in ushering agricultural growth and rural development at large. The present study has included a score of socio-ecological variables predicting women behaviour in the KVK ecosystem and an endeavour has been made to estimate the role and contribution of all these factors, both in quantum and quality terms, in empowering themselves as well as making contribution to the agricultural and rural development. The perceptual variables considered in the study had undergone a data normalisation process so that it can be rendered analysable and interpretative in the multivariate analytical techniques. The study reveals and illicits an array of policy implications as to how we can make the farm women empowered, well directed and a strong contributor to the sustainable development.

References

Acharya, S., Ghosh, A., Mahato, M., Mazumder, D., Ghoshal, S., Biswas, A. & Haque, M. (2020). Socio-Ecological Correlates of Attitude towards KVK Functioning: A Multivariate Analytical Approach. Current Journal of Applied Science and Technology. 39. 23-31.

Biswas, A., Mahato, M., Ghosh, A., Haque, M., Ghoshal, S., Mazumder, D. & Acharya, S.K. (2020). Production Orientation of KVK Adopted Farmers: The Socio-Ecological Estimation and Interpretation. Asian Journal of Agricultural Extension Economics & Sociology. 38. 102-113.

Chakkaravarthy, D.N. & Balakrishnan, T. (2020). Perceived suggestions for improving the effective functioning of Krishi Vigyan Kendra. Test Engineering and Management. 82. 3977-3979.

Mahato, M., Acharya, S., Ghoshal, S., Haque, M. & Ghosh, A. (2020). Market Orientation of KVK Adopter Farmers: The Socio-Ecological and Managerial Interpretation. Advances in Research. 38-46.

Mishra, B., Mishra, R. and Kanungo, A.P. (2005) Women's access to farm information and technology, Journal of Extn. Edn. 9 & 10 (1,2).72-77.

Sailaja, A and Reddy, M.N. (2003) Changing needs or farm women in agriculture. 4(1). 165-175.

Upamanya, G., Sarma, H., Barman, P. & Deka, A. (2020). Evaluation of Training Programmes on Oyster Mushroom Cultivation Conducted by KVK, Barpeta and Constraints Perceived by Mushroom Growers in Barpeta District of Assam, India. International Journal of Current Microbiology and Applied Sciences. 9. 560-568.

5

Constraints Faced by the Rural Women in Participation of Different Off-Farm Activities

Eshant Kumar Sukdeve[1], Hem Prakash Verma[2]

[1]*Department of Agricultural Extension, College of Agriculture Indira Gandhi Krishi Vishwavidyalaya, Raipur, Chhattisgarh*
[2]*Department of Agricultural Extension, College of Agriculture Indira Gandhi Krishi Vishwavidyalaya, Raipur, Chhattisgarh*

Abstract

The study was conducted during the year 2015-16 in the Balod district of the Chhattisgarh State, India. The main constraints faced by the majority of rural women in the participation of different off-farm activities were lack of proper knowledge about different off-farm activities, Poor economic status of rural women. Lack of training facility in rural area for skill development related to different off-farm activities, poor economic status of rural women's and lack of transportation facility in rural area. Some of the major suggestions expressed by the respondents to overcome the constraints related to off-farm activities suggestions obtained from the respondents were training facility should be provided by the govt. for different off-farm activities like business, small industry and enterprises establishment, credit should be available at proper time, government should be started new policy for employment of rural women and transportation facility should be available.

Keywords: Off-farm, Rural women, Participation, Constraints, Suggestions.

Introduction

The total population of India according to recent estimates in 2015 is 1.28 billion (Anonymous, 2011). The population in the age-group of 15-34 years in India increased from 353 million in 2001 to 430 million in 2011. More than 50 per cent of India's current population is below the age of 25 years old and over 65 per cent below the age of 35 years old.

The total population of Chhattisgarh State is 25,545,198 of which male and female are 12,832,895 and 12,712,303. Sex ratio is 991 *i.e.* for each 1000 male, which is above national average of 940. Literacy rate in Chhattisgarh has seen upward trend and is 71.04 per cent (Anonymous, 2011). Out of total population of Chhattisgarh 23.24 per cent people live in urban regions and 76.76 per cent people live in the villages of rural areas.

Women are the most potent segment of the population of a country. The women of today are the hopes of tomorrow. They are the back bone of the country. The socioeconomic development and prosperity of rural areas depends to a considerable extent, on the type of women living in rural areas, because the rural women have abilities to orient themselves to go along the main stream of the development process. They reflect the national potentiality and represent the life blood of a nation. Development of women determines the development of community and country as a whole. So, the future of the country lies in their hands, what they will become, what role they will play in a democratic society and what they will do, will be dependent to a greater extent on the period between their childhood to adulthood.

Materials and Methods

The study was conducted during the year 2015-16 in the Balod district of the Chhattisgarh State. The Chhattisgarh State consists of 27 districts out of which Balod district was selected randomly. Out of total 5 blocks in the district only three blocks namely Gurur, Balod and Doundi-lohara were selected randomly. Three villages were selected randomly from each selected block to take a total of 9 villages in the sample. 15 rural women were Selected randomly from each selected villages. Thus, the total 135 rural women were considered as respondents for this study. The data were collected and analyzed by using appropriate statistical methods like mean, S.D., percentage, correlation and multiple regression analysis etc.

Results and Discussion

Constraints faced by rural women in participation of different off-farm activities

An attempt has been made in this study to identify the constraints which were responsible for the participation of rural women in different off-farm activities.

The data compiled in table 1 shows that the majority (80.74%) of the respondents faced constraints related to different off-farm activities like lack of proper knowledge about different off-farm activities, followed by Poor economic status of rural women(72.59%), Lack of training facility (62.22%), Health problems (47.40), Difficulties to spend time for off-farm works/operations due to domestic work (44.44%), Engagement of rural women's other activities (40.74), Secondary status in decision making (34.81%), Lack of govt. support/subsidy (31.11%), Working under unfavourable condition (29.62%), Burden of family work (28.88%), Lack of transportation facility in rural area (26.66%), Non availability of credit at proper time (20.74%), Being women it is difficult to perform certain off-farm activities (16.29%), Male dominated society which of about only 13.33 per cent of the respondents faced problems high rate of interest with the involvement of off-farm activities.

Table 1: Distribution of the respondents according to their constraints faced by the rural women in various off-farm activities

(n=135)

Particulars	Frequency*	Percentage	Rank
1. Lack of proper knowledge about off farm activities.	109	80.74	I
2. Lack of Govt. support/subsidy.	42	31.11	VIII
3. Non availability of credit at proper time.	28	20.74	XII
4. Poor economic status of rural women.	98	72.59	II
5. Lack of training facility in rural area.	84	62.22	III
6. Lack of transportation facility.	36	26.66	XI
7. Engagement of rural women's other activities.	55	40.74	VI
8. Secondary status in decision making.	47	34.81	VII
9. Difficulties to spend time for off-farm works/ operations due to domestic work.	60	44.44	V
10. Being women it is difficult to perform certain off-farm activities.	22	16.29	XIII
11. Burden of family work.	39	28.88	X
12. Male dominated society.	18	13.33	XIV
13. Health problems.	64	47.40	IV
14. Working under unfavourable condition.	40	29.62	IX

*Data based on multiple responses

Suggestions from the rural women for improving the participation of different off-farm activities

As regards to suggestions given by the respondents to overcome the constraints and for improving the participation of different off-farm activities the findings are presented in table 2.

With regard to off-farm activities the data reveals that the majority of the respondents (75.55%) suggested that the Govt. should be provided training facility for different off-farm activities like business, small industry and enterprises established for improving the participation of different off-farm activities, followed by 65.18 per cent of the respondents suggested that the credit should be available at proper time for off-farm activities, 47.40 per cent of the respondents suggested that the Govt. should be started new policy for employment of rural women and 23.70 per cent of the respondents suggested that transportation facility should be available.

Table 2: Distribution of the respondents according to their suggestions from the rural women to overcome the problems/constraints

(n=135)

Suggestions	Frequency*	Percentage	Rank
1. Govt. should be provided training facility for different off-farm activities like business, small industry and enterprises establishment.	102	75.55	I
2. Credit should be available at proper time.	88	65.18	II
3. Govt. should be started new policy for employment of rural women.	64	47.40	III
4. Transportation facility should be available.	32	23.70	IV

*data based on multiple responses

Conclusion

From the findings of this study, it can be concluded that, if the government take proper initiative for increasing the involvement of the rural women in different off farm activities. If so, then the problem of training in rural area can be solved. In this context, it is necessary that government should plan the programme on the basis of the constraints faced by them.

References

Adekunle, O. A., L. L. Adefalu, F. O. Oladipo, R. S. Adisa and A. D. Fatoye (2009). Constraints to Youths' involvement in Agricultural Production in Kwara State. Nigeria Journalof Agricultural Extension, 13(1) : 102-108.

Angaitkar, A. G., V. B. Janjal, K. N. Barse and V. R. Shedge (2013). Problems faced by rural youths while choosing agriculture as their profession. Agric. Update, 8(4) : 685- 686.

Chauhan, N. M. (2012). Impact and constraints faced by tribal farm women in kitchen gardening. Raj. J. Extn. Edu., 20 : 87-91.

Mandavkar, P. M., P. A. Sawant and R. P. Mahadik (2011). Training needs of Tribals in relation to agricultural occupation. Raj. J. Extn. Edu., 19 : 20-24.

Rokonuzzaman, M. (2013). Training needs of tribal people in carrying out income generating activities. Indian Res. J.Ext. Edu., 13(1) : 77-84.

Singh, D. P. and S. K. Yadav (2014). Knowledge and adoption gap of tribal farmers of Bastar towards rice production technology. American International Journal of Researchin Humanities, Arts and Social Sciences, 5(1) : 54-56.

6

Impact of Training Programs Conducted by KVK on Rural Women

[1]Kiran Pant, and [2]A.K. Sharma

[1]*Program Assistant (Extension), KVK Dehradun, GBPUA&T, Uttarakhand*
[2]*Professor & OIC, KVK Dehradun, GBPUA&T, Pantnagar, Uttarakhand*

Abstract

Skill development among the rural women is a crucial process which equips them to organize themselves to improve their agricultural knowledge and capabilities, making them more independent, giving them decision-making skills, and a better control over their resources. A lot of rural women lack educational qualifications and technical skills, either because of lack of education in the villages or the need to drop out of school due to economic conditions. Vocational skills play a very important role in eradicating poverty by training unskilled farm women and ultimately creating a range of entrepreneurs in the rural areas. Krishi Vigyan Kendras (KVKs) are working all over the nation with the main objective of imparting training and the vision to raise the socio-economic status of the rural masses. Since its inception, KVK Dehradun has been working under the administrative control of G B pant University of Agriculture & Technology, Pantnagar and is sincerely accomplishing its objectives of imparting trainings to the farming community. A study was conducted to understand the "Impact of Training Programs Conducted by KVK on Rural Women" for the upliftment of farm women with the help of scanning the reports and well-structured questionnaire to find out realistic data. The data analyzed to study the profile of the participants reveals that majority of the farm women belonged to the middle age group (35-50 years), were literate up to higher secondary, had income in the range of Rs.5,000-10,000 rupees per month, have medium level of mass media exposure but were not exposed to any of the developmental programs run by the government or other departments. Participation of traineesduring the years 2017-18 was the highest, resulting in maximum participation in both the categories i.e.: male and female, and majority of the beneficiaries were fully skilled after receiving training and demonstration. The extent of knowledge gain was higher where one to one interaction was created. As per the data, the training on cow dung diya making and packaging, and herbal colour making came out as the most interesting areas where the number of participation as well as skill-gain both were observed on the higher side. The findings reveal that KVK is working in the right direction by providing more options for income generating activities in rural areas by imparting vocational/skill trainings to women folk (Anithakumari et al,2021). KVK has had a positive impact on the quality of life of farming community in terms of income, yield, productivity, and above all, their capacity for optimal utilization of the available resources. Demonstration and trainings were the activities where maximum extent of knowledge transfer benefitted the respondents, primarily because of the effectiveness of one-to-one interaction of the trainer with farm women.

Keywords: KVK, Training, Impact

In a country like India where the economy is dominant by agriculture and its allied sectors, nearly 75% of women are engaged in agriculture in comparison with their male counterparts. Among rural female workers, 50% comprises of agricultural laborer and 37% are cultivators. Not only this, around 70% of total farm work is performed by women folk. Majority of agricultural works are done by women farmers (*FAO, 2011*) and thus empowering them will have far-reaching effect on agricultural development. They usually operate multiple livelihood fields, including farming, animal husbandry, and marketing along with doing the household chores and are closely involved in social issues. Among these activities when we look specifically into the agricultural tasks, it is very clear that approximately 70–80% of women are participating in all agriculture activities from raising seedlings, weeding, harvesting and, processing and value addition activities and sometimes their involvement extends to marketing their product. To compete in this fast-growing industry with others, the rural women need to be up-to-date with new technologies to foster the reach of innovations to the farm women. Training is one of the most important means by which farmers and rural women can upgrade their professional and technical skillset. This is an essential component in the area of agriculture where technology is changing day by day and the knowledge of the farming community including women needs to be upgraded at a comparatively quicker pace.

Skill development for rural women is a process whereby rural women can organize themselves to increase their own capabilities towards the work, making them more independent to make their decisions and to have a control over the resources they possess. Rural women either are less educated or are school droppers who lack any type of educational qualification or technical skill. Vocational skills play a very important role in eradication of poverty by converting the unskilled farm women to trained artisans and ultimately enhancing the number of entrepreneurs in the rural areas. After participating in these trainings women are competent to enter the market network system and, that helps them to take the decisions confidently and leads to an overall improvement in women's social and economic status.

An impressive number of organizations have been established to upgrade the skills of the farming community aiming at overall rural development. Krishi Vigyan Kendra provides vocational trainings to the rural women for making them self-dependent which ultimately helps to empower them. Indian Council of Agriculture and Research (ICAR) has established a huge network of over 731 Krishi Vigyan Kendras (KVKs) all over the country with a common objective to conduct technology assessment, refinement, knowledge dissemination and provide critical input support to the farming fraternity in a multi-disciplinary way. These KVKs are working towards upgrading the socio-economic status of the farmwomen in terms of capacity building through training, demonstrations, conducting awareness campaigns and through different sensitization programs. Some of the trainings are specially designed for the women so that they can earn and sustain their family through remunerative enterprises like vermi-compost, nutrition gardening, tailoring, fruit and vegetable processing and value addition, handicraft, bag making, candle making, herbal colour making, artificial jewelry making, donapattal making etc. Adequate and continuous attention to rural women and their participation has a significant impact on economic prosperity at the family and eventually at local, regional, and national levels (Naseri et al., 2020).

Krishi Vigyan Kendra Dehradun working under the administrative control of G B pant University of Ag & Tech, Pantnagar is sincerely accomplishing its objectives including imparting trainings to the farming community. A study was conducted to know the "*Impact of Training Programs Conducted by KVK on Rural Women*" for the upliftment of farmwomen with the help of scanning reports and using well-structured questionnaire to find out relevant and realistic data. The present applied study is a quantitative, non-experimental based on survey, and the statistical population includes rural women and girls who took part in the training programs conducted by KVK during the study period i.e. 2017-2021. The research tool used for conducting the study was a researcher-designed questionnaire.

Training Programs given by KVK to Farmers and Farm Women

KVK generally deal with training programs related to needy areas including both men and women. The types of courses generally include packaging and processing of various crops, (field crops, vegetables crops, oil seed crops, etc.), plant protection techniques, farm planning practices, care and feeding of animals, poultry keeping, marketing of agricultural product, nutrition gardening, health and hygiene, value addition, rural handicrafts, entrepreneurship development among the farm women etc. To impart training in an efficient way, KVKs are equipped with subject matter experts and specialized training groups/individuals. Some trained and established farmers are also invited as practical teachers and to share their experiences during the trainings. The help of agricultural universities, reputed NGOs, various agro-based industries and other state government agencies is also appreciated to link the farming communities with these departments. The programs are problem oriented and field oriented with follow-up measures. "Learning by doing" and "Learning by seeing" are KVK's motto. It gives direct bearing on our agricultural productivity (S.K. Acharya, et al 2019)

Rural women can play a significant role by their effectual and competent involvement in entrepreneurial activities. So, the training program must promote critical analysis for women and encourage them to think independently and challenge gender inequality (Acharya, Sankar & Chatterjee, Riti, 2020).Training programs further intend to cover the have not section of the society including tribes and hill farmers on priority basis. Influential or the early adopters of the technologies play the role of motivators and are encouraged to facilitate new entrepreneurs during the implementation of any program. All the courses and trainings in KVK are tailored according to the needs of the areas served.

Brief Profile of the Farm Women

Majority of the farm women i.e. 46% belonged to the middle age group (35-50 years) followed by young i.e.: 43% and old age groups, i.e: 11% .The data pertaining to education revealed that almost half (48%) of the respondents were having education up to higher secondary followed by up to 12th standard (36%) and higher education (16%). The data gathered reveals that 48% respondent had income in the range of Rs.5,000-10,000 rupees per month whereas 33% respondents' income ranges between Rs.11,000-15,000 per month while 19% of respondents belong to income group of Rs.16,000-20,000 Rupees per month. A good number of respondent (69%) had a medium level of mass media exposure whereas 19% have high level of exposure, followed by low level of media exposure (12%). The respondents were also asked whether they have any awareness about the developmental programs, and it was observed that 67% of the respondents had no exposure about any of the developmental programs while 33% had the exposure for one or two developmental programs.

Table 1: Profile of the participants

S.No.	Aspects	Categories	Frequency	Percentage
1	Age	Young (18-35)	43	43
		Middle (35-50)	46	46
		Old (50 and above)	11	11
2	Education	Up to Higher Secondary	48	48
		Up to 12th	36	36
		Graduation and above	16	16
3	Economic Status	Low (5,000-10,000)	48	48
		Medium (11,000-15,000)	33	33
		High (16,000-20,000)	19	19
4	Mass Media Exposure	Low	12	12
		Medium	69	69
		High	19	19
5	Any exposure towards developmental programs	Yes	67	67
		No	33	33

Year wise Distribution of Trainees

To find out the participation ratio of women and their counterpart men in the training conducted by KVK during the last four years, the data gathered from the previous reports were analyzed. It is very clear from the Graph 1, that the total number of trainings conducted during the year 2017-18 were highest resulting in maximum participation in both the categories i.e. male and female, followed by the numbers during the year 2020-21. Distribution ratio of men and women depicts that it was maximum (63.09%) in the year 2019-20 while was minimum (38.77%) in the year 2018-19. A good number of rural women are coming out of their household chores for participation.

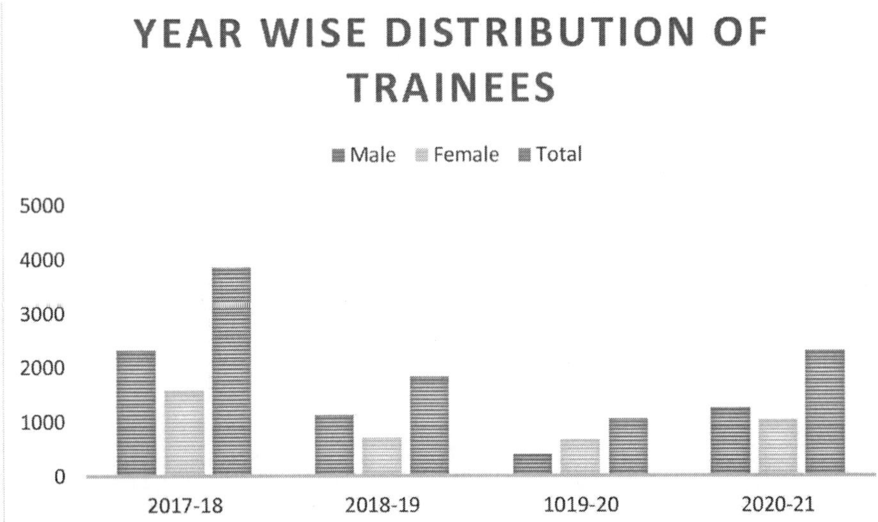

Figure 1: Year wise distribution of the participants

Table 2: Year wise Distribution of Trainees

S.No.	Year	Male	Female	Total
1	2017-18	2311 (59.50)	1573 (40.49)	3884
2	2018-19	1113 (61.22)	705 (38.77)	1818
3	2019-20	389 (36.90)	665 (63.09)	1054
4	2020-21	1242(54.64)	1031(45.35)	2273

Role of KVK in Gain in Knowledge

Krishi Vigyan Kendra is conducting various programs, viz: training, demonstration, goshthies, visits, and diagnostic services for its clientele which lead to the upgradation of subject knowledge. To gain knowledge means to gain an understanding and awareness regarding a particular subject. The role of KVKs in women empowerment through enhancement in knowledge transfer was studied. Tables 2 depicts the status of knowledge gain by the respondents through different programs. Goshthies and trainings were found as the most popular method of knowledge gain where 89% and 88% respectively stated that they gained knowledge followed by demonstration (74%) diagnostic services (71%) and field visits (67%).

This concludes that goshthies and trainings were the programs where maximum extent of knowledge gain was received by the respondents. Demonstration and diagnostic services were also considered as good ways of knowledge gain where almost equal number of respondents revealed them as a good way of gain in knowledge regarding agricultural and its related aspects. When the respondents were asked about the extent of knowledge gain, it was observed that highest (58%) number of respondents agreed upon that they were fully skilled after receiving training followed by demonstration (29%) and field visit (22%). The higher extent of gain in knowledge was observed in the programs where one to one interaction was there.

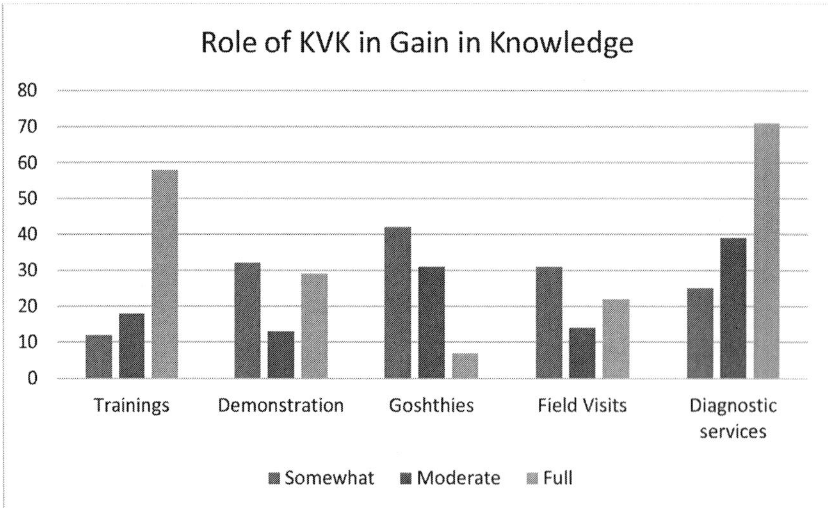

Figure 2: Role of KVK in Gain in Knowledge

Table 2: Role of KVK in Gain in Knowledge

Program	No. of Respondents	Whether KVK helped in Knowledge Gain		Up to what extent		
		Yes	No	Somewhat	Moderate	Full
Trainings	100	88	12	12	18	58
Demonstration	100	74	36	32	13	29
Goshthies	100	89	11	42	31	7
Field Visits	100	67	33	31	14	22
Diagnostic services	100	71	39	25	39	71

Role of KVK in Skill Development

Skill is the ability of a person to perform the task with the help of optimum utilization of resources. It helps in improving the ability of a person so that he/she could utilize skills in a particular field and ultimately get success in his/her professional work. Hence an attempt was made to know the role of KVK in skill development of the women through various programs. Extent of skill gain was studied separately in seven specific training programs which were repeatedly conducted during last four years. Highest number of respondents agreed that KVK did help in gaining skill on cow dung diya making (86%) followed by packaging, herbal colour making, and candle making (81%, 78% and 74% simultaneously). Extent of skill gain was also studied separately in each category and the same pattern was observed i.e. highest was in cow dung diya making (53%) followed by packaging, herbal colour making, and candle making (46%, 38% and 32% simultaneously). This reflects that training on cow dung diya making, packaging and herbal colour making came out as most interesting areas where the number of participation as well as skill gain both were seen on the higher side. A good number of trainees accepted it as income generating activities as well and it is on very high demand within and outside the district too.

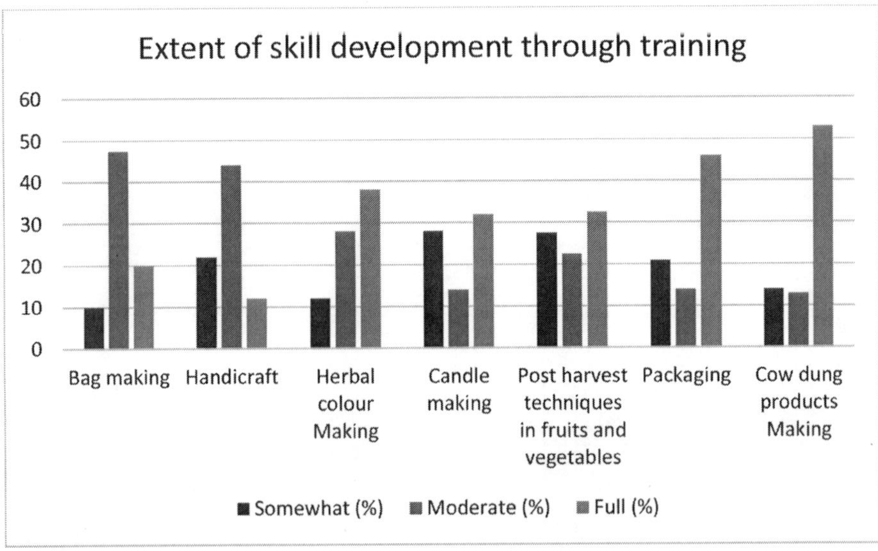

Figure 3: Extent of skill development through training

Table 3: Extent of skill development through trainings

Training Aspect	No. of Respondents	Did KVK help in Skill Gain		Up to what extent		
		Yes (%)	No (%)	Somewhat (%)	Moderate (%)	Full (%)
Bag making	80	62(77.50)	28(35.00)	8(10.00)	38(47.50)	16(20.00)
Handicraft	100	58(58.00)	42(42.00)	22(22.00)	44(44.00)	12(12.00)
Herbal colour Making	100	78(78.00)	32(32.00)	12(12.00)	28(28.00)	38(38.00)
Candle making	100	74(74.00)	36(36.00)	28(28.00)	14(14.00)	32(32.00)
Post harvest techniques in fruits and vegetables	80	66(82.50)	14(17.50)	22(27.50)	18(22.50)	26(32.50)
Packaging	100	81(81.00)	19(19.00)	21(21.00)	14(14.00)	46(46.00)
Cow dung products Making	100	86(86.00)	24(24.00)	14914.00)	13(13.00)	53(53.00)

Conclusion

Trainings are meaningful only if it changes the attitude, motivates, and reflects in actual work situation and it should be continued as per needs. Rural women can play a significant role due to their effectual and competent involvement in entrepreneurial activities. So, the training program must promote critical analysis in women and encourage them to think independently and challenge gender inequality (Acharya, Sankar & Chatterjee, Riti. (2020). The data analyzed to study the profile of respondents reveals that majority of the farm women belonged to the middle age group (35-50 years), literate up to higher secondary, had income in the range of Rs.5,000-10,000 rupees per month, having medium level of mass media exposure and also no exposure towards any of the developmental programs run by government or other departments.

The scanning of report reveals that participation of trainees during the year 2017-18 was highest resulting maximum participation in both the categories i.e.: male and female, followed by during the year 2020-21. The study also explains that a good number of rural women are coming out of their households for participation in the trainings whether it is on agricultural aspects or household aspects.Krishi Vigyan Kendra is continually performing several farm support activities like providing technology dissemination to farmers, training and awareness, farm advisory service, conducting demonstrations to establish production potential of technologies on the farmers' fields. Gain in knowledge means understanding and awareness regarding the subject specific. It reflects performance level of a person may be enhanced by acquiring knowledge through training, exposure to different activities performed by the KVK for enhanced learning. Goshthies and trainings were found as the most popular method of knowledge gain followed by demonstrations, diagnostic services, and field visits. While studying the extent of knowledge gain, it was observed that majority of the beneficiaries were fully skilled after receiving training followed by demonstrations. Extent of knowledge gain was higher where one to one interaction was there.

Extent of skill gain through training, was also studied and explains that cow dung diya making, packaging, herbal colour making, and candle making were the aspects which were on highest skill gained aspects and fullest of skill gain categories. This reflects that training on cow dung diya making, packaging, and herbal colour making came out as most interesting areas where the number of participation as well as skill gain both were seen on higher side. A good number of trainees accepted it as income generating activities as well and it is on very high demand within and outside the district too. The findings reveal that KVK is working in the right direction by providing more options for income generating activities in rural areas by imparting vocational/skill training to women folk (Anithakumari et al, 2021). KVK has a positive impact on the quality of life of farming community in terms of income, yield, productivity, and above all their capacity for optimal utilization of the available resources. Demonstration and trainings were the activities where maximum extent of knowledge gain was received by the respondents, primarily due to one to one interaction of the trainer with farm women.

References

Anithakumari. P1, Mahima Mohan2 and Muralidharan. K3, (2021) Impact of Training Programs for Women Farmer Groups under the Farmer FIRST Program Indian Res. J. Ext. Edu. 21 (2&3), April & July, 2021

Acharya, Sankar & Chatterjee, Riti, (2020). Krishi Vigyan Kendra (KVK) and its Role in the Upliftment of the Farm Women in Indian Agriculture Introduction. 10.21088/ijab.2454.7964.5219.3Naseri S, Amin HC, Poursaeed A, Arayesh MB (2020) Investigating the effective factors on employment-oriented empowerment of rural female heads of households in ilam, Irán. NexoRevistaCientífica 33(02):539–546

FAO (2011). The role of women in Agriculture, ESA. Working Paper No. 11-02

Sankar Kumar Acharya, Riti Chatterjee, (2019) , Krishi Vigyan Kendra (KVK) and its Role in the Upliftment of the Farm Women in Indian Agriculture.78 Indian Journal of Agriculture Business / Volume 5 Number 2 / July – December 2019

7

Status of Successful Self-Help Group In Udham Singh Nagar District, Uttarakhand

Priya Jadon, Bhawna Sharma and Sangita Saini

Department of Home Science, Faculty of Arts, Dayal Bagh Educational Institute, Agra, Uttar Pradesh

Abstract

Self-help groups are an initiative of the central government for women who belong to the same background and are willing to improve their lifestyle. This study undertaken so far has assessed the status of SHGs in Udham Singh Nagar District in Uttarakhand, focusing on the working procedure of SHGs and the involvement of NRLM in the development of SHGs. This study tried to highlight the growth of women after joining the group and the perception of SHG women towards the working process of the National Rural Livelihood Mission in favour of SHGs. The study focuses on ten SHGs from two blocks, Gadarpur and Khatima. These blocks belong to Udham Singh Nagar. Each SHG has 10 members. Thus, the study group included 100 members. The result of the study revealed that women had a positive attitude toward SHGs. Lifestyle of women also changed after joining the SHG. Women are now able to provide a good education to children, and decision-making capacity of women is also enhanced. Presently, women of SHGs have a good position in the family because of their financial dependency.

Keywords: *Self Help Group, Capacity, Women and Development*

Introduction

The key to the advancement of women belong to poor background is self-help groups. Members of society with greater social influence are women. The World Bank estimates that women make up 48.0% of the population in India. Women in India are impacted by a number of economic and social variables. They are essential to any economy. Men and women must be viewed as equal participants in development if a nation is to achieve overall success and positive growth. The nation's economic growth and social advancement depend on the independence of women. Through collective participation for growth, the empowerment of women through SHGs will benefit not just the individual women but also the family and network as a whole.

Self- Help Groups

Self-Help Groups (SHGs) have developed in order to assist underprivileged women in obtaining services and resources like loans. In 1985, the idea of SHG was first introduced in India. Self-help Groups (SHGs) are informal associations of people who come together to find ways to improve their living conditions. SHGs help in improve the lives of the poor and marginalised segments of society. They also foster financial inclusion. The self-help groups play an important role in ruler women life.India's Self Help Group movement has emerged as the world's largest network of community-based organizations.Under this programme, banks were allowed to open savings accounts for Self-Help Groups (SHGs). This was started in the early nineties. Empowering women is crucial to the prosperity and development of the country.

Review of Literature

Singh. P et. al (2017), observedthat of all the groups examined, 24.25 percent were active, 13.50 percent were dormant and 62.00 %were stopped. Individually, members of active organisations have engaged in income-generating activities. The major occupation of SHG members was dairy and the average yearly income per member was Rs. 22,130. The study found that rural development organisations place more of an emphasis on SHG establishment than on monitoring, following up and upgrading knowledge and skills, which led to SHG inactivity or discontinuance. To make SHGs more sustainable, more attention has to be paid to monitoring and follow-up, microfinance, knowledge transfer and capacity building trainings and creating market links.

Gupta. M (2021) Stated that the purpose of study is to determine how non-government organisations (NGOs) might support women's empowerment by providing immediate livelihood options and play a significant role in educating and enabling women to secure a stable income. Through business development and entrepreneurship, which are possible with the right skills and information, women are empowered. Women become more confident as a result of motivation and strong leadership and they strive to accomplish their objectives. They can achieve empowerment by proper knowledge of government policies, networking with clients, affiliations with NGOs and self-help groups (SHGs).

Chunera. A and Bhardwaj. N (2019) through this result suggested that improving the capacity of the SHGs will take a lot of work and assistance. To assist people, improve their competence and mindsets, ongoing trainings on capability building, skill development, personality development, leadership development, accomplishment motivation training, technical training orientation and entrepreneurial skills development are required. For the growth of women's SHGs to be successful both public and private organisations must work in concert with one another.

Singh. P et.al (2017) observed that the government, NGOs and other organisations interested in rural development employ the self-help group (SHG) strategy to undertake development programmes at the community level. SHGs provide regular savings, work prospects, credit access, involvement in local government and a shift in how families make decisions to the group members. The study demonstrates that just 24% of the 400 SHGs examined in six districts of Uttarakhand continued to function after eight years of creation. Overall, it became clear that women's SHGs in Uttarakhand were not particularly successful in raising their standard of living or creating any new businesses.

Objectives

1. To analysis the status of Self -help groups in Uddham Singh Nagar district.
2. To find out changes in the status of living of women after joining the SHG.
3. To find out issues faced by women of SHGs.

Methodology

The present study was based on both primary and secondary data. The study is both qualitative and quantitative. The Secondary data was collected from research papers, government websites and newspapers. Primary data was collected through personal interviews with group members. Khatima and Gadarpur blocks in Udham Singh Nagar Districts were selected for the study.

Udham Singh Nagar is a District of Uttarakhand state in northern India. Rudrapur is the district headquartering. This district consists of nine Tehsils named Bajpur, Gadarpur, Jaspur, Kashipur, Kichha, Khatima, Sitarganj, Nanak Matta. The district is located in the Terai region and is part of Kumaon Division. According to census of 2011 sex ratio in Rudrapur is 902 per 1000 males. Self-help group programme started in Udham Singh Nagar in 2000. The number of women SHGs registered in Khatima and Gadarpur under National rural livelihood mission (NRLM) were 1942 in July 2022. A micro level study of women SHGs in Udham Singh Nagar district is undertaken to find out the status and development of successful self-help group in Udham Singh Nagar. For this purpose, primary data was collected from 100 women who belong to 10 SHGs.

Hypothesis Ho

There is no significance impact of SHGs on the empowerment of woman after joining SHGs among various self- help group members.

Results and Discussions

Table 1: Age of the self-help group members

S.no	Age	No of members	Percentage
1	18-30	48	48
2	31- 40	34	34
3	41- 50	18	18
	Total	100	100

Source: Field survey, 2022

The table 1 inferred that 48% respondents are from 18-30 age group, 34% respondents from 31-40 age group and 18 % from 41- 50 age group.

Table 2: Type of Family of the Self-help Groups

S.no	Type of family	No of members	Percentage
1.	Joint Family	52	52
2.	Nuclear family	48	48
Total		100	100

Source: Feild survey, 2022

From the table no 2, it is inferred that 52 % respondents are from joint family and 48% respondents are from nuclear family.

Table 3: Literacy level of SHG members

S.no	Literacy level	No of Members	Percentage
1.	Illiterate	16	16
2.	Literate	30	30
3.	Primary education	39	39
4.	Secondary education	15	15
	Total	100	100

Source: Field survey, 2022

Table 3, show that 16 % individuals are Illiterate and does not know how to do signature,30% individuals are literate and able to do sign, 39% individuals had Primary education, 15% members are educated up to secondary level.

Table 4: Occupation of SHG members

S.no	Occupation Description	No of Members	Percentage
1.	Pickle making	30	30
2.	Agar Batti making	20	20
3.	Spices making	30	30
4.	Basket making	20	20
	Total	100	100

Source: Field survey, 2022

Table 4 show that 60% individuals involved in pickle making and spices making. 40% involved in agar Batti making and basket making.

Table 5: Reasons of joining SHG

S.no	Reason	No of members	Percentage
1.	To start business	15	15
2.	To promote savings	18	18
3.	To get credit	38	38
4.	To meet household expenses	25	25
5.	To strengthen the leadership quality	4	4
6.	For other reasons	0	0
	Total	100	100

Source: Feild survey, 2022

The rationale for joining the self-help group as an individual, the table 5 shows that 35 % of the respondent's assessment with respect to for to get credit, the second significance for to meet household expenses 25%, and staying to begin business, to promote saving, and others. Surmising Most of the respondents are joining for to get credit.

Table 6: Issues faced by SHGs members

S.no	Issues	No of member	Percentage
1.	Marketing	60	60
2.	Low demand of local product	20	20
3.	Got lower price than other brands	20	20
	Total	100	100

Source: Field survey, 2022

From the table no 6 it is inferred that 60% individuals are facing issues related to unavailability of market. According to 40 % of individuals there was low demand of product and got very lower price than another brands.

Table 7: Support of SHGs from NRLM

S.no	Type of Support	No of member	Percentage
1.	Training	100	100
2.	Centre provided		
3.	Loan		

Source: Field survey, 2022

Table no 7 show that all the respondents got training, loan and centre for work from NRLM.

Table 8: Earning of women per month

S.no	earning	No of members	percentage
1	1000-5000	62	62
2	6000 - 10000	38	38
3	11000-20000	0	0
	Total	100	100

Table no 8 shows that 625 of individuals earn 1000-5000 rupees per month and 38 % of individuals earn 6000 – 10000 rupees per month.

Conclusion

The study's findings supported the favourable effects of self-help groups on empowering women in terms of increased social awareness and engagement, savings practises, income levels, independence in the workforce, asset creation, enhanced decision-making abilities, and better nutrition at the household level due to money. Self-help organisations have made it possible for underprivileged women to access a range of financial services and goods. The idea of self-help groups helped many women gain social acceptance. But Giving education, training, and awareness-building to the group's members needs to be prioritised more. Government officials and NGOs need to be much more enthusiastic about convincing and identifying poor people and SHGs, especially in more vulnerable segments of the general public.

Limitations

- The present study is confined only to the women SHGs of Udham Singh Nagar district, Uttarakhand.
- Only those SHGs were selected which was registered in NRLM.

References

Chunera. A and Bharadwaj. N (2019). Capability Assessment of Women Self Help Groups of Uttarakhand, International Journal of Current Microbiology and Applied Sciences, ISSN: 2319-7706 Volume 8, Number 09.

Gupta. M (2021). Role of NGOs in women empowerment: case studies from Uttarakhand, India, Journal of Enterprising Communities People and Places in the Global Economy, DOI: 10.1108/JEC-04-2020-0066

https://www.census2011.co.in/census/city/26rudrapur.html#:~:text=The%20sex%20ratio%20 of%20 Rudrapur, is%20912 %20per%201000%20boyshttps://data.worldbank.org

Ramakrishna. H and Khaja. MJ (2013) Women empowerment through micro finance: An empirical study of women Self Help Groups in Bellary district, Karnataka, Acme International Journal of Multidisciplinary Research,11(1).

Singh. P et.al (2017). Status of Self-Help Groups in Uttarakhand, International Journal of Basic and Applied Agricultural Research, Vol. 15 (3).

8

Nutrition Sensitive Interventions for Combating Malnutrition in High Hills of Uttarakhand: Gender Centric Approach

***Renu Jethi*[1]*, Pankaj Nautiyal*[2] *and Pratibha Joshi*[3]**

[1]*ICAR-DCFR, Bhimtal, Uttarakhand*
[2]*SMS, KVK-Uttarkashi, Uttarakhand*
[3]*ICAR-IARI, New Delhi*

In Uttarakhand, majority of the population lives in rural areas and primarily depends on agriculture for their livelihood. In Uttarakhand, only 14.02% area is under cultivation. Out of total 7.66 lakh hectare cultivated area in the state, 4.21 lakh hectare area is rainfed i.e. more than 50 % of the cultivable area is rainfed. The average land holding is 0.68 ha that too is divided into many patches. Rainfed farming in Uttarakhand gets affected by frequent drought like situations and excessive rainfall. Food and nutritional security of people living in hill region depends more on favorable climatic conditions as agriculture and livestock farming are the main occupations. Prevailing low level of soil fertility and increased incidences of pest and disease occurrences has adversely affected the crop production and productivity in hills. Low agricultural productivity has emerged out as an important push factor for out migration of people from hill areas. Traditional hill agriculture is practiced on small and fragmented terraced fields with lack of irrigation facilities, which are uneconomical. Women constitute the main workforce in agriculture as large section of able bodied men have migrated from hills in search of livelihood. They also take care of the cattle, collect fuel wood and fodder from forests, often situated at considerable distance from the villages involving four to five hours of walking both ways, and do all household chores. Their life is an unending drudgery of hard work.

As a large population in the hill region of Uttarakhand depends on agriculture, the problem of malnutrition can be better addressed through nutrition sensitive agricultural interventions specifically designed for women in hills. Local nutritional needs can be met out locally by reviving traditional practices along with amalgamation of science and technology knowledge. Focusing on nutrition helps to build the resilience of rural households. Nutrition sensitive approach was promoted at local level with active participation of women who are mainly responsible for food security of the households.

Methodology

Two sites were selected in high hills of Pithoragarh and Uttarkashi for studying dietary pattern and nutritional status of women. A transect walk was planned along with local women to explore food crops grown in the area, socio-economic conditions of the people living in the area by observing, asking, listening and developing and map. Anthropometric measurements of some selected women based of socio-economic condition and their willingness were recorded that includes weight, height, skin fold measurements, blood pressure and pulse rate. The extent of malnutrition in women was assessed by computing the BMI (Body Mass Index). The questionnaire recommended by FAO for MDD-W was adapted with some modification like replacing examples of food items by a list of common food in the local context. Structured questionnaire was used to collect baseline data during initial stage of project. Concept of nutri-garden was introduced in high hill areas to encourage women to cultivate healthy food crops in their backyards. A well planned nutri-garden ensures regular supply of fresh vegetables rich in nutrients. Anthropometric measurements, dietary diversity among women and socio-economic condition laid basis for selection of beneficiaries. There are variety of vegetables that can be grown in a plot of 100 m^2 to 200 m^2 in hilly areas vegetables are rich source of nutritional bio-active compounds. Various postural ergonomics techniques were used for quantification of musculoskeletal issues with traditional method and postural evaluation of farm women while performing various agricultural activities.

Objectives

- To assess nutritional and health status of hill farmwomen in selected production systems
- To design and demonstrate and concept of Nutri-gardens with active participation of farm women in hills;

Approach

Nutrition-sensitive agriculture interventions were implemented in three main areas. Enhancing food productivity for making enough food available and accessible at local level, food diversity was enhanced through production of diverse crops, making food itself more nutritious were the main areas in which women farmers of high hills of Uttarakhand were actively involved.

Results

Dietary Diversity Score (DDS)

In the present study, sixty-three selected women were encouraged to produce seasonal vegetables in nutri-gardens along with mushroom production, bee-keeping and microgreens. Regular trainings and exposure visits were conducted on the improved package of practices for vegetable production along with field monitoring for controlling insect-pest infestations. Women were linked to Need-based SMS service for receiving messages related to insect-pest infestation, improved agronomic practices, improved crop varieties and health and nutrition issues.

Minimum Dietary Diversity for Women of Reproductive age (MDD-W) was calculated before the implementation of the interventions and after interventions with the same set of respondents. Women had very low dietary diversity score (3.9) with standard deviation of 0.9 as 69.8 per cent were found to be consuming less than five food group in their diet. Gitagia*et al* 2019 also reported DDS of 3.78±0.99 among women of reproductive age. After the implementation of nutrition sensitive agricultural intervention, MDD-W score reached to 5.5 with standard deviation 0.8. More than 90 per cent of women achieved minimum dietary diversity and they are more likely to have higher (more adequate) micro-nutrient intake (Table 3). Similar results were reported by Shashikantha*et al* 2016 and Singh *et al* 2020. To address the issues of nutritional deficiencies among women, Helen Keller International implemented homestead food production programs in various developing nations through various interventions like home gardening, animal husbandry along with nutrition education which resulted in significant improvement in dietary diversification and food security (Talukder *et al* 2010).

Table 1: Minimum Dietary Diversity Score before and after nutrition-sensitive agriculture interventions

Range of Minimum Dietary Diversity Score	*Before intervention*	*After intervention*	*Difference*
Less than 5	69.8	7.9	Z value = 9.3*
% and above 5	30.2	92.1	P value = <0.0001
Mean and SD	3.9 ± 0.9	5.5 ± 0.8	

*significant at 0.05 probability level

It is also very important to present the average number of fruit/vegetable groups consumed out of the four groups ("Dark green leafy vegetables", "other vitamin A-rich fruits and vegetables", "other fruits" and "other vegetables"). Before intervention, the consumption of number of fruit/vegetable groups was 1.5 with standard deviation of 0.87 whereas after intervention it got increased to 2.5 with a standard deviation of 0.76 among women of high hills. There was found to be a significant difference at 0.05 level of probability in consumption of number of fruit/vegetable groups after the implementation of interventions.

Field Demonstration of High yielding varieties, crop diversification and Value-addition of Products

In order to address the issue of low productivity, high yielding varieties of wheat (VL *Gehun* 829, VL *Gehun* 953), finger millet (VL *Mandua* 352) and lentil (VL *Masoor* 126) were demonstrated in the project area along with the recommended package of practices. There was 50-60 percent increase in yield of lentil as a result of improved variety and improved method of sowing. Similarly there was 30-32 percent increase in yield as a result of improved variety and improved method of sowing.

Gender mainstreaming and economic empowerment

For economic sufficiency, farm women were provided with seedlings of vegetables initially. Gradually with the progress of project, nurseries were raised by farm women under low-cost poly-tunnels. More than 8 nutritionally rich vegetables were grown per seasons in nutri-gardens. As a result, women were able to save Rs 1500-3300 per season on buying vegetables and surplus were sold to the market for additional income. Women were also trained in other entrepreneurial activities as bee keeping, microgreen production etc. Capacity development of women were also done for scientific bee-keeping practices and are able to fetch Rs 700 for 1 kg of honey in local market. Groups of women were also motivated towards production and consumption of micro-green as they are the most affordable source of micronutrients even for land less poor farm families. Microgreens have potential to overcome mineral malnutrition because of the presence of antioxidant, vitamins, minerals and several other bioactive compounds. They are easy to grow in home as they have potential to provide a range of phytonutrients per gram plant biomass as compared to mature vegetables.

Conclusion

Agriculture based nutrition interventions had positive impact on enhancing dietary diversity and food security by increasing round the year availability of diverse food. Food production at household level has the potential to increase micro-nutrient availability and improve the nutritional status of women in hills. The issue of increasing malnutrition among women is due to a lack of awareness about the right kind of diet required for the proper growth and functioning of the body. Creating awareness about nutrition and health is an important task especially in the rural areas of hills. The climatic conditions of hill region is highly suitable for the production of fruits and vegetables. Micronutrient malnutrition can be overcome by including a variety of fruits and vegetables in the daily diet. Dietary diversity is one of the important dietary strategies that can help in improving both situations of undernourishment and overweight. More emphasis should be given on farming of diverse crops through family farming, nutri-gardens and homestead food production of vegetables, fruits, mushroom local levels. Nutrition-sensitive agriculture interventions are advantageous ways to improve nutrition levels in women with minimum investment.

9

Agricultural Information in Relation with the Tribal Community

Sayan Sagnika Mohapatra and Bishnupriya Mishra

Department of Extension Education, Odisha University of Agriculture and Technology, Bhubaneswar

Abstract

This study was conducted in Keonjhar district of Odisha state. Agri-information management behaviour is to be focussed on the flow of information regarding seeking information on various practice of crops and seeking information on various agri-allied activities . The flow of information is based on the various aspect in which tribal people are most probably involved in transferring information through various methods. Agricultural information interacts with and influences agricultural productivity in a variety of ways. It can help in making decision regarding land, labour, livestock, capital and management. Agricultural productivity can arguably be improved by relevant, reliable and useful information and knowledge. Hence the creation of agricultural information by extension services, research, education programmes and others are managed by agricultural organizations that create information systems to disseminate information to farmers so that farmers can make better decisions in order to take advantage of market opportunities and manage continuous changes in their production systems. All these information seeking behaviours with respect to ICT, the farmers help them to manage in agricultural and other allied activities.

Keywords: *ICT, Agricultural information, Research, Education*

Introduction

Agriculture denotes as the key development in the rise of sedentary human civilization, where by farming of domesticated species creates food surpluses that enables people to survive and showing them the way of earning. Thus our country is enhancing agro-based industries. Industrial agriculture based on large scale monoculture, which is dominated by agricultural output. Hence it is the dynamically economic improvement of our country with the help of agriculture basing on modern information technology. Information technology means the latest version of information facilities like computer, mass media and Internet. Information technology (IT) or Information communication technology (ICT) is a broad subject concerned with technology and other aspects of managing and processing information.

Information technology contributes much to agriculture and rural development. It facilitates rural activities and provides more comfortable and also saves the rural life with equivalent services to those in urban areas such as provision of distance education, tele-medicine, remote public services, remote entertainment etc. IT initiates new agricultural and rural business such as e-commerce, real estate business for satellite offices, rural tourism, and virtual corporation of small-scale farms. It also supports policy-making and evaluation on optimal farm production, disaster management, agro-environmental resource management etc using tools such as geographic information systems (GIS). It also helps in improving farm management and farming technologies by efficient farm management, risk effective information or knowledge transfer etc realizing competitive and sustainable farming with safe products.

Objectives

1. To study the existing agri-allied activities undertaken by the farmers in relation to flow of information
2. To study the source of flow of information for the identified agri-allied activities
3. To analyse the relationship between information seeking sources with some socio-economic and psychological variables

Research Methodology

The study was conducted during 2018 with use of Ex-post facto research design. A research method is a systematic plan for conducting research. Hence research methodology is described as a systematic, theoretical analysis of the methods applied to a field of study. It comprises the theoretical analysis of the body of methods and principles associated with a branch of knowledge. It especially encompasses concepts such as paradigm, theoretical model, phases and quantitative or qualitative techniques. Research methodology is a vital pre-requisite procedure of any research study since it bears the relevant and valid research findings. The research study was followed by both purposive and non probability random sampling method for selection of district, block, grampachyat and respondents. The study was undertaken in two blocks i.e.Ghatagaon and Harichandanpur. From these two blocks it had been selected four grampanchayats and four villages .

Results and Discussion

Source of flow of information for the identified agri-allied activities

Receiving Information	Rice				Blackgram				Greengram			
	Yes		No		Yes		No		Yes		No	
	f	%	f	%	f	%	f	%	f	%	f	%
Cultivation practices	40	33.33	80	66.66	30	25	90	75	35	29.16	85	70.83
Harvesting	31	25.83	89	74.16	30	25	90	75	29	24.16	91	75.83
Storage	35	29.16	85	70.83	38	31.66	82	68.33	30	25	90	75
Processing	42	35	78	65	39	32.5	81	67.5	38	31.66	82	68.33
Packaging	41	34.16	79	65.83	38	31.66	82	68.33	37	30.83	83	69.16
Marketing	55	45.83	65	54.16	55	45.83	65	54.16	53	44.16	67	55.83
Total	244	174.15	476	396.64	230	191.65	490	408.32	222	184.97	498	414.98

It revealed from above table that the respondents were seeking less information on crop practices. The main crop of this area is rice. So it was observed that some respondents were receiving information about cultivation practices 33.33% followed by greengram cultivation practices 29.16% and black gram cultivation practices 25%. Some of the respondents were receiving information on marketing i.e. 45.83% in rice, 45.83% also in blackgram and 44.16% in greengram.

Seeking information on major vegetables practised by the respondents

Receiving information	**Tomato**				**Watermelon**			
	Yes		**No**		**Yes**		**No**	
	f	**%**	**f**	**%**	**F**	**%**	**f**	**%**
Cultivation practices	60	50	60	50	65	54.16	55	45.83
Harvesting	30	25	90	75	45	37.5	75	62.5
Storage	20	16.66	100	83.33	39	32.5	81	67.5
Sorting and grading	25	20.83	95	79.16	35	29.16	85	70.83
Marketing	45	37.5	75	62.5	50	41.66	70	58.33
Total	180	149.99	420	349.99	234	194.98	366	304.99

It revealed from table 2.2.2 that most of the respondents were seeking information on cultivation practices i.e.54.16% in watwrmelon cultivation followed by tomato cultivation i.e. 50%. The information about marketing of these two vegetables were relevant to the respondents i.e.37.5% in tomato and 41.66% in watermelon.

Seeking information on allied activities

Receiving information	**Forest products**				**Receiving information**	**Goatery**			
	Yes		**No**			**Yes**		**No**	
	f	**%**	**f**	**%**		**F**	**%**	**f**	**%**
Collection	33	27.5	87	72.5	Maintainance of shed	21	17.5	99	82.5
Manufacturing	20	16.67	100	83.33	Protection from disease	29	24.16	91	75.83
Marketing	40	33.33	80	66.67	Marketing	35	29.17	85	70.83
Preparing enterprise	26	21.66	94	78.33					
Total	119	99.16	361	300.83	Total	85	70.83	275	229.16

It revealed from above table that most of the respondents are receiving information about marketing i.e. 70.83% in goatery followed by 33.33% in forest products. The respondents have not sought the information on other practices of allied activities. They can't get good opportunity to enhance these allied activities.

Receiving Information on Agri-allied Activities

The graph revealed that the respondents were receiving information on crop practices followed by vegetables and allied activities. The respondents sought less information for their agri-allied activities compared to the individuals of other areas. The tribals had low literacy level, they can't learn new method of practices. So they were facing many problems in cultivation practices as well as marketing . So that the tribal farmers can't get remunerative price of the product and they also much time to do agri-allied practices.

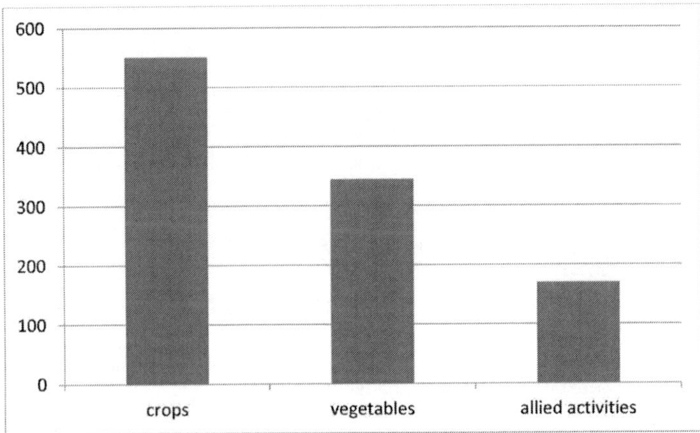

Conclusion

The study observed the relationship of source of information with some socio-economic and psychological variables. Gender,education, annual income, family size, are positively correlated with source of information. Age is negatively correlated with source of information. Among the study it was observed that majority of the respondents were facing social constraints followed by technical constraints and financial constraints.

The study with limited resources, mobility, specific time period due to academic and financial compulsion, definitely unfolded the importance areas of Agri-information management behaviour of tribal farmers in Keonjhar district of Odisha. The study was conducted in particular area and left untouched vast areas which can be identified and evaluated in other studies.

References

Aghazia.M.D(2008) "Trainning needs of onion cultivation in Afghanistan country,"M.Sc(Ag) Thesis,ANGRAU.

Amtul warish(2004) " Utilisation of Information sources by farmers in Arid Rajasthan,"Indian journal of extension education,Vol40(1&2)pp75-77.

Ashok kumar,Jha.S.K and Kumar Aurovind(2008) "Critical analysis of Information Sources and Channels preferred by rapessed-mustard farmers."

Bhagat.G.R,Nain.M.S and N.Rakesh(2004) "Information Sources for Agricultural Technology," Indian Journal Of Extension Education,Vol 40(1&2)pp111-112.

Bose.S and Chowdhury.S (2015) "Farm women access to Farm Information-A study on Dolcpursubdivision,"Journal of Extension Education,VolXX(2)pp74-78.

Chidananda. M(2008) "A study on entrepreneurial behaviour of dryland farmers in Karnataka state,"M.Sc(Ag)Thesis, ANGRAU,Hyderabad.

Daudu.S, Chado.S and Igbashal.A(2009) Agricultural Information Sources Utilised by Farmers in Benue state,Nigeria,Vol5(1)pp39-48,ISSN:0794-5213.

Dwivedi.N.Khan,R.A,Bisen.A.Patel,R.K and Khare.Y.R(2007)Factors associated with adoption of nursery management practices of vegetable crops in Kaymoreplateau,Abstract published in the compendium of 4th Natinal Extension Education Congreessorganised by Society of Extension Education,pp124.

Gawande Ravindra P, Ahire Milind C and Patil Sandip D(2009) "Mass media utilisation pattern and constraints faced by cotton growers,"Journal of Communication studies 27(2).

Kapse.P.S and Chole.R.R(2007) " Information Sources Utilisation Pattern at various stages of innovation decision process,"Marathwada Agricultural Universities,Parbhani Abstract Compendium Fourth

Khan Mohd Ameer et al(2010) "Socio-economic profile and information source utilisation of organised and unorganised poultry farmers of Uttarpradesh-A comparative study,Indian Journal of social research,Vol5*1(1)p*p95-102.

Khurana.G.S(2004) "Information needs of young farmers of Punjab,Abstract Indian Journal of Extension Education,Vol10(1&2)pp23-27.

National Extension Education Congress livelihood security and extension education system. perspectives in collaboration with JNKVV and Society of Extension Education,Agra March(09-11)pp119.

10

Farmers' Attitude Towards the Use of ICT Interventions in Khurda District of Odisha

Shilpa Bahubalendra, B.P. Mohapatra, Adyasha Mohanty and Poka Malini

Department of Extension Education, College of Agriculture, OUAT Bhubaneswar, Odisha

Abstract

Agriculture sector is one among the foremost important sector in India. Keeping in view of the demand for agricultural growth, evolutionary changes have been made in transfer of technology to reach the farming community effectively. The Information and Communication Technology (ICT) is one of these solutions, and has recently unleashed incredible potential to improve agriculture in developing countries specifically. Therein regards, extensive use of information technologies must be promoted to farm level for transfer of technologies. The main target of the study was on the attitude of farmers towards information and communication technology tools in two blocks i.e, Balianta and Balipatna of Khurda district of Odisha. Pre-structured interview schedule was used for data collection from the farmers and appropriate statistical tools were used to analyze the collected data. Proportionate random sampling was used in the selection of one hundred and twenty (120) farmers as the sample of the study. The outcomes confirmed that a half (50.00%) of the respondents opined that they got timely information on inputs through the ICT interventions and also informed that they increased their knowledge on specification of prices. More than half (54.16%) of the respondents told that the available data through the station may not accurately represent the current weather condition on a farm. Majority of the respondents (56.68%) were happy, satisfied. Half (50.00%) of the respondents told that they were getting timely information on the demand of product and subsidies available in the market. An appraisal of the content analysis of the statements show that, the majority of the respondents had more favorable attitude towards ICT tools. So, there is an urgent need to organize effective training programmes and demonstrations for rural youth at village level on ICT tools.

Keywords: Attitude, Farmers, ICT, Knowledge, Training

1. Introduction

Over the years, Agriculture Extension has been at the fore front within the delivery of adequate data to the farming community not just for increasing productivity however additionally to reinforce their standard of living. Keeping in-sight of the demand for agricultural growth, evolutionary modifications had been made in transfer of technology to succeed in the farming community effectively. The Information and Communication Technology (ICT) is one among the required counterparts which made the agriculture extension more realistic and quite interesting.

Developing nations like India are though capable to get self-sufficiency in food production after green revolution however they're in risk to keep this self-sufficiency as they are possibly to face food shortage in near future due to rapid population growth. Thus, the demand of continuous increasing population is often fulfilled with the assistance of ICT as a tool of revamping extension network of the country. ICT is an emerging tool for development of farming communities.

Information and communication technologies have unique feature that provide opportunities to harness them in ways which are extraordinary from traditional media. ICT provides two-way communication between rural communities and development organizations. ICTs also improves the capacity to look for information and increase the number of information available, provide quality information, reduces uncertainty and enhance market participation.

Meera et al. [1] had stated that as a results of the emerging new paradigm of agricultural development, old ways of delivering vital services to citizens are being challenged; traditional societies are additionally being modified into knowledge societies all over the world. Positive attitude of farmers toward information and communication technology will magnify their agriculture productivity and also improve their socio economic conditions.

Information is a valuable input through which farmers adopt new technologies to make the farming more profitable. The application of ICT offers wider possibilities, there by strengthening transfer of technology between research and extension system and further onward transmission to the end users. Due to the decreasing strength of field level functionaries at gross root level it becomes a challenging task to the concerned agencies to reach the farming community with latest technical knowhow. It is quite evident that the advent of ICT revolution which is well utilised by other sectors than agriculture sector. Hence, use of ICT for farm communication plays a vital role in bridging this gap at rural areas hence access to these ICT tools is a crucial requirement for sustainable development of the farming community. It is quite evident that many of the bottlenecks hinder the free flow of information through these tools.

Hence it is time to seriously look into these issues to make future communication system more sustainable.

With this history this study has been formulated with the following objective. To study the attitude of farmers towards information and communication technology tools.

2. Methodology

The present study was conducted in Anantapur district of Andhra Pradesh for the duration of the year 2017-2018. An ex-post facto research design was used in the present investigation. Anantapur district was selected purposively based on largest district in Andhra Pradesh with an area of 19.13 lakh ha. Three divisions Kadiri, Anantapur andKalyanadurg were selected randomly. From each division one mandal were selected by means of using random sampling technique. From each mandal four villages were selected by using random sampling procedure. 34 farmers from Gandlapenta mandal, 47 farmers fromBukkarayasamudram mandal and 39 farmers from Beluguppa mandal were selected from the villages by proportionate random sampling method. Thus a complete of 120 farmers were sampled for the study by proportionate random sampling method who are using ICT tools. Data was collected through a well-structured interview schedule which was developed keeping in view of the objective of the Over the years, Agriculture Extension has been at the fore front within the delivery of adequate data to the farming community not just for increasing productivity however additionally to reinforce their standard of living. Keeping in-sight of the demand for agricultural growth, evolutionary modifications had been made in transfer of technology to succeed in the farming community effectively.

The Information and Communication Technology (ICT) is one among the required counterparts which made the agriculture extension more realistic and quite interesting

Over the years, Agriculture Extension has been at the fore front within the delivery of adequate data to the farming community not just for increasing productivity however additionally to reinforce their standard of living. Keeping in-sight of the demand for agricultural growth, evolutionary modifications had been made in transfer of technology to succeed in the farming community effectively. The Information and Communication Technology (ICT) is one among the required counterparts which made the agriculture extension more realistic and quite interesting.

The economy of the developing country like India predominantly depends on agriculture and development of it emphasizes on the need of related information to be transferred to the farmers. The farmers are also showing interest towards ICT. A positive attitude towards ICT will surely enhance this interest towards harnessing benefits out of it on a sustained basis. As attitude depends upon the socio-personal disposition of an individual, the present study tries to tress out those factors in relation to use of ICT. Attitude is the degree of positive or negative feeling of farmers towards the Information and Communication Technologies (ICTs). The favourable attitude of farmers towards ICT is very much required in obtaining benefit of effective and efficient information support tool which would lead to stronger conviction and efficient extension programme planning in changing agri-rural environment.

Attitude is seen as an evaluative disposition based upon cognitions, affective reactions, behavioral intention and past behaviors and it provides the response of someone's like or dislike towards something (Shih, 2004; Luarn and Lin, 2004).

ICT has tremendous potential in improving the livelihood of farmers via positive impact on factors such as farming efficiency, farm productivity, and farmers' income (Sangeetha et al., 2015).

Furthermore, agricultural services leveraged by ICT have been seen toprovide a greater scope for incorporating the farmers' feedback intothe agricultural knowledge system (Glendenning, 2010). Although the recent developments in ICT have facilitated flow of information to various stakeholders in agriculture, especially farmers; however, factors such as lack of awareness, not enough ICT infrastructure, nonstrategic location of information centres and lack lustre attitude towards ICT use continue to inhibit the potential of ICT for agricultural development (Maningas, 2006).

For increasing the scope of ICT media penetration in agricultural extension services as well as raising the interest of farmers in adopting ICT, there is a need to look closely at factors the constraints the farmers currently face with the use of ICT as well as study their perception regarding the use of ICT in agriculture. This research paper makes an attempt to explore the same.

Materials and Methods

Ex post facto research design was used in the present investigation. The study was undertaken in Balianta and Balipatna blocks of Khordha district with 120 respondents taken randomly for data collection. A total of four villages were covered under the study.From each selected grampanchayat, two villages from each grampanchayat namely Atala and Balabhadrapurvillages of Balianta block and Ariloand Baradapadavillages of Balipatna block were selected randomly on the basis of maximum ICT activities undertaken. The primary data was collected with the help of well-structured and pre-tested interview schedule, designed especially in the light of objectives, whereas secondary data was collected from sources like thesis, journals, literature etc.

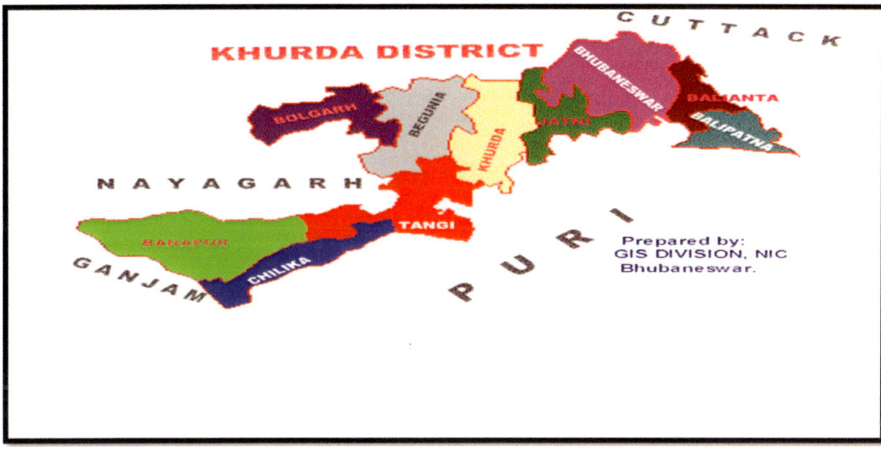

Figure 1: Sample Area

The statistical measures like frequency, percentage, mean, standard deviation, were used to analyze the data to draw tangible inferences.

Results and Discussion

An attempt was made to find out the response of farmers towards a set of statements of attitude towards ICTs use with varying degrees of positive and negative impressions. Data on content analysis of the attitude statements were furnished accordingly on the nature and magnitude of the attitude statements in table as given below.

Table 1: Content analysis of attitude statements on input specification and prices (n=120)

S. no	Statements	SA		A		UD		DA		SDA	
		F	%	F	%	F	%	F	%	F	%
1)	Input Specification and Prices										
a)	Getting timely information of inputs possible through ICT	45	37.50	60	50.00	4	3.33	6	5.00	5	4.16
b)	Correct information through ICT helps to increase the knowledge on the specifications of prices	44	36.66	60	50.00	6	5.00	5	4.16	5	4.16
c)	Farmers are getting remunerative prices of the produce through ICT based market intelligence.	20	16.66	50	41.66	30	25.00	30	25.00	10	8.33
d)	The information is adequate to meet my needs.	18	15.00	55	45.84	6	5.00	28	23.33	13	10.83

S. no	Statements	SA		A		UD		DA		SDA	
		F	%	F	%	F	%	F	%	F	%
e)	This information is influencing my decision making.	19	15.83	44	36.68	40	33.33	15	12.50	2	1.66

SA- Strongly Agree, **A**- Agree, **UD**- Undecided, **DA**- Disagree, **SDA**- Strongly Disagree, F = frequency, % = percentage

A half (50.00%) of the respondents opined that they got timely information on inputs through the ICT interventions and also said that they increased their knowledge on specification of prices. However, 25% of the respondents did not agreed to the statement that farmers were gaining remunerative prices through ICT based market intelligence. Even, the information is adequate to meet the needs for 10.83% of the respondents and it influence the decision making for 36.68% of the respondents.

Table 2: Content analysis of attitude statements onweather-based forecasting of pest and diseases (n=120)

S. no	Statements	SA		A		UD		DA		SDA	
		F	%	F	%	F	%	F	%	F	%
2)	Weather Based Forecasting of Pest and Diseases										
a)	Available data through the station may not accurately represent the current weather condition on a farm perhaps 10-20 miles away.	35	29.16	65	54.16	10	8.33	5	4.16	5	4.16
b)	Weather network provide accurate pest prediction?	5	4.16	15	12.50	20	16.66	60	50.00	20	16.66
c)	ICT based Pest/disease outbreak warning system facilitate to take preventive measures.	34	28.33	55	45.85	4	3.33	25	20.83	2	1.66
d)	The information is adequate to meet our needs.	11	9.16	57	47.52	4	3.33	34	28.33	14	11.66
e)	The information is adequate to improve the crop productivity.	30	25.00	56	46.68	7	5.83	25	20.83	2	1.66
f)	This information is influencing my decision making.	18	15.00	55	45.84	6	5.00	28	23.33	13	10.83

SA- Strongly Agree, **A**- Agree, **UD**- Undecided, **DA**- Disagree, **SDA**- Strongly Disagree, F = frequency, % = percentage

More than half (54.16%) of the respondents opined that the available data through the station may not accurately represent the current weather condition on a farm perhaps 10-20 miles away yet half (50.00%) of the respondents agreed that weather network provide accurate pest prediction and 45% of them opined that this warning system facilitate to take preventive measures. The information is adequate to meet the needs for 47.52% of the respondents and is adequate to improve the crop productivity for 46.68% of the respondents. This is also influencing the decision making of 45.84% of respondents.

Table 3: Content analysis of attitude statements on weather forecasting(n=120)

S. no	**Statements**	**SA**		**A**		**UD**		**DA**		**SDA**	
		F	**%**	**F**	**%**	**F**	**%**	**F**	**%**	**F**	**%**
3)	Weather Forecasting										
a)	Getting timely information on weathers.	45	37.50	60	50.00	4	3.34	6	5.00	5	4.16
b)	I saved my stored grains from getting affected from cyclone by weather forecasting	30	25.00	50	41.68	8	6.66	28	23.33	4	3.33
c)	I am happy and satisfied with the service.	29	24.17	68	56.68	5	4.16	7	5.83	11	9.16
d)	The information is adequate to meet our needs.	18	15.00	55	45.84	6	5.00	28	23.33	13	10.83
e)	Forecasting on weather saves our life and property.	21	17.50	39	32.50	21	17.50	37	30.84	2	1.66
f)	This information is influencing our decision making.	19	15.83	44	36.68	40	33.33	15	12.50	2	1.66

SA- Strongly Agree, **A**- Agree, **UD**- Undecided, **DA**- Disagree, **SDA**- Strongly Disagree, F = frequency, % = percentage

A half of the respondents (50.00%) opined that they are getting timely information on weathers and also saved the stored grains from getting affected from cyclone for 41.68% of the respondents and more than half (56.68%) of the respondents are happy and satisfied. The information is adequate to meet the needs for 45.84% of the respondents and saves life and property for 32.50% of the respondents. This is also influencing the decision making of 36.68% of respondents.

Table 4.3.4: Content analysis of attitude statements on market forecasting(n=120)

S. no.	Statements	SA		A		UD		DA		SDA	
		F	%	F	%	F	%	F	%	F	%
4)	Market Forecasting										
a)	Getting timely information on the demand of our product in the market.	45	37.50	60	50.00	4	3.34	6	5.00	5	4.16
b)	I am getting information on different segments of the market.	16	13.33	44	36.66	10	8.33	34	28.33	16	13.33
c)	This information is helping us to penetrate into the market.	29	24.17	68	56.68	5	4.16	7	5.83	11	9.16
d)	This information helping us to know our competitors in the market and why customers buy from them.	16	13.33	44	36.66	10	8.33	34	28.33	16	13.33
e)	I am wasting my precious time by browsing unnecessary sites	7	5.83	58	48.34	15	12.50	28	23.33	12	10.00

SA- Strongly Agree, **A**- Agree, **UD**- Undecided, **DA**- Disagree, **SDA**- Strongly Disagree, F = frequency, % = percentage

A half (50.00%) of the respondents opined that they are getting timely information on the demand of product and 36.66% of the respondents getting information on different segments in the market. 24.17 % of the respondents opined that this is helping them to penetrate into the market and to know their competitors in the market for 36.66% of the respondents. Also 23.33% of the respondents are not wasting their time by browsing unnecessary sites.

A half (50.00%) of the respondents opined that they are getting timely information on the subsidies available in the market. This is also influencing the decision making of 36.68% of respondents. More than half (56.68%) of the respondents are happy and satisfied and 41.68% of the respondents can access at any time.

Table 4.3.5: Content analysis of attitude statements on subsidies(n=120)

S. no	Statements	SA		A		UD		DA		SDA	
		F	%	F	%	F	%	F	%	F	%
5)	Subsidies										
a)	I am getting timely information on the subsidies available in the market.	45	37.50	60	50.00	4	3.34	6	5.00	5	4.16
b)	This information helps me to take decision wisely.	19	15.83	44	36.68	40	33.33	15	12.50	2	1.66
c)	I am happy satisfied with the service.	29	24.17	68	56.68	5	4.16	7	5.83	11	9.16
d)	I can access at any time	30	25.00	50	41.68	8	6.66	28	23.33	4	3.33

SA- Strongly Agree, **A**- Agree, **UD**- Undecided, **DA**- Disagree, **SDA**- Strongly Disagree, F = frequency, % = percentage

An appraisal of the content analysis of the statements as shown in the above table that, the majority of the respondents had more favourable attitude towards ICT tools. So, there is an urgent need to organize effective training programmes and demonstrations for rural youth at village level on ICT tools.

Conclusion

Due to the lack of credibility in the information provided by the ICT tools, the farmers are not adopting the recommendations given by them. So, the extension agents should make efforts to enhance the credibility of the ICTs by providing adequate, timely and useful information and impart skill in using ICT tools. This enables them to utilize the ICT tools and develop a favourable attitude towards ICTs use. Therefore, there is a need to make available these tools at village level along with creating awareness about the importance of ICT tools and types of services providing by these tools to the farming community. Providing such facilities at village level will further attract them to use these tools for doubling their income. Further providing services in local language and making the tools more users friendly which will result in developing positive attitude towards ICT tools. Variables like education and extension participation were found significant to the attitude of farmers. Hence, there is a strong need to educate the farmers during extension programmes regarding usefulness of ICT tools, type of information provided and authenticity of information to make them aware about these tools.

References

Dhaka, B.L. and Chayal, K.2010. Farmers Experience with ICTs on Transfer of Technology in Changing Agri-rural Environment. Indian Research Journal Extension Education. 10 (3):114-118.

Ganeshkumar, P., PrabhakaraSastry, T., Prasad, T. and Prasad, S.V. 2008. Attitude of Farmers towards RythuMitra Television Programme. Journal of Research, ANGRAU. 36 (2 & 3): 79-81.

Ganeshkumar, P. 2008. Information and Communication Technologies enabled Agricultural Extension System in Andhra Pradesh-A critical analysis. Ph.D. Thesis. Acharya N.G. Ranga Agricultural University, Hyderabad, India.

Glendenning, et.al. (2010). Review of Agricultural Extension in India: Are farmers information needs being met?" DOI: 10.1.1.220.8439.

Kabir, K.H. (2015). Attitude and level of knowledge of farmers on ICT based farming. European Academic research, (10)2, 13117 – 13196.

Luarn, P.andH.Lin, 2004. Towards an understanding of the behavioral intention to use mobile banking. Journal of Computing Human Behavior, 30: 1-9. DOI: 10.1016/j.chb.2004.03.003

Madhu, P. 2010. Awareness and Listening Behavior of the Listeners of Krishi Community Radio station M. Sc. (Ag.) Thesis. University of Agricultural Sciences, Dharwad, Karnataka (India).

Madhubabu, K. 2008. A Study on Training Intervention of Extension Education Institute on the Capacity Building of Functionaries of Developments. Ph. D Thesis. Acharya N.G.Ranga Agricultural University, Hyderabad, India.

Maningas, R. V. (2006). Mainstreaming farmers and intermediaries into information and communications technology (ICT): A strategy towards adopting ICT for rural development and agricultural extension. Computers in Agriculture and Natural Resources, 4th World Congress Conference, USA.

Palaiah, R.K., Bharatesh, S.M., Dechamma, S and Devaraja, K. 2016. Attitude of Farmers about Use of ICT tools in Farm Communication. 21stInternational Academic Conference, Miami. 101-111.

Selviya, V and Reddy, V.G. 2008.Profile of Faculty of Home Science in ANGRAU.Journal of Research ANGRAU.36 (1): 46-48.

Shih, H. P. 2004. Extended technology acceptance model of internet utilization behavior. Journal of Information Management, 41: 719-729. DOI: 10.1016/j.im.2003.08.009,

V. Sangeetha, R.R. Burman, S.K. Dubey, J.P. Sharma, I. Singh. Attitude of agricultural stakeholders on use of Short Message Service (SMS) in transfer of technology. Indian Journal of Extension Education. 2015; 51, 60-65.

Verma, A.P., Ansari, M.A., Rajan, R., Archana, B., Rupan, R and Diksha, P. 2016. Farmers Attitude towards e-choupal: A Critical Investigation in Gonda District of Uttar Pradesh. International Journal of Agriculture Sciences.8 (49):2076- 2078.

11

An In-depth Analysis of Socio-Psychological Attributes of Krishi Vigyan Kendra Officials of Odisha

Smaranika Mohanty[1], Mahamaya Prasad Nayak[2] Rajeeb Kumar Behera[3], Sweta Sahoo[4], Soumya Shakti Dash[5] and Santosh Kumar Rout[6]

[1,3,4,5]Department of Agricultural Extension & Communication, IAS, SOA Bhubaneswar, Odisha
[2]Extension Information, Odisha University & Agricultural Technology Bhubaneswar, Odisha
[6]Siksha 'O' Anusandhan Deemed to be University, IAS, Bhubaneswar

Abstract

The present study was conducted in Odisha state of India in the year 2019. The data were collected from the officials of thirty-three KVKs operating in Odisha spread across 30 districtsthrough the mailed questionnaire method. Fifty percent from each category of KVK manager's i.e.Upper level: Sr. Sc. and Head, Middle Level: Scientists and Lower level: Programme Assistant / Farm managers were selected randomly as sample (125 Nos.) respondents for the study. Survey research design and exploratory research were used for the study and analysis. This study revealed that the scientists and technical staff of KVKs (Krishi Vigyan Kendras)were working in the active age of their life.Majority of the officials were male. Majority of them had agricultural background, education level of post-graduation and above, more than 10 years of experience, adequate knowledge on KVK, medium to higher level of commitment for KVK, medium level of job satisfaction, medium level of organizational stress followed by high level,and medium level of job autonomy. They were using formal meetings and discussion with colleagues and seniors at KVK for acquiring knowledge on KVK mostly. They were participating less in management-oriented trainings as compared to technical trainings, working under moderately favourable level of organizational climate.

Keywords: Krishi Vigyan Kendra Officials, Socio-psychological attributes, Odisha

Introduction

In Puducherry, the first KVK was founded in 1974. In the newly constructed districts and several larger districts, 106 additional KVKs will be established, bringing the total number of KVKs to 645. KVKs are·approved for use by Agricultural Universities, ICAR Institutes, associated Government Departments, and Non-Governmental Organizations (NGOs) engaged in agriculture. The KVK initiative is fully funded by the Government of India. KVK, which is a crucial component of NARS, aims to evaluate location-specific technology modules in agricultural and related companies through technology assessment, refinement, and demonstrations. As a knowledge and resource centre for agricultural technology, KVKs have supported public, private, and non-profit sector efforts aimed at enhancing the district's agricultural economy(1).

The KVK's mandate has changed from vocational training to technology assessment and demonstration for its application and capacity building in a district in preparation for its growing contribution to NARES in India. Though technologies were developed at ICAR Research Centres or State Agricultural Universities, KVK carried out location-specific technology assessments, micro-location refinements on farms with genuine problems, and final technology certification for widespread adoption by farmers(2).

As a lot of works and responsibilities are performed by KVK officials of Odisha, it is becoming very important to study their socio-psychological attributes like which Age, Sex, Education, Professional Experience, Knowledge on KVK, Training attended by respondents, Organizational climate, Level of commitment, Job satisfaction, Organizational Stress, Job Autonomy etc. are involved in these activities.

Research Methodology

Survey & exploratory research design was adopted for the study. The present study has been undertaken in the in the state of Odisha. The state is located in the eastern part of India have unique social, cultural and ecological background which influence the living standard and behavioural patterns of the people in many ways. For the purpose of the study 50% from each category of KVK managers (1. Upper level: Sr. Sc. and Head, 2. Middle Level: Scientists and 3. Lower level: Programme Assistant / Farm managers) were selected randomly as sample i.e. 125 nos. of respondents for the study.

Results & Discussion

The socio-psychological variables are defined on the basis of personal, social, psychological characteristics of the respondents. Variables taken in this study like age, sex, educational background,educational level, professional experience, knowledge on KVKs, source availed for acquiring knowledge on KVK, training programmes attended, types of training attended, organizational climate, level of commitment towards KVK, job satisfaction, organizational stress, job autonomy were taken to study the socio-psychological characteristics. Analysis &discussions are presented below:

Age: After analysis it was found that majority (56.80%) of the respondents were in the middle age group followed by the respondentsin young age group 43.20 % implying that the scientists and technical staff of KVKs are in active age of their life.

Sex: Majority (64.00 %)of respondents were male and rest (36.00 %) were female.

Educational background: Majority of the respondents (67.20%) were from agricultural background followed by computerscience (7.20 %)and agriculture engineering (6.40%) and home science (6.40 %) as their educational qualification. Further (5.6%), (4.8%), and (2.4 %) respondents were found to be from fishery, veterinary, forestry background respectively (Fig.1).

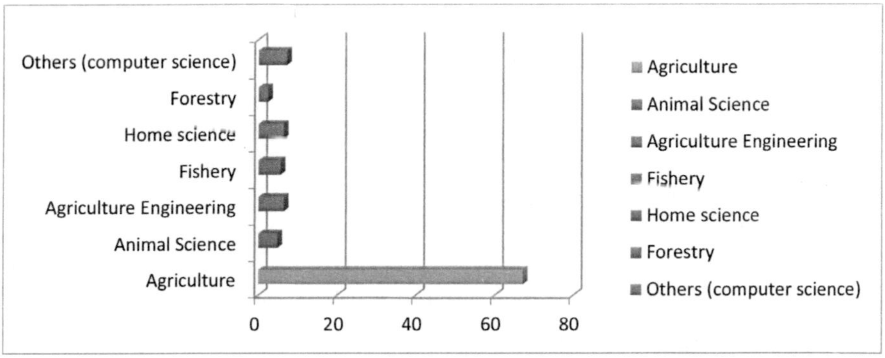

Fig. 1: Distribution of the respondents according to their educational background

Educational level: Majority (92.80 %) of the respondents were having education level of post-graduationand above, and 28.00 % out of this completed PhD. Only7.20% respondents were having qualification up to graduation level.

Professional experience: Majority (41.60 %) respondents were having more than 10 years of experience followed by 36.00%and 22.40% respondents were having less than 5 years and 5 to 10 years of experience respectively. The study signifies majority of the KVK officials were having more than 10 years of experience

Knowledge on KVKs: Majority 98.40% of the respondents were having adequate knowledge on KVK whereas remaining only 1.60 % respondents had expressed to have inadequate knowledge on KVK.

Source availed for acquiring knowledge on KVK: Formal meetings and discussion with colleagues and seniors at KVK was the major source (62.40%) of acquiring knowledge on KVK as opined by of the respondents followed by 54.40% respondents indicating formal training conducted for the purpose as the source of knowledge on KVK. This implies that there are deliberate activities conducted which helps the scientists and technical staff to acquire knowledge on KVK.

Types of training attended: Technology oriented trainings attended by the respondents inside the state showing highest mean score (1.52) followed by technology-oriented training outside the state (MS = 1.22). The respondents were found have relatively less participation in management-oriented trainings as compared to technical trainings. Similar findings are reported by Kumar and Singh (1994), Godara et al. (2006), Tayade et al. (2011) and Bimal(2016).

Organizational climate: Majority (76.00%) of the respondents perceived a moderately favourable level of organizational climate to prevail in KVKs whereas, 16.00%perceived the prevailing organizational climate to be less favourable. In total, 84.00 per cent of the respondents opined a moderate and highly favourable organizational climate prevailing in KVK. Similar findings were reported by Benjamin Schneider (1975), Sandika (2007) and Sandika (2007).

Level of commitment towards KVK: Majority (88.00%) of the respondents have explained to have medium to higher level of commitment for KVK. Only 12.00%were reported to have lower level of commitment. This explains a higher order involvement and commitment of the employees for their own organization.

Job satisfaction: More than half (68.00%) of the respondents were having medium level of job satisfaction. Even though there exists a significant group of respondents showing high level of job satisfaction, still 12.80 %respondents revealed to have low level of job satisfaction. Job satisfaction is directly proportional to their performance in job. These results are satisfied with the study done by Prakash *et al.* (2003), Meena and Singh (2003), Jagdale (2004), Godara*et al.* (2006) and Jyothi (2006).

Organizational stress: Majority (56.00%) of the respondents were found to be under medium level of organizational stress followed by 30(24.00 %) respondents in high organizational stress category. However, 20%of the respondents have expressed to be under low organizational stress category. This indicates KVK staffs in general to tune of 76.00 per cent were not under heavy organizational stress. The study has also attempted to analyze the stress factors in KVKs.The three most important factors perceived as always a source of stress were inconvenience created by colleagues, frequently changing job responsibilities even without the staff's consent, and assigned with works those were not suitable to one's position. Several stress factors with their analysis are represented in the Table:1.

Table 1: Stress factorsof KVK officials in Odisha (n = 125)

Sl. no.	Stress factors	NSS		RSS		SSS		ASS	
		f	%	f	%	f	%	f	%
1	Assigned with works those are not suitable to one's position	48	38.4	10.0	8.0	47	37.6	20	16.0
2	Frequently changing job responsibilities even without my consent	51	40.8	15	12.0	39	31.20	20	16.0
3	Unpredictable authority	55	44.00	07	5.6	44	35.2	19	15.2
4	Having no power to take decision for own activities	37	29.6	19	15.2	52	41.6	17	13.6
5	Not being trusted by authorities	69	55.2	19	15.2	52	41.6	17	13.6
6	Being forced to work with people that one does not like	54	43.2	24	19.2	37	29.6	10	8.0
7	Having no personal life and privacy	51	40.8	22	17.6	33	26.4	19	15.2
8	No respect from superior and co-workers	68	54.4	26	20.8	17	13.6	14	11.2
9	Absurd and bothersome organization rules	44	35.2	27	21.6	37	29.6	17	13.6
10	Inconvenience created by colleagues to work	54	43.2	20	16.0	30	24.0	21	16.8
	Average		42.24		15.12		31.04		13.92

Table 2: Job autonomy of KVK officials of Odisha (n=125)

Sl. no.	Aspect of job autonomy	Allowed always for all the tasks		Allowed sometimes for some tasks		Never allowed	
		f	**%**	**f**	**%**	**f**	**%**
1	In deciding what is to be done	46	36.8	76	60.0	3	2.4
2	In prioritizing the works	61	48.8	61	48.8	3	2.4
3	In deciding the procedure to be followed	43	34.4	77	61.6	5	4.0
4	In deciding the team mates	45	36.0	60	48.0	20	16.0
5	In use of resources	43	34.4	73	58.4	9	7.2

Job autonomy: Majority (63.20&) of the respondents have expressed to avail medium level of autonomy whereas 27 (21.60 %) respondentsavailed high level of autonomy.However, only 19 (15.20 %) respondents have expressed that the job autonomy allowed to them in Krishi Vigyan Kendra's is low. While analyzing different aspects in deciding the job autonomy in KVKs the following information was revealed.48.80%respondents opined that they have been given complete autonomy in prioritizing their works whereas same proportion of the respondents expressed that the level of autonomy is medium for prioritizing their works (Table:2). The results are supported by Leach et al., (2003)and Sao (2018).

References

Christo, B. and Pienaar, J.2006, South Africa Correctional Official Occupational Stress: The Role of Psychological Strengths, Journal of Criminal Justice, 34(1): 73-84

Dkhar. Sao et al.2018.An explorative study on the Organizational climate of KVKs, College of post graduate studies, Central agriculture university, Imphal, Meghalaya Operational Research, 154, 573-584.

https://kvk.icar.gov.in/aboutkvk.aspx

Jyothi, V. 2006.Decision making process of KrishiVigyanKendras (KVKs) in Northern Karnataka. Unpublished Ph.D. thesis, University of Agricultural Sciences,Dharwad,Karnataka, India.

Manshor, A. T., Rodrigue, F. and Chong, S. C.2003, Occupational Stress Among Managers: Malaysian Survey, Journal of Managerial publication, New Delhi, pp. 98-126.

Meena, B.S and B.Singh. 2003. Effect of traits of the trainers on their job performance in KrishiVigyanKendras in Rajasthan. Agricultural Science Digest, 23(3):187-190

Sagi.A and Friedland.N. 2007.The effect of the size and Diversity of Decision sets on Post – Decision Regret,journal of personality and social psychology,93(4);515-24

Sahoo, A. K., Sahu, S., Meher, S. K., Begum, R., Panda, T. C., & Barik, N. C. (2021). The Role of Krishi Vigyan Kendras (KVK) in Strengthening National Agricultural Research Extension System in India. Insights into Economics and Management Vol. 8, 112-122.

Sandika, A.L. 2006. A study on organizational climate perception by Veterinary Officers (VOs) and Veterinary Livestock Inspectors (VLIs) of Department of Animal Husbandry and Veterinary Service, Karnataka. Unpublished M.Sc. thesis, University of Agricultural Sciences, Dharwad, Karnataka.

12

Farm Women's Knowledge About Drudgery Reduction Tools

Sunita Bairwa[1], K. Chayal[2] and S.R. Verma[3] and Asha Dagar[4]

[1,2]Department of H.Sc. University of Rajasthan, Jaipur, Rajasthan
[3](M&E), CCS Haryana Agricultural University, Hisar, Haryana
[4]Department of Home Science , University of Rajasthan, Jaipur, Rajasthan

Abstract

Agriculture and related fields contribute more than 20.2 percent of India's GDP (NSO 2021). It is the largest un - organized area, with such a large number of farm women making a living from it. They are essential in most agricultural tasks such as land preparation, seed treatment, sowing, planting, weeding, intercultural operation, crop irrigation, application of manures and fertilisers, crop protection, harvesting, threshing, storage, and so on. The majority of these tasks are tedious and time-consuming when carried out manually or with existing tools, resulting in drudgery, physical and mental fatigue, monotony, hardship, exploitation, pain, economic stress, and so on.

Therefore, this study was conducted in Rajasthan's Alwar district to investigate farm women's knowledge of drudgery reduction tools. The research was carried out in two villages chosen at random from panchyat samiti Laxmangaarh. From each village, 25 farm women were selected randomly. As a result, the study's sample consisted of 50 farm women. Five drudgery-reduction tools were chosen for research: a serrated sickle, a wheel hoe, a fertiliser broadcaster, a maize sheller, and a kudali. The majority of farm women were familiar with the serrated sickle and Kudali, drudgery-reducing tools; however, knowledge of the wheel hoe was inadequate. Further knowledge about fertilizer broadcaster among the farm women was good; level of knowledge of farm women about maize seller tool was medium. It also can be concluded that overall knowledge of respondents was 52.96 MPS which is neither high nor low. This study suggested that the required knowledge and skills should be provided to rural women regarding improved drudgery reduction tools.

Introduction

Agriculture is a major sector in Indian economy. It is one of the biggest production sector contributing more than 20 percent of GDP for India's economy. Its is the largest provider of employment in Rural India. Over 70 percent household depends on agriculture. Its provides raw materials to industries and market for industrial product(Kekane, 2013).

Therefore, agriculture is a main sector for the planning of social and economic development of the country. Rapid growth of agriculture is essential not only to achieve self reliance at national level but also for household food security and to bring about equity in distribution of income and wealth resulting in rapid reduction in poverty line (Shrivastava *et al.* 2010). Agriculture is considered as the biggest unorganized sector where large numbers of women are working for their livelihood. The farm women comprise a quarter of the word'spopulation . Women also make up 41 percent of the word's agriculture labour force, a ratio which rises to 49 percent for low income countries (ILO,2018).

Women play an important role towards the growth of the agricultural sector and its other associated areas. They perform multifarious tasks in agriculture and its allied fields. They are playing an important role in most of the farm operation like land preparation, seed treatment, sowing, planting, weeding, intercultural operation, irrigation of crop, application of manures and fertilizers, protection of crop, harvesting, threshing, storage etc. There is hardly any activity in agricultural production, except ploughing in which women are not actively involved. In some of the farm activities like processing and storage, women are dominant so strongly those men workers numerically insignificant (Patil and Babus 2018).Naidu *et al.* (2020) found that 63 percent of women are engaged in sowing of seeds and transplanting followed by in harvesting operation (61%), wedding (59%), winnowing (27%), manure and fertilizer application (25%) on regular basis.

Most of these tasks performed by women are tedious as well as time consuming. As most of these operations are done manually or by using traditional tools, they are slow and cause considerable fatigue and drudgery. Drudgery is generally conceived as physical and mental strain, agony, monotony and hardship experienced by human beings (Momin, 2009).

Women's condition in this regard is alarming, as they are hindered by illiteracy, poor health, unemployment, and a lack of technological expertise and skill. As a result, women's needs for comfortable work participation continue to be ignored. Drudgery reduction may result in women working with improved productivity capacity and health. The application of simple, scientific, and appropriate technologies can bring about the desired change in the life of rural women, which is full of drudgery.

There is an increasing number of available technologies that can increase women's productivity and income in agriculture, but these technologies have not reached women. As a result, women must gain technological empowerment in agricultural sectors. This can be accomplished by improving their knowledge and skills in these technologies.

Research Methodology

The current study was carried out in the Rajasthan district of Alwar. The Alwar district was chosen on purposely. There are fourteen panchayat samities in Alwar district, one of which, Laxmangarh, was chosen on purpose. One village panchayat Khohara-Malawali was selected at random from the selected panchayat samiti, and two villages, Khohara and Malawali, were chosen at random from this village panchayat. 25 farm women were taken randomly from each village. Drudgery-reduction tools used in the study included a serrated sickle, a wheel hoe, a fertiliser broadcaster, a maize sheller, and a kudali.

Result and Discussion

The present investigation was conducted with one of the objective "To find out level of Knowledge about selected drudgery reducing tools. The results are shown in the following tables. To obtain an overview of respondents' knowledge status about selected drudgery reducing tools, they were classified into three categories, namely low, medium, and high knowledge level, based on the calculated mean and standard deviation of the knowledge score obtained.

Table 1: Distribution of respondents according to knowledge about serrated sickle N=50

Level of Knowledge		Frequency	Percent
Low	0-5	19	38.00
Medium	6-10	6	12.00
High	11-15	25	50.00

The results in table 01 show that the majority of respondents (50%) had a high level of knowledge about serrated sickle. While only 12% of farm women were found to have a medium level of knowledge. A sizable proportion of respondents (38%) reported having little knowledge of the serrated sickle. It is possible to conclude that the majority of farm women were familiar with this drudgery-reducing tool.

Table 2: Distribution of respondents according to knowledge about wheel hoe N=50

Level of Knowledge		Frequency	Percent
Low	0-5	31	62.00
Medium	6-10	5	10.00
High	11-15	14	28.00

Table 02 indicates that majority of respondents'(62 percent) possessed low level knowledge regarding wheel hoe equipment. While 28 percent of the farm women were found to have high level, knowledge followed by 10 percent of respondents having medium level of knowledge about this intercultural operation tool. These results show that knowledge of farm women about wheel hoe was unsatisfactory.

Findings are in line with the findings of Kumari *et al.* (2019) and Kumar *et al.* (2018)

Table 3: Distribution of respondents according to knowledge about fertilizer broadcaster N=50

Level of Knowledge		Frequency	Percent
Low	0-4	20	40.00
Medium	5-9	3	6.00
High	10-14	27	54.00

The data displayed in table 03 shows that majority numbers of respondents (54%) were having high level knowledge regarding fertilizer broadcaster equipment whereas, only 6 percent of respondents had medium knowledge level. It was noted that 40 percent of the respondents were found in low level knowledge category. Inferences can be taken that knowledge about fertilizer broadcaster among the farm women was good.

Table 4: Distribution of respondents according to knowledge about maize seller N=50

Level of Knowledge		Frequency	Percent
Low	0-4	24	48.00
Medium	5-8	4	8.00
High	9-13	22	44.00

The figures depicted in table 04 indicates that 48 percent of respondents were having low level knowledge regarding maize seller equipment followed by 08 percent farm women having medium level knowledge. While a significant number of respondents (44%) were found to have high level knowledge about maize seller. From the above table it can be concluded that level of knowledge of farm women about maize seller tool was medium.

Table 5: Distribution of respondents according to knowledge about kudali N=50

Level of Knowledge		Frequency	Percent
Low	0-4	15	30.00
Medium	5-9	5	10.00
High	10-14	30	60.00

Table 05 depicts that majority of respondents (60%) were having high level knowledge about kudali tool followed by 10 percent respondents having medium level of knowledge. The remaining respondents were classified as having a low level of knowledge. According to these figures, farm women had a high level of knowledge about the drudgery-reduction tool-Kudali.

Table 6: Distribution of respondents according to overall knowledge about selected tools N=50

Level of Knowledge		Frequency	Percent
Low	0-23	12	24.00
Medium	24-47	21	42.00
High	48-71	17	34.00

The results presented in table 06 shows that 42 percent of the respondents have medium level of knowledge (overall) about all selected drudgery reduction equipments. Whereas, 34 percent of the respondent were in high level of knowledge category about all selected tools. The table also depicts that 24 percent of farm women had low level of awareness about the drudgery reduction tools. Therefore conclusion can be drawn that majority of the farm women have good knowledge about the selected tools.

Table 7: Mean score, mean percent scoreand standard deviations of knowledge N=50

S. No.	Equipments	Total score	Mean score	Mean Percent Score	Standard Deviation
1.	Serrated Sickle	15	8.68	57.87	6.97
2.	Wheel hoe	15	5.18	34.53	6.75
3.	Fertilizer board caster	14	8.08	57.71	6.71
4.	Maize seller	13	6.36	48.92	6.25
5.	Kudali	14	9.30	66.43	6.26
	Overall	71	37.60	52.96	21.52

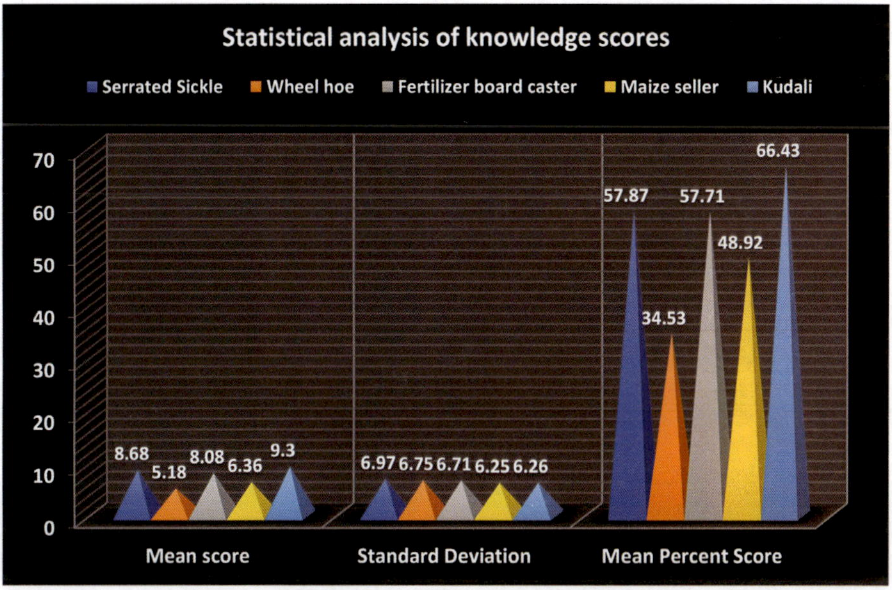

Figure 1: Graphic presentation of statistical data of knowledge about tools

For the purpose of the comparison of knowledge of farm women about different drudgery reduction equipment mean percent score and standard deviation was calculated and presented in the table 07 and in figure 01.These results shows that farm women have highest knowledge about the tool Kudali (66.43) followed by serrated sickle (57.87), fertilizer broadcaster (57.71), maize seller (48.92). It was also found that farm women having lowest knowledge about wheel hoe equipment. From the above results it also can be concluded that over all knowledge of respondents was 52.96 MPS which is neither high nor low. The results of the study are in the line of Kumari*et al.*(2019) and Kumari *et al.* (2017).

Conclusion

It can be concluded that the majority of farm women were familiar with the serrated sickle and Kudali, drudgery-reducing tools; however, farm women's knowledge of the wheel hoe was inadequate. Furthermore, it was determined that farm women's knowledge of fertiliser broadcaster was good, while their knowledge of maize seller tool was medium. The tool Kudali was found to have the high knowledge, followed by serrated sickle, fertiliser broadcaster, and maize seller, in that order, and wheel hoe equipment had the least knowledge. It can also be concluded that respondents' overall knowledge was 52.96 MPS, which is neither high nor low. *As a result, the study recommended that rural women's knowledge and skill in the proper use of those tools, about which they have less knowledge and are less skilled in handling them, be increased through training and educational programmes.*

References

ILO (2018). Rural women at work, bridging the gaps. Gender, Equality and Diversity & ILOAIDS Branch. file:///Users/lkvatta/Desktop/wcms_619691.pdf

Kekane, M.A. (2013). Indian Agriculture- Status, Importance and Role in Indian Economy. International Journal of Agriculture and Food Science Technology, 4(4), 343-34. Research India Publications.

Kumar, S., Srivastava, A.K., Mishra, S.B. and Choudhar, R.C. (2018). Reducing Drudgery of Farmwomen through Appropriate Farm Implements in Uttar Pradesh, India. International Journal of Agriculture Sciences, 10, (7), 5761-5764.

Kumari, A., Yadav, R., Tripathi, S.and Kunwar, N. (2017). Knowledge and awareness of farm women about selected drudgery reducing practices in dairy farming. International Journal of Veterinary Sciences and Animal Husbandry, 2(6): 38-41

Kumari, A.R., Dubey, S. K., Gautam, U. S. (2019). Assessment of awareness and use about drudgery Reducing Improved Farm Tools and Implements by Farm Women of Uttar Pradesh. Indian Journal of Extension Education, 55(1),157-160.

Momin, J.W. (2009). http://www.shvoong. com/social-sciences/1883032-women-drudgeryagrarian- society.

Naidu, D. C., Bhagya, K. L. and Roy, G.S. (2020). Participation of Farm Women in Agricultural Operations in Srikakulam District of Andhra Pradesh, India. International Journal of Current Microbiology and Applied Sciences, 9(6): 500-503.

NSO (2021) https://www.pib.gov.in/PressReleasePage.aspx?PRID=1741942

Patil, B. and Babus, V., S. (2018). Role of women in agriculture. International Journal of Applied Research, 4(12): 109-114

Shrivastava, M.P.Sahay, Nilima, Vidhyarthi, V.P. and Singh S.P. 2010. Book on Second green revolution Vs Rainbow revolution published by Deep and Deep publication, New Delhi. Available on the link https://books.google.co.in/books?id=ehqiBfxY-AEC&pg=PA285&dq=

13

Delphi Experiment Forecasts to Reform Policies for Agricultural Extension:A Sustainable Development Perspective

Bhalerao Amol Kamalakar*

(Agri. Extension), Training and Education Centre, ICAR-Indian Veterinary Research Institute, Pune - 411005, Maharashtra

Abstract

Agricultural extension policies have shifted focus time to time to cater the needs of farmers. As per the farmers need, extension education transformed a lot from mere transfer of technology domain to emancipation of an individuals. However, in consideration to the challenges witnessed by Indian agriculture, the agricultural extension polices need reforms to promote sustainable development. In order to rethink about reforms, the most appropriate approach would be to devise policies driven by sustainable development philosophy.

In the context of sustainable development, the policymakers need to be informed about what are the real-time challenges, where to invest on priority, how best we can stabilize the socio-economic conditions of farmers. To generate forecasts to answer above stated research questions, a systematic Delphi experiment was conducted with three iterations wherein 41 experts participated. This experiment was confined for the North Eastern region of India (NER), so senior level scientists and the Heads of KVKs were invited to participate in this experiment.

The findings of the Delphi reveal that changes in climate have already adversely affected farming, livelihoods, and sustainability in NER. Organic farming, conservation practices, efficient irrigation and scientific approaches to farming, such as Integrated Farming Systems, are the most advocated policy options to combat the challenges identified by the experts. Concerning the institutions, experts forecast that the Krishi Vigyan Kendras must play a pivotal role in implementing policies to promote sustainable and climate-resilient agriculture. These findings are based on the consensus of participating experts derived from policy Delphi method.

Keywords: Sustainable Development, Policy Reforms, Delphi Experiment Forecasts, Sustainable Agriculture, Consensus for Policy Reforms

Introduction

Climate change is the long-term alteration of temperature and meteorological conditions in the Earth's atmosphere. It is caused by unsustainable human activities such as burning fossil fuels, deforestation, and rampant industrialization which released massive quantities of greenhouse gases into the atmosphere. These gases trap heat from the sun and cause the planet to warm, resulting in various impacts on the environment and human society. Climate change is likely to disrupt the monsoon patterns and lead to more extreme weather events such as droughts and floods. This will have a significant impact on crop yields and food security in India, which is already facing a growing population and increasing demand for food. Additionally, rising temperatures and changing rainfall patterns may also lead to the spread of new pests and diseases, further exacerbating the challenges faced by Indian farmers.

Additionally, climate change impacts are disproportionately affecting the livelihoods of smallholder and resource-poor farmers, who are more vulnerable to weather-related risks and have fewer resources to adapt. The Indian government and private sector are implementing various measures to help farmers adapt to changing conditions, but more needs to be done to ensure the long-term sustainability of the sector. The agricultural sector in India is particularly vulnerable to the impacts of climate change, as it employs a large portion of the population and is a major contributor to the country's economy. The Indian government and various organizations are working to develop strategies and technologies to adapt to the changing climate and mitigate the effects on the agricultural sector. Time to time the Indian government have implemented various policies and programs to improve the productivity and profitability of the agricultural sector, including the transfer of technology and extension services.

Agricultural extension and transfer of technology are critical to improving the performance of the agricultural sector by providing farmers with access to the latest knowledge and techniques. Extension services are designed to deliver information and advice to farmers on a wide range of topics, including crop selection and management, soil health, water conservation, and pest control. These services are typically provided by government agencies, non-government organizations, and private companies. As, transfer of technology refers to the process of introducing new and improved agricultural technologies to farmers, this includes new seed varieties, irrigation systems, farming equipment, and other inputs that can help to increase crop yields, improve water and nutrient management, and practices that reduce the attacks of pests and diseases. Agricultural extension services in India are provided by various government agencies, such as the Indian Council of Agricultural Research (ICAR) and the State Agricultural Universities, a vibrant network of *Krishi Vigyan Kendras* as well as private organizations. These services include providing information on new technologies and farming practices, as well as offering need-based trainings and demonstrations to farmers.

In addition, the Indian government has initiated several schemes and programs to provide farmers with the necessary resources and support to improve their agricultural practices and increase their incomes. Hence agricultural extension plays a crucial role by providing farmers with the knowledge and tools needed to improve productivity, profitability and sustainability, and help them to adapt to the changing climate conditions.

However, the traditional approach of providing extension services and transferring technology has not been adequate to address the complex challenges faced by farmers, such as climate change, water scarcity, and soil degradation. So, to address the challenges being faced by Indian agriculture and to promote sustainable development, the agricultural extension policies need to be reformed. The best approach, as suggested by literature review, for reforms would be to devise policies driven by sustainable development philosophy.

Sustainable development philosophy prioritizes the needs of current generations while preserving resources and opportunities for future generations. In the context of Indian agriculture, this means implementing policies that not only increase crop yields and farmers' incomes but also protect the environment and conserve natural resources. A sustainable development philosophy, which prioritizes economic growth, social well-being, and environmental protection, can also provide a framework for devising policies that promote sustainable agriculture in India. This approach would involve considering the long-term impacts of agricultural policies on farmers, the environment, and the broader society.

One key aspect of this approach would be to prioritize smallholder and resource poor farmers, who are often marginalized and have limited access to extension services and technology. Policies shall be re-designed to empower them by providing the necessary resources, information, and training to adapt to changing conditions and improve their livelihoods. Another key aspect of sustainable development in Indian agriculture is promoting the use of climate-resilient crop varieties, farming practices via awareness campaign and trainings on how to adapt to changing weather patterns. To facilitate it, the government and private sector needs to invest in research and development to create appropriate agricultural technologies and farming practices that consider the unique challenges faced by Indian agriculture. Additionally, sustainable development in Indian agriculture also requires the participation and empowerment of various stakeholders (such as scientists, change agents, extension professionals, farmers etc.) in decision-making process, as they are the ones who are directly impacted by the policies and programs.

Therefore, the government and policymakers need to be systematically informed by answering following questions.

i. What are the real-time challenges for the sustainable development of Indian agricultural sector?

ii. Where to invest available funding on priority to promote sustainable agriculture?

iii. Which kind of technological interventions are needed to stabilize the socio-economic conditions of smallholder / resource poor farmers?

So, to answer the above stated questions, we deployed a Delphi methodology experiment to generate forecasts from experts.

Methodology

Research Locale

The North Eastern Region of India consists of eight states (Fig. 1) namely Arunachal Pradesh, Assam, Manipur, Meghalaya, Mizoram, Nagaland, Tripura, and Sikkim (stretched across 89°42'E to 97°25'E longitude and 22°N to 29°25'N latitude). These states are a home of more than 160 tribes and farming is characterized as resource poor subsistence farming. Considering the ongoing impacts of climate change in the North Eastern Region of India, this locale was purposively selected for conducting Delphi experiment.

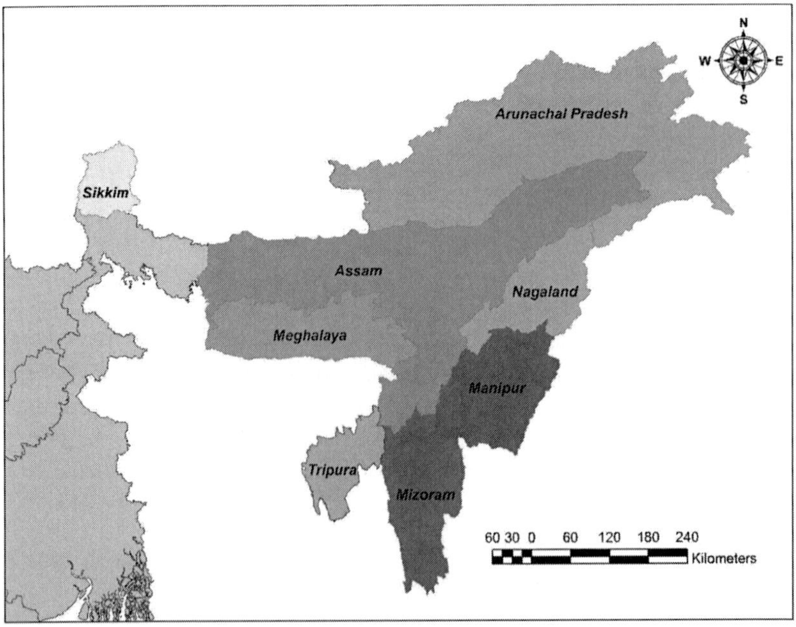

Figure 1: Map of North Eastern Region of India indicating eight constituent states

Study Design

The objective of this scientific investigation is to obtain forecasts on policy reforms for sustainable development of agricultural sector with controlled feedback. This method, called the Delphi technique, allows experts to test divergence, to form a consensus and to prioritize adaptation options. It includes four unique features such as anonymity, iteration, controlled feedback of individual judgments, and statistical aggregation of group responses (Rowe & Wright, 2001; Skulmoski, Hartman, & Krahn, 2007).

The anonymous treatment of individual judgments removes a common obstacle in collaborative decision-making (Adler & Ziglio, 1996; Mitroff & Turoff, 1975; Rowe & Wright, 2001; Somerville, 2007). Through several iterations, participants experience the aggregated perspective of the entire group and have an opportunity to revise their judgment. The objective of iterations is to gradually produce a consensus among participants (Ilbery, Maye, Kneafsey, Jenkins, & Walkley, 2004). The controlled feedback aims at reducing noise from data deficiencies and other influences (Dalkey, Brown, & Cochran, 1969; Hsu & Sandford, 2007). Controlled feedback helps to diminish various group research biases. Moreover, it also maintains the focus towards the purpose of the study by generating insights based on clear understanding among participants (Hsu & Sandford, 2007).

Delphi Modification

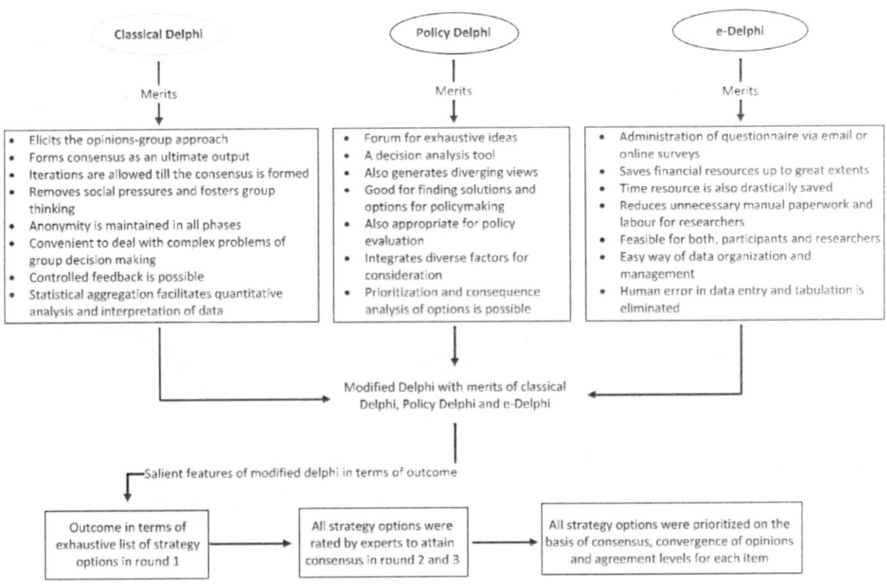

Figure 2: Modified Delphi approach derived from classical, policy and e-Delphi – schematic representation

The application of the Delphi technique in scientific studies includes several variations(Hasson & Keeney, 2011; Klenk & Hickey, 2011; Lund, Banta, & Bunting, 2014; Mukherjee et al., 2015). Van Zolingen and Klaassen (2003) distinguish them as classical, policy, and decision Delphi. Hasson and Keeney (2011) categorize different popular approaches into Modified Delphi, Real-time Delphi, e-Delphi, Technological Delphi, Argument Delphi, Online Delphi, and Disaggregative Delphi. With each variation of the Delphi design having particular strengths and weaknesses, researchers can choose the design best suited for a particular application.

In this study, we use features from the classical Delphi but modify the process of data collection. Particularly, we use electronic communication (email) and online questionnaires rather than postal mails and printed questionnaires. As we aim at forecasting adequate strategies for policymaking, we adopted also features from a Policy Delphi. A Policy Delphi is a variation to generate innovative solutions for complex problems such as climate change and sustainable development (Mukherjee et al., 2015). Policy Delphi has been applied to generate a broad range of ideas, strategies, and solutions (Hasson & Keeney, 2011; Turoff, 1970). In nutshell, our approach combines the elements from Classical Delphi, e-Delphi, and Policy Delphi. We use a structured, iterative process to understand the converging and diverging opinions of experts (Frewer et al., 2011) about the most relevant agricultural adaption strategies in North East Region of India.

Questionnaire Development and Administration

To develop a questionnaire for a Delphi experiment, identification of the specific research question is necessary. Then, researcher must determine the types of questions that will be included in the questionnaire, such as open-ended or closed-ended questions. After the questionnaire development, we pilot tested it to ensure that the questions are clear, easy to understand, and that they elicit the desired information. The pilot test was conducted with a small group of people who were identified from The North Eastern Region of India. After making any necessary revisions, the questionnaire was to the participants in the Delphi study. All necessary instructions were given to the participating experts. Finally, the data collected from the questionnaires we analyzed to identify patterns and trends, and to generate insights into the research questions.

Data Collection Approach

To collect data according to the objectives of this investigation, we used a modified Delphi approach which is a mixture of classical Delphi, policy Delphi, and e-Delphi. We utilized google form service to create online surveys for saving financial resources. This service facilitated quicker submission and retrieval of data which also saved a great amount of time. After designing the questionnaire, only the link was communicated to all experts through email. In this process, no group emails were sent to maintain the privacy and anonymity of experts. Multiple follow up emails (i.e., two to three) were also sent as reminders to experts for timely submission of responses. Sometimes personal telephonic calls were also made to experts who delayed. Most of the experts delayed response submission due to the work overload in their respective offices. Frequent electricity supply interruptions and breakdown of internet services in North East Region of India also caused delays.

Participants and Participation

To address questions in the methodology section, we selected experts from agriculture-related institutions in the North East Region of India. Most of the experts are affiliated with the Indian Council of Agricultural Research (ICAR), which has a wide network of research institutes, research centers, directorates, and Farm Science Centers (Krishi Vigyan Kendras or KVKs) across India. For this study, we recruited experts from ICAR's research institutes who are actively working in the North East Region of India and have at least five years of experience in the agriculture and related fields such as animal husbandry, beekeeping, silkworm rearing, poultry farming, etc. Although ICAR institutes typically operate at the national or regional level, we also wanted to include perspectives from the grass-root level. To do this, we included professionals from district-level KVKs. We included the heads of KVKs as they have a deep understanding of local farming communities, their farming practices, and specific issues in the district. Therefore, the scientific opinions of the heads of KVKs are important in the strategy formulation process.

128 experts who were professionally active in the Northeastern region of India were pre-identified for this investigation. Invitations to participate were sent to all 128 pre-identified experts via email. Of those experts, 41 (32.3%) accepted the invitation and participated in Round 1 of the Delphi investigation. However, only 35 (85.3%) of those experts had the required 5 years of work experience in the Northeastern region of India. The remaining 6 (14.63%) experts were eliminated from Round 1 analysis along with their inputs. The Round 2 questionnaire was sent to all 35 valid experts via email, and 32 (91.4%) of them participated and their responses were considered for analysis. Similarly, the Round 3 questionnaire was sent to all 32 experts via email, and 31 (93.7%) of them participated and their responses were considered for analysis. The iteration of Delphi rounds was terminated as consensus was reached at significant levels.

Qualitative Data Analysis

Both qualitative and quantitative methods were used to analyze Delphi survey data, with open-ended questions in pre-testing and first round questionnaires. Nvivo software was employed to analyze the qualitative data and content analysis was used to extract insights from similar ideas, terms and concepts, including expert suggestions.

Quantitative Data Analysis

To understand the degree of agreement and measure the consensus among experts, we used several descriptive statistics as well as inferential statistics tools.

In the first phase, we calculated the mean of Likert scores to find out average positions of all response items (Figure 1). However, we estimated consensus among experts using Inter Quartile Range (IQR) and Quartile Deviation (QD) (De Vet, Brug, De Nooijer, Dijkstra, & De Vries, 2004; Heiko, 2012). IQR is a popular measure of dispersion and widely used in Delphi investigations to estimate the consensus. Usually, IQR value less than 1 indicates that 50 % of expert ratings fall within 1 point on the scale(De Vet et al., 2004). De Vet et al. (2004) also established that consensus is obtained in case of 7-point rating scale if IQR is 1 or less than 1. In general, Delphi researchers (Raskin, 1994) used this criterion of IQR less than or equal to 1, to determine consensus for 5 points Likert scale. In another study,Heiko and Darkow (2010) used IQR less than or equal to 2 criteria for 10 points Likert scale. Above arguments implies that use of IQR less than or equal to 1, for 7-point rating scale provides a much robust and reliable consensus.

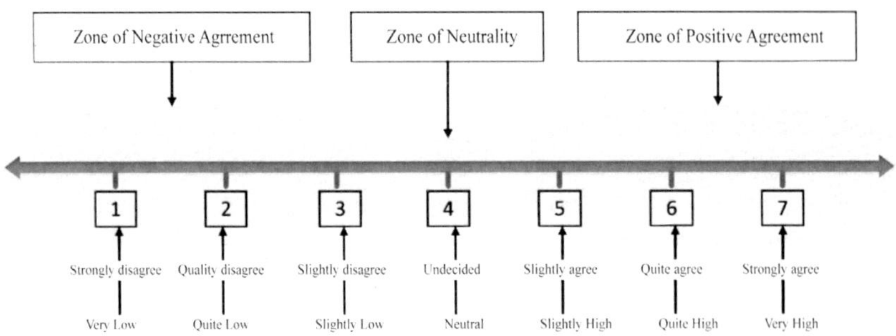

Figure. 3: Likert like scale used to explore expert agreement and disagreement

In the second phase of analysis, to explore the opinion convergence in the zone of positive agreement, we tested all items with a median, which represents the middle value of the data set. As values 5, 6 and 7 on Likert like scale indicate the zone of positive agreement (Fig. 3), the median more than 6 signified higher convergence in this zone. Similarly, Mode value represented the most popular vote in data series and used to triangulate the results with a median. If the majority of participants, as thumb rule say 51 %, vote for a single point, then consensus is assumed.

In the final phase of analysis, we summed Likert scores of "Strongly agree" "Quite agree" "Slightly agree" positions to find out the agreement scores for all items (Mcilfatrick & Keeney, 2003). Likewise, also totaled Likert scores of "Strongly disagree" "Quite disagree" "Slightly disagree" positions to observe the disagreement score. Besides, we established the reliability of the internal consistency of opinion ratings using Cronbach's (α) alpha (Bland &Altman, 1997; Cronbach, 1951). We performed a statistical analysis of quantitative data using IBM SPSS Statistics Software (Version-24).

Ethical Considerations

To preserve the anonymity of all experts, the research study only contacted them individually through email and avoided sending group emails. Participants were informed of the confidentiality and anonymity of the data and the study's objectives were clearly communicated in the emails. It was emphasized that the data would be used solely for research. All communication with experts during the data collection process was conducted in a respectful and professional manner.

Results and Discussion

This section discusses the findings of Delphi experiment to answer the research questions posed in the introduction section.

Real-time Challenges for the Sustainable Development

Table 1 indicates key challenges that are impediments for sustainable development of Indian Agriculture, specifically in the context of North East Region of India. These challenges are derived after the consensus among experts at the end of third iteration of Delphi experiment. High-intensity rainfall and floods first and foremost challenge for sustainable agriculture in the North East Region of India. Heavy rainfall causes soil erosion, wash away crops, and damage irrigation systems. Flooding contaminates soil and water, making them unsuitable for growing crops. Additionally, floods in North East Region of India disrupt transportation and distribution networks, make it difficult for farmers to access markets and sell their products (Chetia, Ahmed, Singh, & Feroze, 2015; Pal, Biswas, Mondal, & Pramanik, 2016). These challenges lead to decreased crop yields and financial losses for farmers, making it impossible for them to continue farming sustainably. To mitigate these challenges, farmers may need to adopt new practices such as terracing, crop rotation, tunnel farming, protected farming and more efficient irrigation systems.

Similarly, access to high-quality seed is essential for achieving high crop yields and maintaining the health of the soil. However, many farmers in North East Region of India have limited access to good quality seed, which can lead to reduced crop yields and increased vulnerability to crop pests and diseases. Additionally, the use of poor-quality seed can also lead to the loss of genetic diversity, which is essential for maintaining the resilience of agricultural systems. The use of low-quality seed subsequently leads to a greater dependence on chemical inputs, such as pesticides and fertilizers, which can be harmful to the environment and human health. Inadequate seed supply also increases the risk of crop failures, which can have severe economic and social consequences for farmers and their communities. To overcome this challenge, farmers across India need access to high-quality seed that is adapted to local conditions which is produced and distributed in a sustainable manner. This can be achieved by promoting seed production and distribution systems that are community-based and that involve farmers. Additionally, public and private sector organizations can play a key role in developing and distributing high-quality seed varieties that are adapted to local conditions.

Table 1: The real-time challenges for the sustainable development of Indian agricultural sector

Challenges	Median	Mode	IQR	QD	Agreement
High-intensity rainfall and floods	6	7	1	0.5	100
Inadequate seeds, poor quality seeds	6	6	1	0.5	100
Weak as well as unstable marketing set up	6	6	1	0.5	100
Low extent of value addition in farming	6	6	1	0.5	100
Rainfall dependent farming	7	7	1	0.5	96.77
Less farm mechanization	6	6	1	0.5	96.77
Trivial farm income	6	6	1	0.5	93.55
Unscientific and resource poor livestock farming	6	6	1	0.5	93.55
Policy apathy and low scheme penetration	6	6	1	0.5	90.32
Drought management	6	6	1	0.5	90.32
Temperature increase and warming	6	6	1	0.5	90.32
High labor cost and energy shortages	6	6	1	0.5	90.32
Unscientific and exploitative farming	6	6	1	0.5	90.32

Further, lack of access to stable and profitable markets for agricultural products can make it difficult for farmers to make a living and to invest in sustainable farming practices. Without a reliable market for their products, farmers may be forced to sell at low prices, which is not even sufficient to cover their production costs and make a profit. Additionally, weak marketing systems hinders farmers access to information about prices, demand, and other market conditions. As a result, farmers are unable to make informed decisions about what to grow and when to sell. In addition, weak and unstable marketing makes supply chain inefficient and blocks farmers' access to new markets and consumers. Furthermore, this can also lead to the dominance of middlemen which limits the bargaining power of small farmers. hence, to overcome this challenge, it is important to develop and strengthen marketing systems that are inclusive, transparent, and efficient. This can be achieved by developing market infrastructure, strengthening farmer organizations (such as Farmer Producer Organizations i.e. FPOs), and promoting collective action among farmers. Additionally, government policies and programs must play a key role in promoting fair trade, and providing technical and financial assistance to help farmers access new markets and to improve their bargaining power.

Moreover, Low value addition is also a significant challenge for sustainable agriculture in India. Value addition refers to the process of increasing the value of a product by adding value to it through activities such as processing, packaging, and branding. Low value addition paralyzes farmers to compete in the global market and to achieve a fair price for their products. Additionally, low value addition hinders to access new markets and consumers, which limits farmers income-earning potential. The lack of value addition in Indian agriculture also results in a high proportion of agricultural produce being perishable and therefore, a large percentage of the produce is lost due to lack of proper storage and transportation facilities. This not only leads to financial losses for farmers but also results in food insecurity and wastage of resources. Therefore, farmers need to be provided with the necessary resources and support to add value to their products. This can be attained by promoting the development of small-scale processing and packaging facilities, as well as by providing training and technical assistance to farmers on how to add value to their products. Additionally, government policies and programs for promoting value addition in agriculture must provide financial and technical assistance to farmers to create an enabling environment for agribusinesses to flourish.

With respect to rainfall dependency of farming, a large portion of Indian agriculture is dependent on monsoon rains, which can be unpredictable and unreliable. This does not allow farmers to systematically plan and manage their crops. This dependency also leads to crop failures and decreased yields. Additionally, the over-reliance on monsoon rains often triggers lack of water management and conservation practices, which can further exacerbate the impacts of droughts and floods. Moreover, the over-reliance on monsoon hinders the adaptation to changing climate conditions, such as rising temperatures and changes in precipitation patterns. Ultimately, farmers face increased vulnerability to pests, diseases and extreme weather events as they could not grow crops that are well-suited to local conditions. To overcome this challenge, farmers across India need to be provided with the necessary resources and support to develop and adopt more sustainable water management and irrigation practices. This can be achieved by promoting the development of small-scale irrigation systems, as well as by providing training and technical assistance to farmers on how to conserve water and manage their crops in a sustainable manner.

Furthermore, less farm mechanization in India hinders sustainable agriculture for several reasons. One of the main reasons is low productivity. Without the use of modern machinery and equipment, farmers in North East Region of India often have to rely on manual labor, which is time-consuming and less efficient (K. Dikshit & J. K. Dikshit, 2014a, 2014b; K. R. Dikshit & J. K. Dikshit, 2014). This leads to lower crop yields and reduced income for farmers. This also leads to a lack of modernization, as farmers are less likely to adopt modern farming techniques and technologies, which can improve efficiency and sustainability. Another major reason is high labor costs. The reliance on manual labor also increases labor costs, which can be a significant burden for small farmers in India. This is particularly problematic in the context of labor scarcity and increasing labor costs, as it can make farming less profitable and less sustainable in the long term. In addition, soil degradation is another major issue caused by the lack of farm mechanization. Overuse of manual labor can lead to soil degradation, which can reduce crop yields and make it more difficult to grow crops in the future. This can also have a negative impact on the environment, as it can lead to erosion and other forms of land degradation. Finally, less farm mechanization in North East Region of India also leads to dependence on human labor, which is becoming scarce and expensive, and this can also be a problem for sustainable agriculture. As the population increases and more people move to urban areas, there will be fewer people available to work in agriculture, making it more difficult to sustainably produce food for the growing population. So, increasing farm mechanization in India would help to improve efficiency, productivity, and sustainability in agriculture.

Besides above, additional challenges with less agreement are trivial farm income, Unscientific and resource poor livestock farming, Policy apathy and low scheme penetration, Drought management, temperature increase and overall warming. Unscientific and resource poor livestock farming lead to overgrazing, soil degradation, and increased greenhouse gas emissions. This type of farming is often characterized by the use of low-yielding breeds, poor nutrition, and inadequate veterinary care, which can reduce the productivity and sustainability of livestock production. And policy apathy and low scheme penetration is another challenge for sustainable farming in India. Government policies and schemes are crucial for promoting sustainable farming practices, but their implementation is often hindered by lack of political will, lack of coordination among government agencies, and poor implementation at the grassroots level. This can limit the ability of farmers to access the support and resources they need to adopt sustainable practices. Further, some experts feel drought management is a challenge for sustainable farming in India as it affects crop productivity, water availability, and soil health. Climate change is making droughts more frequent and severe, which can lead to reduced crop yields and increased water stress. This can also exacerbate land degradation and desertification, making it more difficult to sustainably produce food. And same is reported by majority of experts as temperature increase and warming may affect crop growth and productivity. Rising temperatures can lead to water stress, heat stress, and increased pest and disease pressure, all of which can reduce crop yields and make it more difficult to sustainably produce food. This along with other climate change impacts such as extreme weather events, irregular rainfall pattern and soil erosion can impede sustainable agriculture in India.

Technological Interventions Needed to Stabilize the Socio-economic Conditions of Smallholder / Resource Poor Farmers

The hilly farming terrain in North East Region of India demands certain technological interventions to provide an assured income to smallholder farmers. Table 2 indicates some of those interventions forecasted by the experts which not only stabilize the socio-economic conditions of smallholder / resource poor farmers but also aid in fostering sustainable agriculture.

Firstly, scientific crop and livestock farming support sustainable agriculture by utilizing advanced technologies and research-based practices to optimize yields, improve the efficiency of resource use, and reduce negative impacts on the environment (Jat et al., 2015; Lal, 2013; Misra, 2014; Nandi, Gowdru, Bokelmann, & Dias, 2015). This can include the use of precision agriculture techniques, such as precision planting and nutrient management, as well as the development of resilient crop varieties and sustainable livestock breeding practices. Additionally, scientific farming can support sustainable agriculture by incorporating principles of agroecology, which emphasizes the interdependence of plants, animals, and the environment, and by promoting soil health and biodiversity. With scientific techniques, smallholder farmers can increase crop and livestock productivity, leading to higher income from sales of these products. By using scientific techniques, smallholder farmers can produce higher quality and value-added products such as organic or certified products, leading to higher income from sales.

Secondly, smallholder farmers can diversify their production, for example by adding new crops or livestock species, leading to higher income from sales (Khatri-Chhetri, Aryal, Sapkota, & Khurana, 2016; Nandi et al., 2015). Scientific approach of diversification reduces the amount of resources needed to produce their crops and livestock, leading to reduced costs and higher profit margins. And by producing high-quality, value-added products, smallholder farmers may be able to access new and more lucrative markets, leading to higher income from sales. Most importantly, these practices diminish the environmental impact of their farming practices, making them eligible for eco-labeling and premium prices.

Thirdly, adequate input supply at the right time is crucial for farmers to plant, grow, and care for their crops and livestock, leading to higher yields and higher income. Quality inputs such as seeds, fertilizers, and pesticides are essential to ensure healthy growth of crops and livestock. Poor quality inputs can lead to poor yields, disease and pest outbreaks, and reduced income. With timely and quality input supply, farmers can generate income from sales, which can be used to cover expenses, including input costs and invest in future farming activities, leading to financial stability. Further, smallholder farmers can improve their livelihood by increasing their income, food security, and overall well-being by receiving right inputs at right time. Moreover, timely and quality input supply can support sustainable agriculture by reducing the environmental impact of farming practices, promoting soil health and biodiversity, and ultimately leading to sustainable livelihoods for smallholder farmers.

Fourthly, Indigenous traditional knowledge provides smallholder farmers with a deep understanding of the local environment and its specific conditions, including climate, soil types, and water resources. This knowledge enables farmers to adapt their farming practices to the local context, increasing the chances of successful crop and livestock production (Ingty, 2017; Maiti et al., 2014; Sharma & Dhakal,

2011; Sharma & Rai, 2012). Further, Indigenous traditional knowledge has been developed and passed down over generations, based on the principles of agroecology and conservation of biodiversity. This knowledge enables smallholder farmers to grow crops and raise livestock in a way that is sustainable and preserves the local ecosystem. Consequently, it enables smallholder farmers to grow a variety of crops that are suited to the local environment and can provide food security for their families and communities (Saha, Sundriyal, & Sundriyal, 2014). It is also observed that indigenous knowledge is closely tied to the culture and identity of the local community. So, by promoting the use of indigenous knowledge, farmers preserve their cultural heritage and promote the identity of their community.

Table 2: Technological interventions are needed to stabilize the socio-economic conditions of smallholder / resource poor farmers

Technological interventions	Median	Mode	IQR	QD	Agreement
Scientific crop farming and livestock rearing	6	6	1	0.5	100
Diversification of farming	6	6	0	0	96.77
Quality input supply in a Timely manner	6	6	1	0.5	96.77
Use of indigenous traditional knowledge	6	6	1	0.5	96.77
Integrated farming system (IFS)	6	7	1	0.5	93.55
Soil, water and resource conservation	6	6	0	0	93.55
Co-operative and e-marketing	6	6	1	0.5	93.55
Price assurance and appropriation	6	7	1	0.5	93.55
Post-harvest value addition	6	6	1	0.5	93.55
Insurance and credit	6	6	1	0.5	93.55
Community approach for farming and awareness creation	6	5	1	0.5	90.32
Formulation and execution of national climate plan	6	6	1	0.5	90.32
Socially acceptable and efficient technologies	6	6	1	0.5	90.32
Weather advisory and ICTs	6	6	1	0.5	90.32

Fifthly, experts recommended the promotion of Integrated Farming Systems (IFS) to stabilize the assured income for resource poor as well as small holder farmers. IFS allow farmers to use resources such as land, water, and labor more efficiently by combining multiple farming practices such as crop cultivation, animal husbandry, agroforestry, and aquaculture in an integrated way (Dar et al., 2018). As a result, improved yields and reduced costs, increasing farmers' income and financial stability are possible. Similarly, IFS encourage the conservation of biodiversity by integrating different crops, livestock, and other organisms in a single system (Suh, 2014). This helps to improve soil health, enhance the resilience of the ecosystem, and reduce the negative impact of farming on the environment. Further, resilience of smallholder farmers to the impacts of climate change is attainable by promoting sustainable farming practices in IFS that conserve resources and enhance the resilience of the ecosystem, reducing the vulnerability of farmers to climate change. Food security and IFS are closely linked as by producing a diverse range of crops and livestock, which can provide a steady supply of food for the farmers and their families and communities. Most importantly, IFS diversify the income of smallholder farmers by providing multiple products and services, such as crops, livestock, non-timber forest products, and fish, which can be sold in local markets. So ultimately, sixth intervention proposed by experts is soil, water and resource conservation are also attainable by promoting IFS in the North East Region of India.

Figure 4: An overview of Integrated Farming Model with several components of farming (Digital sketch by author)

Seventhly, experts have forecasted the marketing sector reforms to stabilize the socio-economic conditions of farmers in the North East Region of India. Co-operatives can help farmers of North East Region of India to pool their resources and negotiate better prices for their products. E-marketing can help them to access new markets and sell their products at a fair price. Together, these can help to stabilize the income and financial situation of smallholder farmers. in addition, price assurance and price appropriation can help farmers to secure a fair price for their products. By providing price assurance, farmers can be protected from price fluctuations and market volatility. Price appropriation can help farmers to receive a fair price for their products by ensuring that they are not exploited by middlemen or traders. Besides, insurance and credit facilities will empower farmers to manage risks and invest in their farming activities. Insurance can protect farmers from financial losses due to crop failures or natural disasters. Whereas credit can help farmers to invest in inputs, equipment, and infrastructure, which can improve their yields and income.

Finally, experts have stressed on Community approach for farming and awareness creation as well as use of Socially acceptable and efficient technologies. These dimensions are compatible with earlier recommendation of Indigenous traditional knowledge which is even today a socially compatible community farming activity i.e. *jhooom* farming in North East Region of India. though the jhoom farming area is decreasing day by day, experts still recommended Formulation and execution of national climate plan as well as use of Weather advisory and ICTs for effectively managing the hazardous impacts of climate related calamities in North East region of India.

Areas of Priority Investment to Promote Sustainable Agriculture

Table 3 provides the key priority area for investment in order to boost the sustainable development in the North East Region of India. first of all, efficient natural resource management and utilization is crucial for the sustainable agriculture in North East India because the region is characterized by poor soil quality, limited water resources, and a high dependence on rain-fed agriculture. Proper management of these resources is pivotal to enhance the crop yields and income for farmers, as well as improved food security for the region. Additionally, investing for efficient utilization of natural resources will help to conserve the region's biodiversity and reduce the negative impacts of agriculture on the environment. Overall, prioritizing investment in natural resource management and utilization is essential for the long-term sustainability and prosperity of the agricultural sector in Northeast India.

Further, the first investment area calls for second logical recommendation i.e. investment for Integrated farming systems in the North East Region of India. Integrated farming systems (IFS) are a holistic approach to agriculture that combines multiple farming practices in order to increase efficiency, productivity, and sustainability. IFS prioritize investment in Northeast India's sustainable agriculture because they offer several benefits. IFS often include the use of crop rotation, intercropping, and the integration of livestock into farming systems, which can help to improve soil quality and fertility. By integrating practices such as rainwater harvesting, irrigation, and conservation agriculture, IFS aids a lot to better manage and utilize limited water resources in the region. IFS modules often include the use of integrated pest management techniques, which reduce the dependence on chemical pesticides and improve crop health. As a result, IFS enable farmers to diversify their income streams and reduce their dependence on a single crop or market, which can improve their overall economic stability. And this is the third recommendation forecasted by experts that priority investment is needed for crop diversification. Therefore, experts made crucial recommendation that priority investment in integrated farming systems is essential for achieving sustainable agriculture in Northeast India, by increasing efficiency, resilience, and income while protecting the environment.

Further, the experts feel that credit and insurance are important tools for achieving sustainable agriculture in Northeast India. Priority investment in credit can help to provide farmers with the necessary financial resources to improve their operations. Smallholder farmers in Northeast India often have limited access to credit, which can impede their ability to invest in new technologies and practices that can improve their crop yields and income. Similarly, crop / livestock insurance will help farmers to manage the risks associated with agriculture, such as crop failure due to weather or pest damage. Priority investment in insurance is essential to protect farmers from financial losses and provide them with the security they need to invest in their operations. As access to credit and insurance can help farmers to improve their livelihoods by increasing their income and providing them with greater financial stability, priority investment is needed.

In addition, experts recommended investment for Solar energy infrastructure as its crucial for achieving sustainable agriculture in Northeast India. One of the main reasons for this is that the region has high solar radiation levels, making it an ideal location for solar energy production. This can provide farmers with a reliable source of energy for powering irrigation systems, crop drying, and other farm operations, thus reducing dependence on fossil fuels and decreasing the environmental impact of agriculture. Additionally, investment in solar energy infrastructure will increase the efficiency of agricultural operations, by providing farmers with the energy they need to power precision farming technologies such as precision irrigation, crop monitoring and mapping. It will subsequently optimize crop yields, reduce waste and improve the overall efficiency of the agricultural sector. Furthermore, solar energy infrastructure will also be used to power rural electrification projects in remote areas, where grid-based electricity is not available. Importantly, investment in solar energy infrastructure will also create employment opportunities and support local economies, by providing jobs in the construction, maintenance, and operation of solar energy systems.

Furthermore, experts recommended investment for germplasm conservation and its improvement. As the North East region of India is home to a rich diversity of plant and animal species, many of which have been traditionally used by local farmers for food, medicine, and other purposes. These traditional knowledge and local varieties are adapted to the specific agro-ecological conditions of the region, making them well-suited to local farming systems. Priority investment in germplasm conservation is essential to preserve valuable genetic diversity and ensure that it remains available for future generations of farmers. consequently, it will improve the resilience of the agricultural sector to changing weather conditions, pests, and diseases. It can also ensure that the cultural heritage of the region is preserved and valued, while also promoting the participation and empowerment of local communities in the development of sustainable agricultural systems.

Moreover, experts forecasted that, providing intelligence-based market reforms, investment for community seed banks and nurseries, and investment for nutritive feeding to livestock is also essential. Intelligence-based market reforms will ensure that farmers in the region have access to fair and profitable markets for their products. By providing farmers with accurate and timely information about prices, market trends, and other relevant data, these reforms will be beneficial to improve their bargaining power and increase their income. Community seed banks and nurseries will provide disease free high-quality seed and planting material for farmers in the region. By providing farmers with access to diverse and locally adapted seed varieties, these institutions can help to improve crop yields and reduce the dependence on external seed sources. Similarly, nutritive feeding to livestock is important for improving the health and productivity of animals in the region. With appropriate feed and nutrition to livestock, farmers can improve the quality of milk and meat, which will eventually increase their income and improve food security in the region.

Table 3: Areas that demand Investment on priority to promote sustainable agriculture

Investment priorities	**Median**	**Mode**	**IQR**	**QD**	**Agreement**
Efficient natural resource management and utilization	6	6	0	0	96.77
Integrated farming system	6	6	1	0.5	96.77
Crop diversification	6	6	1	0.5	96.77
Credit and insurance	6	6	1	0.5	96.77
Post-harvest value addition	6	6	1	0.5	96.77
Solar energy infrastructure	6	6	1	0.5	93.55
Germplasm conservation and improvement using ITKs	6	6	1	0.5	93.55
Research for climate resilient policy	6	6	0	0	93.55
Intelligence based market reforms	6	6	0	0	93.55
Community seed banks and nurseries	6	6	0	0	93.55
Nutritive feeding to livestock	6	6	1	0.5	93.55
Organic and conservation farming	6	6	1	0.5	90.32
Awareness and trainings for enhancing resilience	6	6	1	0.5	90.32
Accurate forecasting by institutional convergence	6	6	1	0.5	90.32
Climate resilient housing for livestock	6	6	1	0.5	90.32
Low carbon farming and carbon sequestration	6	6	1	0.5	83.87

Besides, investment is needed to promote organic and conservation farming practices. These practices are crucial for improving soil health and reducing dependence on chemical inputs. It will boost yields, and ultimately help to maintain the long-term productivity of the land and reduce the negative impacts of intensive farming practices. Additionally, farmers need to be educated on the use of new technologies such as weather forecasting, precision agriculture, and climate-smart farming practices through awareness and trainings to enhance their resilience to climate change. Accurate forecasting by institutional convergence is also important to help farmers make better decisions about planting and harvesting. This coordination helps to ensure that accurate and timely information is available to farmers, which is critical for the success of sustainable agriculture. Climate resilient housing for livestock is also necessary to protect animals from extreme weather conditions and improve their health and productivity. further, low carbon farming and carbon sequestration are also critical investments to reduce greenhouse gas emissions and mitigate the effects of climate change (Pachauri et al., 2014). This helps to reduce the amount of carbon dioxide in the atmosphere, which can help to mitigate the effects of climate change. Therefore, priority investments are necessary to improve the long-term sustainability and resilience of agricultural systems in the region, which can lead to increased food security and economic growth.

Conclusion

This study signifies that experts have noticed adversities caused by climate change in North East Region of India. The key challenges in North East Region of India also highlight the interlinking of climatic catastrophes with existing socio-economic an environmental adversity. The exhaustive list of forecasted recommendations symbolizes that massive amount of strategic investment is required in almost all facets of the development of North East Region of India. Expert judgments are unanimously inclined towards re-thinking and re-shaping existing agricultural sector policies towards climate-sensitive strategies for the sustainable development in North East Region of India.

Multidisciplinary expert panel recurrently advocated Integrated Farming system, scientific organic farming, germplasm / biodiversity conservation, marketing reforms and promotion of community farming approaches as most plausible reforms to shape up future agricultural extension policies for the North East Region of India to promote the sustainable development. The demand driven, commodity or market driven extension approaches might not be very effective in climate change scenario, whereas sustainability driven agricultural extension policies will be much more applicable for sustainable development.

References

Adler, M., & Ziglio, E. (1996). Gazing into the oracle: The Delphi method and its application to social policy and public health: Jessica Kingsley Publishers.

Bland, J. M., & Altman, D. G. (1997). Statistics notes: Cronbach's alpha. Bmj, 314(7080), 572.

Chetia, S. K., Ahmed, T., Singh, R., & Feroze, S. (2015). Impact of Floods on Rice Based Farming in Assam: A Gender Study. Journal of Eco-friendly Agriculture, 10(1), 43-46.

Cronbach, L. J. (1951). Coefficient alpha and the internal structure of tests. psychometrika, 16(3), 297-334.

Dalkey, N. C., Brown, B. B., & Cochran, S. (1969). The Delphi method: An experimental study of group opinion (Vol. 3): Rand Corporation Santa Monica, CA.

Dar, N. A., Lone, B. A., Alaie, B. A., Dar, Z. A., Gulzafar, Bahar, F. A., . . . Singh, K. N. (2018). Integrated Farming Systems for Sustainable Agriculture. In R. S. Sengar & A. Singh (Eds.), Eco-friendly Agro-biological Techniques for Enhancing Crop Productivity (pp. 111-127). Singapore: Springer Singapore.

De Vet, E., Brug, J., De Nooijer, J., Dijkstra, A., & De Vries, N. K. (2004). Determinants of forward stage transitions: a Delphi study. Health education research, 20(2), 195-205.

Dikshit, K., & Dikshit, J. K. (2014a). Agriculture in North-East India: Past and Present. In North-East India: Land, People and Economy (pp. 587-637): Springer.

Dikshit, K., & Dikshit, J. K. (2014b). People of North-East India. In North-East India: Land, People and Economy (pp. 285-420): Springer.

Dikshit, K. R., & Dikshit, J. K. (2014). Socio-Economic Development of the North-East: An Assessment. In North-East India: Land, People and Economy (pp. 723-749): Springer.

Frewer, L., Fischer, A., Wentholt, M., Marvin, H., Ooms, B., Coles, D., & Rowe, G. (2011). The use of Delphi methodology in agrifood policy development: some lessons learned. Technological Forecasting and Social Change, 78(9), 1514-1525.

Hasson, F., & Keeney, S. (2011). Enhancing rigour in the Delphi technique research. Technological Forecasting and Social Change, 78(9), 1695-1704.

Heiko, A. (2012). Consensus measurement in Delphi studies: review and implications for future quality assurance. Technological Forecasting and Social Change, 79(8), 1525-1536.

Heiko, A., & Darkow, I.-L. (2010). Scenarios for the logistics services industry: a Delphi-based analysis for 2025. International Journal of Production Economics, 127(1), 46-59.

Hsu, C.-C., & Sandford, B. A. (2007). The Delphi technique: making sense of consensus. Practical assessment, research & evaluation, 12(10), 1-8.

Ilbery, B., Maye, D., Kneafsey, M., Jenkins, T., & Walkley, C. (2004). Forecasting food supply chain developments in lagging rural regions: evidence from the UK. Journal of Rural Studies, 20(3), 331-344.

Ingty, T. (2017). High mountain communities and climate change: adaptation, traditional ecological knowledge, and institutions. Climatic change, 145(1), 41-55. doi:10.1007/s10584-017-2080-3

Jat, L. K., Singh, Y., Meena, S. K., Meena, S. K., Parihar, M., Jatav, H., . . . Meena, V. S. (2015). Does integrated nutrient management enhance agricultural productivity. J Pure Appl Microbiol, 9(2), 1211-1221.

Khatri-Chhetri, A., Aryal, J. P., Sapkota, T. B., & Khurana, R. (2016). Economic benefits of climate-smart agricultural practices to smallholder farmers in the Indo-Gangetic Plains of India. Current Science, 110(7), 1251-1256. Retrieved from <Go to ISI>://WOS:000373347000028

Klenk, N. L., & Hickey, G. M. (2011). A virtual and anonymous, deliberative and analytic participation process for planning and evaluation: The Concept Mapping Policy Delphi. International Journal of Forecasting, 27(1), 152-165.

Lal, R. (2013). Food security in a changing climate. Ecohydrology & Hydrobiology, 13(1), 8-21.

Lund, S., Banta, G. T., & Bunting, S. W. (2014). Applying stakeholder Delphi techniques for planning sustainable use of aquatic resources: experiences from upland China, India and Vietnam. Sustainability of Water Quality and Ecology, 3, 14-24.

Maiti, S., Jha, S. K., Garai, S., Nag, A., Chakravarty, R., Kadian, K., . . . Upadhayay, R. (2014). Adapting to climate change: Traditional coping mechanism followed by the Brokpa pastoral nomads of Arunachal Pradesh, India.

Mcilfatrick, S. J., & Keeney, S. (2003). Identifying cancer nursing research priorities using the Delphi technique. Journal of advanced nursing, 42(6), 629-636.

Misra, A. K. (2014). Climate change and challenges of water and food security. International Journal of Sustainable Built Environment, 3(1), 153-165.

Mitroff, I. I., & Turoff, M. (1975). Philosophical and methodological foundations of Delphi. The Delphi method: Techniques and applications, 17-35.

Mukherjee, N., Huge, J., Sutherland, W. J., McNeill, J., Van Opstal, M., Dahdouh□Guebas, F., & Koedam, N. (2015). The Delphi technique in ecology and biological conservation: applications and guidelines. Methods in Ecology and Evolution, 6(9), 1097-1109.

Nandi, R., Gowdru, N. V., Bokelmann, W., & Dias, G. (2015). Smallholder organic farmer's attitudes, objectives and barriers towards production of organic fruits and vegetables in India: A multivariate analysis. Emirates Journal of Food and Agriculture, 396-406.

Pachauri, R. K., Allen, M. R., Barros, V. R., Broome, J., Cramer, W., Christ, R., . . . Dasgupta, P. (2014). Climate change 2014: synthesis report. Contribution of Working Groups I, II and III to the fifth assessment report of the Intergovernmental Panel on Climate Change: IPCC.

Pal, R., Biswas, S. S., Mondal, B., & Pramanik, M. K. (2016). Landslides and Floods in the Tista Basin (Darjeeling and Jalpaiguri Districts): Historical Evidence, Causes and Consequences. J. Ind. Geophys. Union (April 2016), 20(2), 209-215.

Raskin, M. S. (1994). The Delphi study in field instruction revisited: Expert consensus on issues and research priorities. Journal of Social Work Education, 30(1), 75-89.

Rowe, G., & Wright, G. (2001). Expert Opinions in Forecasting: The Role of the Delphi Technique. In J. S. Armstrong (Ed.), Principles of Forecasting: A Handbook for Researchers and Practitioners (pp. 125-144). Boston, MA: Springer US.

Saha, D., Sundriyal, M., & Sundriyal, R. (2014). Diversity of food composition and nutritive analysis of edible wild plants in a multi-ethnic tribal land, Northeast India: an important facet for food supply. Indian Journal of Traditional Knowledge, 13(4), 698-705. Retrieved from <Go to ISI>://WOS:000345056800010

Sharma, G., & Dhakal, T. (2011). Opportunities and challenges of the globally important traditional agriculture heritage systems of the Sikkim Himalaya. Biodiversity of Sikkim: exploring and conserving a global hotspot. IPR Department, Government of Sikkim, Gangtok, 411-440.

Sharma, G., & Rai, L. K. (2012). Climate change and sustainability of agrodiversity in traditional farming of the Sikkim Himalaya. Climate Change in Sikkim Patterns, Impacts and Initiatives. Information and Public Relations Department, Government of Sikkim, Gangtok.

Skulmoski, G. J., Hartman, F. T., & Krahn, J. (2007). The Delphi Method for Graduate Research. Journal of Information Technology Education, 6.

Somerville, J. A. (2007). Effective use of the Delphi process in research: Its characteristics, strengths and limitations. Critical factors affecting the meaningful assessment of student learning outcomes: a Delphi study of the opinions of community college personnel. PhD Dissertation, Oregon State University.

Suh, J. (2014). Theory and reality of integrated rice–duck farming in Asian developing countries: A systematic review and SWOT analysis. Agricultural Systems, 125, 74-81.

Turoff, M. (1970). The design of a policy Delphi. Technological Forecasting and Social Change, 2(2), 149-171.

Van Zolingen, S. J., & Klaassen, C. A. (2003). Selection processes in a Delphi study about key qualifications in senior secondary vocational education. Technological Forecasting and Social Change, 70(4), 317-340.

14

Evolving Institutionalization Process of Agricultural Socialization: The Role of ATMA and Its Contribution

Swaraj Preetam Das[1*], B.K. Mohanty[2] and Saumyesh Acharya[3]

[1,2]Department of Agricultural Extension & Communication, Faculty of Agricultural Sciences (IAS), S'O'A Deemed to be University, Bhubaneswar Odisha

[3]Department of Agricultural Extension, Institute of Agriculture Visva-Bharati University, Sriniketan, West Bengal, India

Abstract

Agricultural Technology Management Agency (ATMA) scheme was launched to refine farmers' skill and expertise, provide better understanding, and achieve enhanced proficiency in a number of developmental programs and promote agrarian development. The objective of the present study was to assess the impact of socio-economic variables on development of farmers through various ATMA activities. For this purpose, 100 respondents were selected from Karanjia, Jashipur, Bangiriposi, Baripada and Shamakhunta blocks of Mayurbhanj, Odisha respectively. A pre-tested structured interview schedule was prepared. The data were collected through personal interview, focus group discussion and direct observation methods. Further the data was analysed through different statistical tools; correlation coefficient, regression analysis and path analysis. The results of the correlation coefficient revealed that following socio-economic variables i.e., age, major family occupation, education, land holding size, family size, social participation, communication media used, average annual family income had a significant impact on development of the farmers through ATMA activities. Regression analysis revealed that 12 independent variables collectively contributed 49.4% variance in the dependent variable development of farmers through ATMA activities. It is evident that the independent variables i.e., age, education, social participation, communication media used and average annual family income have significant contribution on the dependent variable development of farmers through ATMA activities. Path analysis concluded that the finding family size of the respondents had exhibited highest total direct effect followed by caste and housing pattern. Similarly, the variable education exhibited highest indirect effect. This study provides inputs for making future policy decisions.

Keywords: Adoption, ATMA, development, family size, involvement, social participation.

Introduction

Agriculture is vital to the Indian economy in terms of job creation, with nearly half of the Indian population relying on agriculture and related activities for a living. A large proportion of the population, particularly in rural areas, is illiterate. The rural population, which is illiterate and ignorant, is unable to cope with the complexities of agricultural technology management. As a result, vocational training and technology dissemination are few viable options for increasing production efficiency and achieving rapid and sustainable agricultural development. The State of Odisha derives more than a quarter of its revenue from agriculture. More than 40% of the people in the state of Odisha depend on farming as their primary source of income and way of life, making up more than 53% of the state's workforce. whereas agricultural technology has developed into a logical extension of the traditions, conventions, knowledge, and skills that date back to the early days of human civilization and have continued to evolve over time. It was not just agriculture that produced food or other biological means; rather, it essentially presented a wide range of traditional values to make it abundantly evident that "TECHNOLOGY and TECHNIQUE" are the cultural means to satisfy the biological as well as social demand for life. To provide the necessary acceleration to the agricultural development, Agricultural Technology Management Agency (ATMA) Scheme was launched during 2005-06 in order to enrich farmer's skills, knowledge and competency several developmental programmes/projects, in agriculture and allied sectors. In the state Odisha, a district level autonomous institute has taken the main responsibility of all technology dissemination activities at district level for development of farmers and farm women through its bottom –up approach. During 2010 the reforms made in ATMA project and farm persons were given due importance to its multifarious activities like training, demonstration, information support, input supply, market-oriented facilities, linking credit and exposure visit etc. These multi-purpose activities provide ample opportunities for the farm communities and makes the resources accessible for them. Additionally, ATMA served as a link between individuals in need and those with the means necessary for undergoing rapid technological change. It is also critical to evaluate various stakeholders of ATMA in order to enhance both its organisational and functional approach. This includes various parameters such as efficiency of skilled professionals of ATMA, availability of resources and feedback from farmer beneficiaries among other factors.

Mass contact had significant and positive relationship with the Block Supervisors' perceptions of on-the-job training. (Hoque and Koichi, 2007) As high as 40 per cent of total respondents and 38.7 and 41.4 per cent of respondents of Agriculture and allied departments belonged to high level of coordination in planning. Whereas, 44.2 per cent of total respondents from agriculture (50.0%) and allied department (48.3%) had medium level of coordination in organizing. Further, 46.7 per cent of total respondents from agriculture (53.2%) and allied departments (39.7%) belonged to high level of coordination in execution. However, with respect to overall coordination in ATMA, 37.5 per cent of total respondents in agriculture (37.1%) and allied departments (41.3%) belonged to medium to high level of coordination process in ATMA, respectively. Moreover, coordination in organizing and coordination in execution was associated closely with coordination process of ATMA. (Kumar and Eswarappa, 2010) The mass media exposure had positive and highly significant correlation with perception of the farmer toward technical capability of the public extension personnel. (Chand, 2012) The majority of the respondents had medium risk bearing capability and the findings revealed that majority 79 per cent of the beneficiaries had medium level of knowledge about programme scheme, 46 per cent beneficiaries had no knowledge about training programme in ATMA. About 50 per cent beneficiaries had medium level of knowledge about demonstration programme, 36 per cent had no knowledge about visit schedule in ATMA and 76 percent had low level of knowledge about others programme in ATMA. (Sahu et al., 2013) The indigenously developed concept of ATMA for innovative transfer of technology in an integrated manner can be adopted in the state and national policy and implemented with full governmental support after comparing it with pre-intervention and post-intervention scenarios. (Singh et al., 2013) Agricultural Technology Management Agency (ATMA) has now become the most important institutional mechanism at district level for implementation of agricultural extension reforms. Capacity building of farmers through organization of training is one of the most important strategies for implementation of ATMA. A majority of trainees perceived that knowledge and skills were enhanced as a result of training. The overall 16 effectiveness of training was found to be 54.6 per cent which came under medium effectiveness category. (Singh, 2014)

Materials and Methods

The present study was conducted in Mayurbhanj district of Odisha. Hundred (100) respondents in total were selected from 5 different blocks of Mayurbhanj district following both purposive and random sampling method. Appropriate statistical tools have been used to carry out the study viz, Correlation coefficient, Multiple regression analysis and Path analysis. Appropriate operationalization and measurement of the variables have helped the researcher to land upon the accurate conclusions. Therefore, the selected variables for this study had been operationalized and measured in the following manner: 1) Independent variables 2) Dependent variables. Independent variables selected for the study were age (x1), major family occupation (x2), education (x3), land holding size (x4), family size (x5), family type (x6), housing pattern (x7), social participation (x8), communication media used 0.07(x9), average annual family income (x10). Dependent variable selected for the study was Development of farmers through ATMA activities (y).

Results and Discussion

The subjective information is measured utilizing explicit numerical methodology. Then data analysis i.e. Co-efficient of correlation, multiple regression analysis, and path analysis has been done to evaluate the information.

Table 1: Co-efficient of Co-relation : Socio-economic variables vs development of farmers through ATMA activities (y):

Sl. no	Independent Variables	'r' value
1	Age (x_1)	-.244*
2	Major Family Occupation (x_2)	.269*
3	Education (x_3)	.319**
4	Land Holding Size (x_4)	.233*
5	Family Size (x_5)	-.259*
6	Family Type (x_6)	-.153
7	Housing Pattern (x_7)	.196
8	Social Participation (x_8)	.242*
9	Communication Media Used (x_9)	.324**
10	Avg Annual Family Income (x_{10})	.307*

The above table indicated that the following socio-economic variables i.e., age, major family occupation, education, land holding size, family size, social participation, communication media used, average annual family income had a significant impact on development of the farmers through ATMA activities. The negative co-relation of the variable age implies that younger respondents are more involved in developmental work through ATMA activities. This may be due to the fact that young farmers are willing to do more innovative work in association with ATMA. The variable major family occupation has positive impact on development of farmers through ATMA activities which implies that when more number of family members are involved in farm activities, higher impact on the development through various activities of ATMA can be observed. Similarly, the independent variable education & land holding size have a positive impact on development of farmer through ATMA activities. The variable family size has a negative impact on development of farmers through ATMA activities. The variable social participation has a positive impact with the dependent variable. This may be due to the fact that higher is the mobility of famers and involvement with fellow farmers such as cooperative society, panchayat society, social organization, cultural organization, religious organization, educational organization & training exposure more on the development of farmers through ATMA activities. The variables communication media used & average annual family income have positive impact on development of farmers through ATMA activities due to exposure towards various information sources and multiple sources of income respectively.

Table 2: Multiple Regression analysis : Socio-economic variables vs development of farmers through ATMA activities(y)

Model		Unstandardized Coefficients		Standardized Coefficients	T	Sig.
		B	Std. Error	Beta		
1	(Constant)	325.137	14.657		22.183	.000
	Gender	1.283	2.487	.051	.516	.607
	Age	-2.128	1.782	-.128	-5.194	.006**
	Caste	1.582	1.057	.167	1.496	.138
	Major Family Occupation	1.301	.749	.169	1.736	.086
	Education	-.171	.919	-.026	-9.186	.013*
	Land Holding Size	.932	2.056	.049	.453	.652
	Family Size	-4.062	2.936	-.188	-1.383	.170
	Family Type	.168	2.247	.010	.075	.941
	Hosing Pattern	3.707	2.491	.161	1.488	.140
	Social Participation	.292	.238	.143	6.226	.023*
	Communication Media Used	.259	.266	.136	7,974	.003**
	Avg Annual Family Income	1.099	1.082	.129	8.017	.012*
a. Dependent Variable: Development of farmers through ATMA activities						

Model	R	R Square	Adjusted R Square	Std. Error of the Estimate
1	.703[a]	.494	.250	9.34314
a. Predictors: (Constant), Avg Annual Family Income, Family Size, Major Family Occupation, Housing Pattern, Age, Social Participation, Land Holding Size, Family Type, Communication Media Used, Education				

Table-2 presents the full model of regression analysis. All the 10 independent variables collectively contributed 49.4% variance in the dependent variable development of farmers through ATMA activities. It is evident that the independent variables age, Education, Social participation, Communication Media used, Average annual family income have significant contribution on the dependent variable development of farmers through ATMA activities.

Table 3: Path Analysis: Decomposition of Total Effect into Direct, Indirect and Residual Effect: Socio-economic variables vs development of farmers through ATMA activities(y)

S. no	Variables	To. Effect	Total Direct Effect	Total Indirect Effect	Substantial Effect		
					I	II	III
1	Age	-.244	-.128	-0.116	0.157_{X8}	-0.139_{X2}	-0.059_{X9}
2	Major Family Occupation	.269	.169	0.100	-0.242_{X2}	0.086_{X5}	-0.041_{X4}
3	Education	.319	-.026	0.345	-0.144_{X6}	0.105_{X7}	-0.069_{X8}
4	Land Holding Size	.233	.049	0.184	-0.206_{X7}	0.167_{X11}	0.088_{X3}
5	Family Size	-.259	-.188	-0.071	-0.223_{X5}	-0.172_{X8}	-0.074_{X6}
6	Family Type	-.153	.010	-0.163	-0.081_{X4}	0.077_{X7}	-0.062_{X5}
7	Housing Pattern	.196	.161	0.035	-0.195_{X11}	0.114_{X5}	-0.023_{X3}
8	Social Participation	.242	.143	0.099	-0.256_{X5}	0.186_{X3}	-0.042_{X9}
9	Communication Media Used	.324	.136	0.188	-0.144_{X6}	0.105_{X10}	-0.042_{X12}
10	Avg Annual Family Income	.307	.129	0.178	-0.206_{X7}	0.167_{X9}	0.088_{X13}

Residual Effect: 0.022

Highest Indirect Effect: 0.345

Table-3 presents the path analysis and reveals that the family size of the respondents had exhibited highest total direct effect followed by housing pattern. Similarly, the variable education exhibited highest indirect effect. Other variables having significant effect are communication media used, land holdings size annual income and family type. The variable education had association with family size, social participation, family occupation, housing pattern and family type. Therefore, it can be concluded that the variable education along with family size, social participation, occupation, housing pattern and family type could able to influence various aspects of development through ATMA activities.

Hence the extension officer involved in process of implementation of ATMA activities may use all these variables resulting for better adoption activities for the development of the respondents. The residual effect being 0.022 showed that 2.2% of the variation in this relation could not be explained.

Conclusion

Agricultural Technology Management Agency (ATMA) has been instrumental in bringing significant changes in bringing revolution in farming society of our country. Over the years, ATMA has evolved its structure and function in order to serve the needs of various stakeholders involved in the production chain of agricultural commodities. From availing various timely inputs to assisting the farmers in providing them need based training, ATMA has proven to be an absolute gamechanger. As policy formulation takes place at district level, ATMA has strengthened the farming ecosystem at local level. The present study came up with a strong revelation in eliciting the fact that family size, education, communication media used, land holding size, annual income and family type are of immense importance to serve the emerging needs of the participating farmers and beyond. Keeping these variables in mind, there is also a gradual need to gear up the managerial reforms among ATMA professionals, proper resource mobilisation, timely sharing of information as well as precise decision making to attain the goals of farmer beneficiaries. Due to this, it is necessary to assess ATMA functionaries across India in order to determine local needs and ensure the organization's seamless operation. This also opens the way for the development of pertinent policies in future and deal with the future challenges in Indian agriculture.

References

Chand, S. (2012). Perception of the farmer toward technical capability of the public extension personnel in middle Gujarat, Unpublished M.Sc. (Agri.) thesis, AAU, Anand.

Hoque, M.J. and Koichi, U. (2007). Determinants of block supervisors' perception of On-the-Job training and Off-the-Job training in Bangladesh, Indian Journal of Extension Education.43(3&4): 22-26.

Kumar, K. A. &Eswarappa, G. (2010). Critical analysis of co-ordination process among different stakeholders of developmental departments involved in planning, organising and execution of agricultural technology management agency ATMA programme in Chittoor District of Andhra Pradesh, Mysore Journal of Agricultural Sciences, 44 (4) : 877-882.

Sahu, B.P., Chaturvedi M.K. &Yadaw K.N. (2013). Knowledge level of ATMA beneficiaries towards the activities of Agricultural Technology Management Agency ATMA. Agriculture Update, 7 (3&4) : 319-322.

Singh, D.K. (2014). Effectiveness of training programmes under Agricultural Technology Management Agency in Bihar, Indian Research Journal of Extension Education, 14 (1): 93-95.

Singh, K., Meena, M.S., Singh, R.K.P., Kumar, A. & Kumar, U. (2013). Agricultural Technology Management Agency ATMA: A Study of its Impact in Pilot Districts in Bihar, India.

15

Relationship Between Profile and Adoption Components of Wadi Project by the Tribal Beneficiaries

***R.P. Khule*[1] *and G.K. Waman*[2]**

[1&2]*(Agril. Extension), Associate Dean (LAE) Office, MPKV, Rahuri Maharashtra*

Abstract

The Wadi project mainly involves development of orchard (locally known as Wadi) on degraded lands. Mango, Cashew, Guava, Custard apple, Aonla, Lemon, Sapota and Drumstick are the major fruit crops that are cultivated by the tribal through Wadi project. The present study was examined the adoption components of Wadiproject and the relationship between profile and adoption components of Wadi project among tribal beneficiaries. The study was conducted in Navsari, Valsad and The Dangs districts purposively, because the Wadi project was started in 1982 in these three districts with maximum number of beneficiaries of the project. The proportionate random sampling method was used for selection of 200 tribal beneficiaries. The tribal beneficiaries possessed medium to high extent of adoption components of Wadi project. The component availability of land was come out as the most adopted, followed by raising of forestry plants and women development, respectively. While, health programme was perceived as the last adopted components of Wadi project. The education, annual income, source of information, social participation, risk orientation, economic motivation, scientific orientation, extension contact, innovativeness, aspiration, cohesiveness, management orientation, communication skill, distance from market found highly significant association with adoption components of Wadi project by the tribal beneficiaries. The land holding, farming experience, occupations was significantly associated with the adoption components of Wadi project by the tribal beneficiaries.

Keywords: *Profile, Adoption components, Wadi project, Tribal beneficiaries.*

Introduction

The *Wadi* project mainly involves development of orchard (locally known as *Wadi*) on degraded lands. Mango, Cashew, Guava, Custard apple, Aonla, Lemon, Sapota and Drumstick are the major fruit crops that are cultivated by the tribal through

Wadi project. *Wadi* is a comprehensive project for improving living standards and socio-economic condition of tribal families in remote areas. Acceptance of new idea is not a unit act but a complex process which involves a series of thoughts and actions. In other words, it is known as mental process through which an individual passes from first hearing about an innovation to final adoption. The components of *Wadi* project so far developed in recent has been for increasing the agriculture and horticulture production and productivity hence, it requires proper adoption by the tribal beneficiaries for production. Unless the components of *Wadi* project are properly adopted to larger extent by the tribal beneficiaries, the desired target cannot be achieved. The study was undertaken on following objectives.

Objectives

1. To study extent of adoption of components of *Wadi* project.
2. To study component wise adoption of *Wadi* project by the tribal beneficiaries.
3. To study relationship between profile of the tribal beneficiaries and adoption components.

Methodology

The present study was examined the socio-economic consequences occurred among beneficiaries as a result of *Wadi* project. The study was conducted in Navsari, Valsad and The Dangs districts purposively, because the *Wadi* project was started in 1982 in these three districts with maximum number of beneficiaries of the project. A list of all tribal beneficiaries of *Wadi* project were collected from the BAIF-DHRUVA office of Village Lachhakadi, P.O.Gangpur, Tal.Vansda, Dist. Navsari, Gujarat. The proportionate random sampling method was used for selection of 200 tribal beneficiaries from the 55 villages of 4 talukas of three districts of South Gujarat for the present investigation.The questionnaire developed on the components of *Wadi* project was administered to the same selected tribal beneficiaries. The relationship was worked out by using the correlation of co-efficient (r).

Results and Discussion

Extent of Adoption of Components of *Wadi* Project

The data in table 1 shows that majority (77.00 per cent) of the tribal beneficiaries had medium extent of adoption regarding components of *Wadi* project, followed by 15.00 and 08.00 per cent had low and high extent of adoption respectively. In general, data showed that the majority (85.00 per cent) of tribal beneficiaries possessed medium to high extent of adoption. This infers that scientific information about the *Wadi* crops and various extension methods had motivated and enhance them for adoption of the components of the *Wadi* project.

Table 1: Distribution of tribal beneficiaries according to their extent of adoption (n=200)

Sr. No.	Extent of adoption	Frequency	Percentage
1.	Low extent of adoption	30	15.00
2.	Medium extent of adoption	154	77.00
3.	High extent of adoption	16	08.00
	Total	**200**	**100.00**

The study result was in line of the findings of Sasane *et al.* (2008 & 2009), Sasane *et al.* (2010) andKhule *et al.* (2011).

Component wise adoption of *Wadi* project by the tribal beneficiaries

The data presented in table 2 reveals that adoption found in respect of availability of land was 96.75 per cent falling in first rank, followed by raising of forestry plants (93.25 per cent) and women development (85.33 per cent) were having second and third rank, respectively. The health programme fell in to the last rank having 74.00 per cent adoption. The average adoption of all the 12 components of *Wadi* project was 81.78 per cent.

Table 2: Component wise adoption of Wadi project by the tribal beneficiaries (n=200)

Sr.No.	Components of *Wadi* project	Average score of adoption	Maximum score	Adoption (%)	Aspect wise rank
1.	Availability of land	5.805	6.00	96.75	I
2.	Raising of fruit crops	40.800	48.00	85.00	IV
3.	Raising of forestry plants	5.595	6.00	93.25	II
4.	Fencing	7.280	9.00	80.89	VI
5.	Water resource development	4.585	6.00	76.41	X
6.	Soil conservation measures	7.425	9.00	82.50	V
7.	Health programme	4.440	6.00	74.00	XII
8.	Women development	10.240	12.00	85.33	III
9.	Support to landless	4.585	6.00	76.42	IX
10.	Processing	4.590	6.00	76.50	VIII
11.	Marketing	4.555	6.00	75.92	XI
12.	Peoples organizations	7.055	9.00	78.39	VII
	Average	8.912	-	81.78	

It could be inferred from above discussion that the component availability of land was come out as the most adopted, followed by raising of forestry plants and women development, respectively. While, health programme was perceived as the last adopted component of *Wadi* project. The probable reason may be that the tribal beneficiaries might have understood and realized the importance of availability of land for adoption of *Wadi* project. The reason of least adoption of health programme might be that they have not aware about the importance of the health programme.

This finding was partially supported by those of Sasane *et al.* (2008 & 2009), Sasane *et al.* (2010) andKhule *et al.* (2011).

Relationship between profile of the tribal beneficiaries and adoption components of *Wadi* project

Table 3: Relationship between profile of the tribal beneficiaries and adoption components of *Wadi* project (n=200)

Sr. No.	Independent variable	Adoption ('r' value)
1.	Age	-0.411**
2.	Education	0.194**
3.	Family size	0.0495^{NS}
4.	Land holding	0.147*
5.	Farming experience	0.146*
6.	Occupation	0.142*
7.	Annual income	0.182**
8.	Source of information	0.224**
9.	Social participation	0.304**
10.	Risk orientation	0.359**
11.	Economic motivation	0.331**
12.	Scientific orientation	0.434**
13.	Extension contact	0.233**
14.	Innovativeness	0.229**
15.	Training acquired	0.0435^{NS}
16.	Aspiration	0.431**
17.	Cohesiveness	0.673**
18.	Management orientation	0.593**
19.	Communication skill	0.695**
20.	Distance from market	0.554**

* Significant at 5 per cent ** Significant at 1 per cent ^{NS}Non significant

The data evident in table 3 revealed that the education (0.194**), annual income(0.182**), source of information (0.224**), social participation (0.304**), risk orientation (0.359**), economic motivation (0.331**), scientific orientation (0.434**), extension contact (0.233**), innovativeness (0.229**), aspiration (0.431**), cohesiveness (0.673**), management orientation (0.593**), communication skill (0.695**), distance from market (0.554**) found highly significant association with adoption components of *Wadi* project by the tribal beneficiaries. Further, the land holding (0.147*), farming experience (0.146*), occupation (0.142*) were significantly associated with the adoption components of *Wadi* project by the tribal beneficiaries. The age (-0.411**) were negative highly significant association with the adoption components of *Wadi* project by the tribal beneficiaries. On the other hand, family size (0.0495^{NS}), training acquired

(0.0435^{NS}) were non-significant association with the adoption components of *Wadi* project by the tribal beneficiaries.

This finding is in conformity with those of Sasane *et al.* (2008 & 2009), Sasane *et al.* (2010) and Khule *et al.* (2011),

Conclusion

The tribal beneficiaries possessed medium to high extent of adoption components of *Wadi* project. The component availability of land was come out as the most adopted, followed by raising of forestry plants and women development, respectively. While, health programme was perceived as the last adopted components of *Wadi* project. The education, annual income, source of information, social participation, risk orientation, economic motivation, scientific orientation, extension contact, innovativeness, aspiration, cohesiveness, management orientation, communication skill, distance from market found highly significant association with adoption components of *Wadi* project by the tribal beneficiaries. The land holding, farming experience, occupations was significantly associated with the adoption components of *Wadi* project by the tribal beneficiaries.

References

Khule, R.P., Lanjewar, D.M., Jagdale, U.D. & Sasane, G.K. (2010). Adoption of low cost crop cultivation technology by the paddy growers. Agriculture Update, 6 (1), 51

Sasane, G.K., Khule, R.P. & Jagdale, U.D. (2008). Knowledge and Adoption of Sugarcane Water Management Practices by the Farmers. Asian Journal of Extension Education, 27, 141-145.

Sasane, G.K., Khule, R.P. & Jagdale, U.D. (2010). Knowledge and Adoption of Food Grain Storage Practices of the Farmers. Asian Journal of Extension Education, 28, 92-95.

16

Digital Innovations and ICT Initiatives by NGOs for Rural Development in India

Akanksha Singh

Department of Agricultural Extension and Communication, Sardar Vallabh Bhai Patel University of Agriculture and Technology Modipuram, Meerut Uttar Pradesh

Abstract

Mahatma Gandhi once said "India lives in its villages" and looking at the present scenario from the social, economic and political perspectives this statement seems valid even today. Hence Rural Development which is concerned with economic growth and social justice, improvement in the living standard of the rural people by providing adequate and quality social services and minimum basic needs becomes essential. Non-government organizations with their advantage of non-rigid, locality specific, felt need-based, beneficiary oriented and committed nature of service have established multitude of roles which can effect rural development. A number of NGOs have been playing a vital role in rural community development, to enable them to lead a dignified life by building their capabilities, strengthening community institutions and mobilising resources. The present era is being characterized by "Revolution in information technology" which is popularly known as Information Communication Technology (ICT), including gadgets which processes to access, retrieve, store, organize, manipulate, produce, present and exchange information by electronic and automated means. So ICT refers to tools that facilitate production, transmission and processing of information. In developing countries like India, ICT has always been an important factor behind development. Even in areas where there is hardly any development, ICT has been considered to change some part of it. It has showed traces in the smallest of villages. In the present era of digitalization ICT plays an important role in rural development. Various ICT initiatives have been undertaken by NGOs like DIGITAL GREEN-Video Kheti, Sampark.net, Oracle Smart Community Interface, ICT for Poor, RAIN etc. These ICT initiatives by NGOs focuses on transforming agriculture and other allied activities in to a sustainable & vibrant activity, which will improve rural people' livelihood and bring respect to their profession and. works towards uplifting the socio-economic status of the rural people.

Keywords: *NGO, ICT, Digital Green, Rural Development, RAIN*

Introduction

Mahatma Gandhi once said "*India lives in its villages" and looking at the present scenario* from the social, economical and political perspectives this statement seems valid even today. According to the World Bank collection of development indicators,around 65.33 percent of total population of India was reported to be Rural population in year 2019. Rural population refers to people living in rural areas as defined by National Statistical offices. Since more than half of the total population of India resides in rural areas Rural Development which is concerned with economic growth and social justice, improvement in the living standard of the rural people by providing adequate and quality social services and minimum basic needs becomes essential.

Superficially, rural development seems to be a simple task but, in reality, it is not. Rural development is based on sustainable development with available resources. Post- Independence era has seen many rural development programmes through different five-year plans. Alleviating poverty, employment generation, more opportunities for generating income, and infrastructure facilities are emphasized through the policies and programmes of the government. Along with this, the panchayat raj institutions have also been initiated by the government to strengthen the democracy at grass roots level. But in India, the scope of development is not narrow but very wide, as it includes not just the economic development but the growth on social front, quality of life, empowerment, women and child development, education and awareness of its citizens. The task of development is so huge and complicated that just implementing government plans is not sufficient to fix the problem. To achieve this, a holistic vision and collaborative efforts was needed which involves various departments, agencies and even NGOs. Owing to such a great need, the number of NGOs in India is increasing rapidly and, large number of NGOs are working for improving the livelihood of rural people.

NGO and Rural Development in India

The term "non-governmental organization" was first coined in 1945, when the United Nations (UN) was created. The UN, itself an intergovernmental organization, made it possible for certain approved specialized international non state agencies i.e., non-governmental organizations to be awarded observer status at its assemblies and in some of its meetings. Later the term became used more widely. Today, according to the UN, any kind of private organization that is independent from government control can be termed an "NGO", provided it is not-for-profit, non-criminal and not simply an opposition political party.

Since ancient times, social service has been an integral part of Indian culture. Soon after Independence, a number of NGOs had emerged in India. These NGOs have been playing a vital role in rural community development, to enable them to lead a dignified life by building their capabilities, strengthening community institutions and mobilising resources. Non-government organizations with their advantage of non- rigid, locality specific, felt need-based, beneficiary oriented and committed nature of service have established multitude of roles which effects rural development. The Government , because of its command over and access to much greater resources and working under a broader institutional framework, has the potential for designing and implementing suitable programmes to reach much wider sections of the poor, on the other hand Non Governmental Organisations, have several advantages over government organizations in the provision of assistance to the rural poor, their close contact with the poor, willingness to spend more time on the various experiments and their easy acceptability by society, has given them an edge over the government.

NGO's in recent days has emerged as an important element of contemporary rural society and act as a link between in the grass roots and civil society. The voluntary sector has contributed significantly in reducing poverty, deprivation or discrimination and exclusion through awareness, social mobilization, service delivery and training. They are effective non political link between government and common people. Misra and Rajeeb (2008) through their study reported that the role of NGO is not to carry out development work in isolation , but through a " Participatory approach" through working with the people. NGO's are essentially grassroots institutions that work closely with local communities. Non Governmental Organizations play an important role in the efficient and cost effective provision of services, in mobilizing communities and groups and thereby providing effective channels of communication between the government and the people , and the implementation of programmes requiring innovative approaches.

ICT initiatives by NGOs for Rural Development

The present era is being characterized by "Revolution in information technology" which is popularly known as Information Communication Technology (ICT), including gadgets which processes to access, retrieve, store, organize, manipulate, produce, present and exchange information by electronic and automated means. So ICT refers to tools that facilitate production, transmission and processing of information. In developing countries like India, ICT has always been an important factor behind development. Even in areas where there is hardly any development, ICT has been considered to change some part of it. It has showed traces in the smallest of villages. Information and Communication Technologies (ICTs) play a key role in development & Economic growth of Rural India. Political, Cultural, Socio-economic Developmental &Behavioral decisions today rests on the ability to access, gather, analyze and utilize Information and Knowledge. ICT is the conduits that transmit information and knowledge to individual to widen their choices for Economic and social empowerment (Ghosh, 2001)

In the present era of digitalization ICT plays an important role in rural development. It gives the information about various schemes and services to the rural people. It highlights the kind of services essential for improving the lifestyle of rural people. The market updates, jobs, product information, travel information, land records, tribal information, grievances are also updated using ICT.

Since digital technologies are key to make rural communities more attractive, smart and sustainable, reduces problems related to remoteness and improves access to services various ICT initiatives has been undertaken by NGOs for rural development. The ICT initiatives by NGOs focuses on transforming agriculture and other allied activities into sustainable& vibrant activity, which will improve rural people' livelihood and bring respect to their profession and works towards uplifting the socio-economic status of the rural people. NGOs are working for rural development and incorporation of digital innovation enhances the pace of rural transformation.

Keeping the above facts in mind I have selected five NGOs namely Digital Green, CARD, PRADAN, DHAN Foundation and Gram Vikas and have tried to compile all the digital innovations and ICT initiatives taken by them for development of rural people. The secondary data was used for this purpose which has been collected from various published sources , reports and websites.

Digital Green

Digital Green is a global development organization that empowers smallholder farmers to lift themselves out of poverty by harnessing the collective power of technology and grassroots-level partnerships. Digital Green is an independent NGO that focuses on training farmers to make and show short videos where they record their problems, share solutions and highlight success stories. It was first conceived as a project in Microsoft Research India›s Technology for Emerging Markets in 2006 by Rikin Gandhi and his colleagues. It later branched out as an independent NGO in 2008. Digital Green›s approach is primarily a technology-enabled means of behavior change communication, which is cost-effective, scalable, and brings together researchers, development practitioners, and rural communities to produce and share locally relevant information through videos. They use a customized data collecting service COCO (connect online and connect offline) method. This eliminates the hurdles which come with lack of connectivity in rural areas. The data can be updated and used offline as well. They have built an analytics dashboard on CoCo database. This is an open source platform and can be used by anyone interested in making an impact with tech. With a database of over 5000 videos, Digital Green plans on implementing recommendation engines based predictive analytics models. Since 2012, it has partnered with the Government of India's National Rural Livelihoods Mission (NRLM) and its state-level counterparts to train over 12,500 frontline workers to use the approach for promotion of sustainable agriculture & nutrition practices in over 15,200 villages across India

Video Kheti – A Video Library

VideoKheti is a responsive web application that provides access to video library through an easy-to-navigate interface for low-literate farmers and extension workers on any device. The navigational layer uses graphics, audio, and touch interaction to make simple choices in crop, crop phase and activity to reach the desired video.

LOOP Mobile App

Loop is a human-mediated mobile phone application that is increasing smallholder farmers' income and reducing their costs by improving their access to markets and market information and increasing their negotiating power. Digital Green nurtures these village-level entrepreneurs to recruit farmers, assess daily produce volumes, determine which nearby market offers the best price, arrange transport based on volume, and sell farmers' produce directly to wholesale buyers. They record volumes and sales on the Loop mobile app, which sends receipts to farmers via text messages. After completing transactions on behalf of all farmers, the entrepreneurs return to the villages to deliver same-day payment and earn their commission. In Bihar in January 2016, more than 3000 farmers in 110 villages have used Loop to sell 6000 tons of vegetables, recording more than $1 million in cash transactions.(getloop.com information retrived on 10 dec 2020).

-*Samvad Mobile Vaani* - Interactive Voice Response (IVR)

Samvad (dialogue in Hindi) Mobile Vaani is an Integrated Voice Response (IVR)-based platform that allows communities to listen to relevant health, nutrition and family planning messages broadcast by the Samvad Project by just giving a missed call and receiving a call back from the system. Digital Green's USAID funded project Samvad aims to improve Family Planning, Maternal Child Health and Nutrition Outcomes across 6 states namely; Bihar, Chhattisgarh, Jharkhand, Odisha, Uttarakhand and Assam through the flagship human-mediated video-based approach. The messages on *Samvad Mobile Vaani* are a combination of drama, testimonials, infotainment directed to behaviour change and action. A unique element of the solution is that the platform allows two-way communication wherein the audience can hear the pre-recorded thematic messages and also record their own messages in response to what they are listening to and can also listen to messages recorded by others like them. These include their feedback, experiences, queries, additional and alternative best practices. The IVR messages reach men and women irrespective of the type of phone they own at zero cost. (Ritika Pandey, Project Director, Samvad, Digital Green)

-Farm Stack: A digital knowledge sharing platform to raise farmer incomes Farm Stack is an open-source, interoperable data platform helping organizations and farmers to make sense of ever-increasing data about food and agriculture to solve problems, make better decisions, and preserve privacy, especially for farmers.

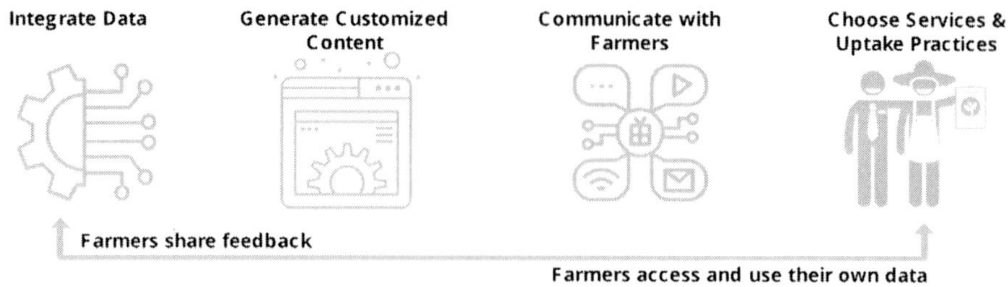

Figure 1. Systematic representation of working of FarmStack
(*Source*: http://www.farmstack.digitalgreen.org)

Training Courseware- A mobile-based courseware for refresher trainings of Agriculture Extension Worker

Digital Green's mobile-based learning platform enables frontline workers to access training courseware on a smartphone or tablet at any time. Pre and post-tests are integrated in the platform, allowing for assessment of learning outcomes. This ensures that frontline workers are always well-prepared when reaching out to farmers.According to Shilpa Rastogi &Ronali Pradhan of Digital Green ,video-based mobile courseware was developed keeping in mind the training needs of frontline workers is made available in 6 languages. The android app developed by Microsoft Research hosts content related to agriculture and nutrition with quizzes to help keep track of the learning curve.The android application has in-built quizzes that test the frontline workers' knowledge before and after they have completed watching the videos in a course. The scores are recorded and keep track of the individual frontline workers' proficiency. A certificate issued after completing a course (on gaining a minimum of 70% marks) incentivizes the frontline workers.

Figure 2: Systematic representation of working of training courseware
(*Source*: http://www.digitalgreen.org)

Seekho-Training App

Digital Green has developed a mobile Android application that guides trainers on the use of this video courseware and includes a means of assessing front line workers and monitoring their progress.

Training Dashboard

The mobile assessment app records the performance of frontline workers and sends it to the training analytics dashboard. The dashboard allows trainers to look across their peers, and frontline workers they have trained, to identify where persistent gaps in their skills exist. The dashboard also enables Digital Green to target follow-up training to frontline workers based on areas in which they're less proficient.

Kisan Diary

This is an Android app developed by Digital Green for farmers to capture, analyse and control their financial data and manage their farming and household expenses better. This first-of-its-kind personal utility app made specifically for farmers provides a free, safe and secure service in 4 languages. The app provides crop and time-based profit/loss analysis. It has seen that over 1800 farmers have downloaded it. It Capture, Retrieve and Analyse sales and expense records as per farmers crop and time.

Mobile Platforms for Rural Social Networks

With support from the Ford Foundation, Digital Green implemented a series of pilots to better understand opportunities and challenges related to rural internet connectivity and social media. In 15 villages in the Mandla district of Madhya Pradesh, Digital Green conducted an internet access pilot, in which 15 farmers were chosen to receive mobile projector phones and SIM cards so that they could select, screen, and lead discussions about community-produced videos. During the pilot, the farmers screened 26 videos during 433 disseminations, which engaged approximately 1,888 viewers and resulted in 400 new practice adoptions.

Farmerbook

Digital Green experimented with the use of social media by developing new platforms to facilitate learning and sharing of agricultural knowledge (through Farmerbook, a Facebook for farmers). Farmerbook enabled Digital Green and partner organizations to learn about video-viewing and practice adoptions of the individuals within their groups, and to identify the most influential farmers within a given area.Farmerbook is an open-access platform that displays detailed timeline-based activities of farmers we have worked with, along with villages plotted on Google Maps. The application stimulates healthy competition among partners, village facilitators, and community members through the sharing of performance data and community feedback.

Wonder Village

Digital Green has developed Facebook game, WonderVillage, to connect urban populations with the agricultural lifestyle. it allows external audiences to learn about and engage in issues related to rural development. Players set up a simulated village economy. Placed in a resource-constrained setting, players pursue quests like setting up small farms of paddy and maize and supplying raw materials to farmers' markets.

Centre for Agriculture and Rural Development (CARD)

Established and registered in the year 2000 under the Societies Registration Act 1860, Centre for Agriculture and Rural Development is a premier Non-Government Organization, playing a vital role in the national efforts of developing India through agricultural transformation. CARD is committed towards reaching out to all parts of the rural society, especially the farming community. CARD also participates actively towards improving the quality of life of rural masses by addressing technical, economic policy issues related to the development of agriculture and rural society. RAIN is a remarkable ICT initiative undertaken by CARD.

The Rural Agri-Innovation Network (RAIN)

The Rural Agri-Innovation Network dedicated to CARD's effort to showcase a remarkable progress taken place in Agriculture space in rural India. RAIN is created with a mandate of knowledge sharing platform; has continued to run its activities through a participatory approach and demand driven agenda. RAIN is focused on transforming agriculture in to a sustainable & vibrant activity, which will improve farmers' livelihood and bring respect to the farming profession, especially in India. The strategic focus areas of RAIN platform of CARD is-Documenting Innovation ,Model Farm, Success Stories, Knowledge Sharing Platform, Strategic Networking & Alliances. Also CARD under its RAIN platform has launched a series of "Learning by Sharing Success Program" to demonstrate the success stories of the farming space in country and reward the progressive farmers thereon.

Professional Assistance for Development Action (PRADAN)

PRADAN was founded in 1983 as an India-based Non-Government Organisation. Currently it is one of the largest rural livelihood promoting NGOs in India. It is a voluntary organisation registered under the Societies Registration Act of India. In 1983, Vijay Mahajan and Deep Joshi set up PRADAN, inspired by the belief that well-educated people with empathy towards the poor must work at the grassroots to remove mass poverty. PRADAN aspires to bring sustainable changes in the lives of marginalized populations in some of the most disadvantaged pockets of the country. To fulfil its vision it created sampark.net.

Sampark.net

'Sampark.net - The Rural Connect', is PRADAN knowledge management portal. It is an attempt to more effectively reach out to the larger development fraternity to trigger stimulating thoughts and initiate innovative action. Sampark means 'connect' and Net comes from the word 'network'. As the name suggests, it is an open portal aimed at building connections. It is a platform for all of us to get into engaging discussions, and unravel complex developmental problems. It is envisaged as a bridge for free exchange of ideas with the external world, bringing fresh perspectives into our routine debates. It is also a sound-box which will help amplify our voices in the din of the mainstream discourses. Knowledge created out of this churning may also set new directions to the development frameworks and policies.

Gram Vikas

Gram Vikas is an Indian non-governmental organisation based in Odisha, and founded in 1979. Gram Vikas works towards enabling rural communities to lead a dignified life. This is done by building capabilities of village communities, strengthening community institutions and mobilising resources. Gram Vikas works on six focus areas which are water, livelihoods, sanitation & hygiene, habitat & technologies, village institutions and education. One of the unique effort of Gram Vikas is creation of Oracle Smart Community Interface.

Oracle Smart Community Interface

The Oracle Smart Community Interface helps in bridging the Market and Health Services Gap in Tribal Areas of Odisha'. This ICT project addresses information separation between small producers and markets on one hand and maternal and child health service (ICDS) users and service providers on the other. It supplements earlier interventions in horticulture and health-related services and is implemented in remote locations of Ganjam and Kalahandi districts in Odisha.The initiative aims to establish a 'smart community interface' to overcome information separation between the market and farmers to aid better livelihoods for small producers, and health service providers and women for improved health of the women and children. It has led to Setting up of hardware systems and software systems for Oracle Smart Community Interface. under this initiative 500 farmers identified from 2 districts receive local mandi prices on their mobile phone 4times a week and video on best agricultural practices 6 times a month.

Development of Humane Action (DHAN) Foundation

Development of Humane Action (DHAN) Foundation, a professional development organisation, was initiated on October 2, 1997. It is a premier development institution in the country, working for the upliftment of the poor and the disadvantaged segment of the communityDHAN Foundation has taken up 'ICT for the poor' as a theme to experiment, develop and implement socially-relevant ICT programmes through ICT based people's organisations built at the grassroots.

ICT for Poor

DHAN Foundation conscious of technological changes in this era and the poor being left out in this process has initiated the "Information & Communication Technology (ICT) for Poor" Programme. DHAN Foundation has taken up 'ICT for the poor' as a theme to experiment, develop and implement socially-relevant ICT programmes. DHAN Foundation established the Community Resource Centers(CRCs) under its Information & Communication Technology Programme(ICT) for poor programme. The CRCs are the grassroots level centre at Panchayat/village level and each CRCs are reaching out to about 5-6 villages. These centers are managed by trained women from the local community. Each CRC is equipped with two computers, internet connection, printer, UPS, public address system, mini-library, interactive multimedia content in DVDs, web/digital camera, etc. The villagers contribute a rent free building which is accessible to all people particularly women irrespective of caste, sex, etc. The CRCs are benefiting community through e-services catering to the needs of school going children, youth, women and farmers.

Conclusion

To deliver various policies and services for development of rural people ICT plays a vital role. ICT provides the better services at lower cost in all the rural area.ICT has the greater potential in providing the information for developing the means of livlihood. It also helps to enhance the change in the society. It supports the information acquisition in local language as and when required. Digital technologies have the potential to revolutionise rural society by helping NGOs to work more precisely, efficiently and sustainably. Thus the ICT initiatives by NGOs focuses on transforming agriculture and other allied activities into sustainable& vibrant activity, which will improve rural people' livelihood and bring respect to their profession and works towards uplifting the socio-economic status of the rural people.

References

Bhaskar, Indu., &Geethakutty, P.S. (2001). Role of Non-Governmental Organisations in Rural Development: A case study. Journal of Tropical Agriculture, Vol .39 , pp. 52-54.

Biswas, Soutik (2013). Digital Indians: Rikin Gandhi helps farmers with video. BBC News, Delhi. Retrieved December 10, 2020.

CARD, www.card.org.inRetrieved 2020-12-08

Chandrashekar, H.M., & Lokesh, M.U. (2009). Role of SHGs in Socio-Economic change of Vulnerable Poor. International Ngo journal, 4(4):127-131.

Dhan Foundation, www.dhan.org Retrieved 2020-12-06.

Digital Green. www.digitalgreen.org. Retrieved 2020-12-06.

Driving Impact through Tech - A story on how technology is changing the livelihoods of farmers in India | Social App Hub. www.socialapphub.com. Retrieved 2020-11-06.

Gandhi, Rikin; Veeraraghavan, Rajesh; Toyama, Kentaro; Ramprasad, Vanaja (2009). Digital Green: Participatory Video and Mediated Instruction for Agricultural Extension. Information Technologies & International Development, 5(1):1–1.

GramVikas, www.gramvikas.orgRetrieved 2020-12-06.

Indu Bhaskar and Geethakutty. (2001). Role of Non-Governmental Organisations in rural development: A case study. Journal of Tropical Agriculture, 39:52-54.

Pradan, www.pradan.net Retrieved 2020-12-06.

Ramakrishna, H. (2013). The Emerging Role of NGOS in Rural development of India: an assessment. International journal of social Science & Interdisciplinary Research, 2(4).

Siddaraju, V.G. (2011).Role of Non Governmental Organizations in promoting sustainable agriculture development in Karnataka. International NGO Journal, 6(2):057-061. doi:10.5897/NGOJ 10.043.

17

NGOs Role in Rural Development and Agriculture in India

Aswani Kumar, Ambrish Kumar Verma, Satendra Kumar and Satyaveer Singh

Department of Agricultural Extension, C.S. Azad University of Agriculture & Technology, Kanpur, Uttar Pradesh

Abstract

Role of non-governmental organizations (NGOs) in rural development the main role of NGOs in agriculture and rural area development are; catalyze rural population, build models experiment, supplement government efforts and organizing rural poor. So we can conclude that NGOs play very important roles in agriculture and rural development which is lead to economic development of any economy. India is a highly populated and predominantly rural country,non-government organizations and voluntary action have been part of the historical legacy..Since ancient times, social service has been an integral part of Indian culture. Soon after Independence, a number of NGOs had emerged in India. Mahatma Gandhi even pleaded to dissolve the Indian National Congress and transform it to a Lok Seva Sangh (Public Service Organization). Though his plea was rejected, but the followers of Mahatma Gandhi started many voluntary agencies to work on various social as well as economic issues of the country. In this way the NGOs can bring the awareness among the poor rural people. It is now the need of the society as well as the nation to make these rural area and people competent to be aware about their fundamental rights. NGO's are the only organizations that could make the rural area developed and agriculture development means providing assistance to the crop producer with the help of various agricultural resources. Providing protection assistance in the research sphere employing latest techniques controlling pests and facilitating diversity all fall within purview of agriculture development. As economic reform and liberalization saw the government vacating several areas to let private sector entrepreneurship flourish and contribute to the high growth rate of the economy in recent years, similar paradigm shifting in needed to transform NGOs from their dependence on aid and grants from within and outside from transforming the rural scenario in the country this is sought to be achieved for the NGOs through engaging them in microfinance , micro insurance and micro entrepreneurship activates for over all development of the rural areas and to per mote the welfare of the people of rural India.

Keywords: *Rural development, NGOs, Agriculture development, Economy, Growth rate* etc.

Introduction

The term, "non-governmental organization" or NGO, came into use in 1945 because of the need for the UN to differentiate in its Charter between participation rights for intergovernmental specialized agencies and those for international private organizations. At the UN, virtually all types of private bodies can be recognized as NGOs .They only have to be independent from government control, not seeking to challenge governments either as a political party or by a narrow focus on human rights, non-profit-making and non-criminal (Ali Mostashari, 2005). On the contrary India is one the poorest countries housing one third of the total world poor population. 43% of Indian children are malnourished, a third ofthe world's total. Over 25% of Indians are illiterate, and more than 20 million children are out of school (Asheesh Bisht et al., 2013). India is a highly populated and predominantly rural country. It is classified as a country of medium human development, ranked 128th in the United Nations Development Programme's Human Development Index 2007. Life expectancy is 63.7 years at birth, and four fifths of the population lives on less than $2/day. The adult literacy rate is 61 percentages. India has a long history of civil society based on the concepts of *daana* (giving) and *seva* (service). Numerous organizations were established during this period, including the Friend-in-Need Society (1858), Prathana Samaj (1864), Satya Shodhan Samaj (1873), Arya Samaj (1875), the National Council for Women in India (1875), and the Indian National Conference (1887). Mahatma Gandhi's return to India in 1916 shifted the focus of development activities to economic self sufficiency. His Swedish movement, which advocated economic self- sufficiency through small-scale local production, swept through the country. Gandhi identified the root of India's problem as the poverty of the rural masses and held that the only way to bring the nation to prosperity was to develop the villages 'self-reliance based on locally available resources. He also believed that voluntary action, decentralized to *gram panchayats* (village councils), was the ideal way to stimulate India's development. Gandhi reinvigorated civil society in India by stressing that political freedom must be accompanied by social responsibility.Voluntary organization is not a new phenomenon in our country. Voluntary effort has always been an important part of our culture and social tradition Voluntary organization can play a crucial role in rural development by supplementing government efforts as they are close to the minds and hearts of the rural people. They have their roots in the people and can respond to the needs and aspirations of the community very effectively. They can experiment new approaches to rural development (Dhillon and Hansra, 1995). Non-Government Organizations (NGOs) and Voluntary action have been part of the historical legacy. In early 20th century, several voluntary efforts were started in the fields of education, health etc.. The plan document states: "A nation-wide network of NGOs will be created. In order to facilitate the working of this network, three schemes relating to the creation/replication/ multiplication and consultancy development have been worked out by the planning commission (Malik, 1995). The widespread belief that NGOs are more successful in reaching the poor in poverty reduction also resulted in rapid growth of funding for NGOs by government and external donors. As far as the government funding is concerned, there are over 200 government schemes initiated by the central and state governments through which NGOs can have direct access to resources for rural development (Reddy and Rajasekhar, 1996).

Objectives of NGOs in india

To promote Information collaborations and Constructive communication between NGOs, to develop effective partnership with eachother To networking for the access, sharing and dissemination of information collaboration and partnership s between NGOs themselves and with other organizations.

- To electronic networking to strengthen community organizations by boosting its knowledge base and its ability to share information and experiences with strategic allies and other partners in relevant field.
- To capacity building of grass root level NGOs, social workers through free online resources and information on a single platform.
- To serve non-profit organizations, charities, grassroots and community groups, educational and research institutions.
- To sharing of ideas among NGOs and development agents.
- To avail free, easy and instant access to information toNGO.
- To help NGOs in their fund-raising efforts.
- To provide more options before Funding Agencies to select the right choice of NGO to support and work.
- To provide information regarding NGO Regisreation and NGO Consultancy.
- To provide Volunteers to NGOs and collaborate between NGOs and Volunteers.

NGOs in India

Since ancient times, social service has been an integral part of Indian culture. Soon after Independence, a number of NGOs had emerged in India. Mahatma Gandhi even pleaded to dissolve the Indian National Congress and transform it to a Lok Seva Sangh (Public Service Organization). Though his plea was rejected, but the followers of Mahatma Gandhi started many voluntary agencies to work on various social as well as economic issues of the country. This was the first phase of NGOs in India. The second phase of NGO development started in 1960 when it was felt that just the government programs were not sufficient to complete the task of development in rural areas. Many groups were formed whose role was to work at grass root levels. Moreover, favourable state policies had drastically affected the formation of NGOs and the irro lesat that time. Over the years, the role of NGO sinrural development of India increased. At present too, their role significantly changes with the change in the policies of the government through different plans.

In the sixth five-year plan (1980-1985), a new role for NGOs in the rural development had been identified by the government. In the seventh five-year plan (1985-1990), the Indian government envisaged an active role of NGOs in developing self-reliant communities. These groups were supposed to show how the village resources along with human resource, skill, local knowledge that is greatly underutilized could be used for their own development. As NGOs were working in close connection with local people so bringing such a change was not a tough task for them. Owing to this, in the eighth five-year plan, more importance to NGOs for rural development in India had been given. Under this scheme, a nation-wide NGO network had been created. The role of these agencies was the rural development at a low cost. In the ninth five-year plan, it has been proposed that NGOs would play a significant role in the development on the public-private partnership model. More scope has been provided to NGOs by the government for rural development through the agricultural development policies as well as their implementation mechanisms. As with every five-year plan, the role of NGOs in the rural development of India is growing, so NGOs are now attracting professionals from different fields. NGOs act as planners and implementers of developmental plans. They help in mobilizing the local resources to be used for development. NGOs help in building a self-reliant and sustainable society. These agencies play the role of mediator between people and government. NGOs are actually the facilitator of development, education andprofessionalization.

Top 10 NGO in India

1. Help Age India - Rank 1
2. Sargam Sanstha - Rank 2
3. Smile Foundation - Rank 3
4. Give India
5. Lepra Society
6. Udaan
7. Deepalaya
8. Uday Foundation
9. Karmayog

Various Role and Functions of NGOs

Voluntary action stimulated and promoted by voluntary agencies engaged in development play a significant role at the grass roots level in the Indian social milieu; the success of rural development depends upon the active participation of the people through Self Help Organizations. The various roles of NGOs are described below for better understanding (Neelima Kumari,2013).

1. Catalyze Rural Population
2. Build Models and Experiment
3. Supplement Government Efforts
4. Organizing Rural Poor
5. Educate the Rural People
6. Provide Training
7. Disseminate Information
8. Mobilize of Resources
9. Promote Rural Leadership
10. Represent the Rural People
11. Act as Innovators
12. Ensure People's Participation
13. Promote Appropriate Technology
14. Activate the Rural Delivery System

The widespread belief that NGOs are more successful in reaching the poor in poverty reduction also resulted in rapid growth of funding for NGOs by government and external donors. As far as the government funding is concerned, there are over 200 government schemes initiated by the central and state governments through which NGOs can have direct access to resources for rural development (Reddy and Rajasekhar, 1996).

Roles of NGOs in Agriculture and Rural Development

The essence of non-governmental organizations remains the same: to provide basic services to those who need them. Many NGOs have demonstrated an ability to reach poor people, work in inaccessible areas, innovate, or in other ways achieve things better than by official agencies. Many NGOs have close links with poor communities. Some are membership organizations of poor or vulnerable people; others are skilled at participatory approaches. Their resources are largely additional; they complement the development effort of others, and they can help to make the development process more accountable, transparent and participatory. They not only "fill in the gaps" but they also act as a response to failures in the public and private sectors in providing basic services. Mirroring the support given to northern NGOs, official funding of southern NGOs has taken two forms: the funding of initiatives put forward by southern NGOs, and the utilization of the services of southern NGOs to help donors achieve their own aid objectives. Donor funding of southern NGOs has received a mixed reception from recipient governments. Clear hostility from many non- democratic regimes has been part of more general opposition to any initiatives to support organizations beyond the control of the state. But even in democratic countries, governments have often resisted moves seen as diverting significant amounts of official aid to non-state controlled initiatives, especially where NGO projects have not been integrated with particular line ministry programs. However, and in spite of a likely expansion and deepening of the reverse agenda, NGOs are likely to maintain their wariness of too close and extensive an alignment with donors. (Newman C. Enyioko, 2012)

Conclusion

By scanning the entire responses of objectively conducted study, it is pertinent to draw a conclusion that the NGO's under study have taken enormous efforts for agriculture and rural development in India and to provide basic services to those who need them. Since ancient times, social service has been an integral part of Indian culture. Soon after Independence, a number of NGOs had emerged in India. This present study also concludes about origin of NGOs and objectives of NGOs and their various roles and functions in agriculture and rural development of area. The main role of NGOs in agriculture and rural area development are; catalyze rural population, build models experiment, supplement government efforts and organizing rural poor. So we can conclude that NGOs play very important roles in agriculture and rural development which is lead to economic development of any economy.

References

Ali Mostashari (2005). An Introduction to Non-Governmental Organizations (NGO) Management Iranian Studies Group at MIT June 2005.

Asheesh Bisht, Amit Joshi and Hitesh Pant (2013). Role of NGO's For Eradication of Rural Poverty Through Women Empowerment Case study Reference of MATREE Tonk District of Rajasthan International Journal of Trend in Economics Managment & Technology (IJTEMT) ; Vol. II, Issue. I, February 2013.

Dhillon, D.S. and Hansra, B.S. (1995). Role of Voluntary Organisations in Rural Development. Kurukshetra18(5): 10-13.

Neelima Kumari (2013). The role of NGOs in socio-economic development of Jharkhand An overview International Journal of Business and Management Invention 2319 – 8028, 2013 PP.19-35.

Newman C. Enyioko, (2012). Role of non-governmental organizations (NGOs) in rural development: a study of the six listed NGOs in rivers state, Nigeria thesis Nov 2012.

Reddy, N.L., and D. Rajasekhar (1996). Development Programmes and NGOs: A guide on Central Government for NGOs in India. Bangalore: BCO and NOVIB

18

Developing Communication Tools for Creating Awareness of Covid-19 Among Communities in Kerala

Athira Krishnan, I.[1], Gayathri, K.V.[2], Sithara Balan, V.[3] and Aparna Khanna[4]

[1&2]PG Scholars, Govt. College for Women, Thiruvanathapuram, Kerala [3]Assistant Professor, Department of Home Science, Govt. College for Women, Thiruvanathapuram, Kerala, [4]Associate Professor, Department of Development Communication & Extension, Lady Irwin College University of Delhi

---

Abstract

COVID-19 pandemic had severe impact among the lives of the people. Due to the lack of proper awareness and the passive attitude of the people, the numbers of cases are increasing, rapidly. In this situation, creating consciousness among the people is highly inevitable. Hence the present study on "Developing communication tools for creating awareness of COVID-19 among communities in Kerala" aims to create awareness about COVID-19 among communities with major focus on vulnerable population. Hundred samples were selected for the study which comprises of fifty children (6-12 years) and fifty older adults (60-80 years). The objectives of the study were to develop and design games for creating awareness about the pandemic and to check the effectiveness of the games used. Online survey and telephonic interview were the techniques carried out to collect the data regarding the sampling. Different games like Puzzles, Word search, Varnamala, Bingo, Find the differences were introduced to the samples, along with many daily tasks and the and responses were positive. Significant difference was found in the level of awareness about the COVID 19 among children. The level of accuracy of the Knowledge regarding the pandemic was also found to be increased.

Keywords: Digital awareness, Communication tools, Covid-19, Games for change, Kerala

Introduction

Corona viruses are a huge family of viruses that cause many chronic diseases. A novel corona virus is an advanced strain of virus that has not been formerly identified in humans. Accepted recommendations to avert the spread of disease include regular hand washing, covering of mouth and nose when coughing and sneezing, avoid close contact (social distancing) with anyone showing symptoms,etc. Since majority of the people, especially children are sitting inside their home the usage of digital platforms are also increased dramatically.

COVID-19 and the associated lockdown in the country, has made a significant impact on the life of the population, across boundaries. Roughly saying, it has been near to a year since the foremost case of corona virus was reported in India, which was reported on January 30th 2020. Today that is, on 10th December 2020, 69,383,534 corona cases were reported, 1,578,707 deaths and 48,155,526 cases were recovered globally. In India 9,768,643 corona cases were reported, Of which 141,817 deaths and 9,253,388 cases were recovered. Whereas in Kerala, 2562 deaths, 658633 confirmed cases and 596593 recovered cases were reported. The only fluctuations are in the statistics, but the conditions are the same in each day. Of this study, we aim to create awareness on COVID-19 through interactive and pleasure seeking games. This helps to make them aware about the severity of the pandemic and also improves their psychological health. Lack of awareness and the passive attitude of people is the major factor of this situation. Children and the older adults are the major two communities more convenient to disseminate the information and at the same time they are the most vulnerable groups to the corona virus. Children are more enthusiastic and older people are more precaution. By providing the same games for both age groups, the interaction between both becomes closer and thereby improves the mental well-being of both in a family. So that the games were most effective in all respects. Games are activities usually undertaken for fun and enjoyment, but it not only meant for fun but also gives edutainment. Games are an effective tool to disseminate information by entertaining way.

Significance of the Study

Wider acceptance of virtual services is one of the major changes happened during COVID-19 pandemic. Even though, the digital technology could enhance public-health education and communication, the whooping majority are unaware about the perspectives of COVID-19. Absence of mindfulness and deceiving data about the pandemic may accelerate the spike in the number of cases, and further to increase in mortality rates. Hence, it is quintessential to provide general awareness to the public among the pandemic and to bring behaviour change communication effectively. So, it is imperative to make every people conscious about the gravity of the pandemic. This could be made possible through focused interventions targeting the vulnerable sections of the population, who can be easily affected by the Pandemic. Such groups include children and the older adults. Daily updates on COVID-19, and the guidelines are made available to the public through various social media platforms, and mass medias. Keeping all these in view, a focussed intervention with the effective use of games as a medium of channelising the information, was designed, were both the children and the adult can enjoy and bring desirable changes alike. The present study entitled 'Developing communication tools for creating awareness of COVID-19 among communities in Kerala' aimed to disseminate information regarding all the aspects of COVID-19 and its management, through games.

Objectives of the Study

1. To assess the socio-demographic background of the samples.
2. To develop and disseminate games for providing awareness about COVID-19 among the samples.
3. To check the effectiveness of the games used in creating awareness among the samples.

Study Design

Thiruvananthapuram, the capital city of the State of Kerala was selected as the locale of the study. Considering the extend of vulnerability of the population, young Children (6-12 years of age) and older adults above 60 years of age were identified as the samples for the study. A sample size of hundred which comprises of fifty samples each from the age category were selected using random sampling method. An inclusion and exclusion criteria were also set for the study. Prior to the study the informed assent of participants was taken for making them understand about the purpose of the study and to agree to become the part of the study. Separate consent was obtained from the parents as well as children and from the older adults as well. The participation of the samples was purely voluntary, and they could withdraw from the study at any point. The socio- demographic background and the existing level of awareness about the five different areas of COVID-19 among the samples were collected using an online Questionnaire. The five areas projected about the COVID-19 Pandemic are Symptoms of the disease, treatment for the disease, misconceptions on COVID-19, nutritional recommendations and Post COVID recovery.

Communication games like Crosswords, Aksharamala and Varnamalas were developed and Puzzles, Crosswords, Jumbled words, Find the difference and Bingo were adapted and translated from the original versions developed by Aparna Khanna (2020). The samples were also given daily challenges and tasks in between the games to create interest, sustain curiosity and attention throughout the period. A period of six weeks was set for the intervention. The content of the games was collected from the official websites of Arogyakeralam, WHO and

UNICEF (from the period of September to November 2020) for better authenticity of the data. Those who joined the survey was further added to a WhatsApp group. Games, tasks, and instructions on how to proceed with the task and play the games were clearly provided in the group. The responses of each sample were collected and analyzed further. All the games were field tested prior to the study. The effectiveness of the games was checked after the intervention programme using a checklist.

Major Findings of the Study

Profile of the Samples

Fifty older adults and fifty children were selected as the samples for the study. The Figures numbered 1 and 2 shows the information regarding the socio-demographic profile of the samples such as age, gender and mode of play for both children and older adults.

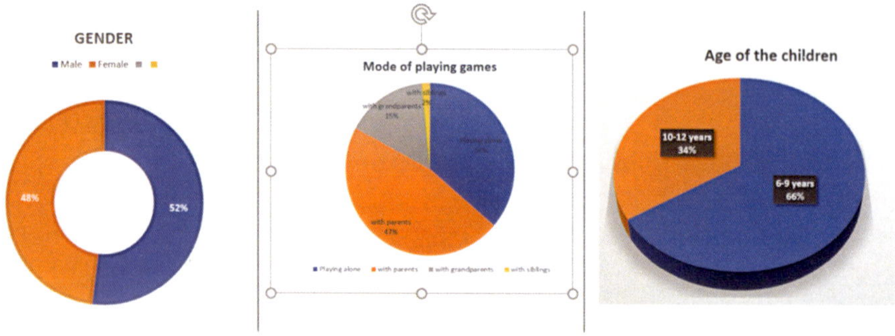

Fig. 1: Socio- demographic profile of the children

In the case of children, the sample belongs to the age group of six to twelve were taken for the study in which sixty six percent of them belongs to the age group of 6-9 years of age and tjirty four percent of them belongs to the age group of 10-12 years of age. Fifty two percent of the sample were boys, and forty eight percent were girls. Regarding the mode of play adopted by the samples, forty seven percent played with the help of parents, thirty six percent played alone, fifteen percent played with grand parents and a mere two percent played with their siblings. Through this, the study also indirectly aimed to create an intergenerational linkage between the children and their grandparents, which may help to improve the relations and to bridge the gap between the two generations.

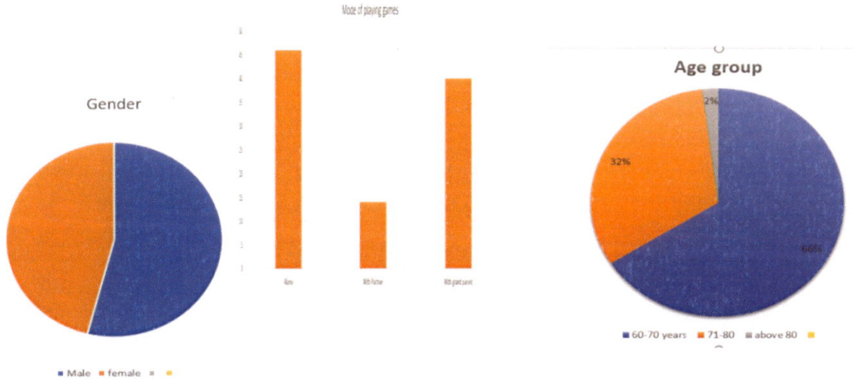

Fig. 2: Profile of the older adults

Older adults from the age of sixty to eighty years and above were taken for the study in which sixty six percent of them belongs to the age group of sixty to seventy and thirty two percent belongs to the age group of seventy to eighty years of age and two percent belongs to the age group of above eighty years. Fifty four percent of the samples were male, and forty six percent were female. In response to the mode of play, forty percent of the older adult sample were played with the help of grandchildren, fourteen percent played with their partner and remaining fouty six percent of the sample played the games alone.

From the hundred samples, ten samples from each categories who have done all the games and tasks actively were selected for assessing the efficacy of the tools used. A telephonic interview was done to collect the data regarding the same. A checklist was used for understanding the acceptance of the games played, the easiness, previous experiences and the effectiveness of the developed games in enhancing the awareness level of the samples and to know about the suggestions.

Table 3: Checklist of the children

	Category	**No.**	**%**
Ease of playing the game	Easy and it was ok	5	50%
	Very comfortable and easy	5	50%
Ease of understanding the instructions	Easy and it was ok	1	10%
	Very comfortable and easy	9	90%
Appeal of the game	Good	2	20%
	Very good	8	80%
Previous experiences	Yes	0	0%
	No	10	100%
Level of awareness on COVID-19 after playing the games	Increased	9	90%
	No change	1	10%

From the checklist of children 50 percent samples responded as the games are very easy and the other half of the sample responded as the games are easy. When asked about the easiness of understanding the instructions 90 percent of the samples responded as very easy and 10 percent responded as easy. 80 percent of the selected samples responded about the appealing of the games as very good and 20 percent as good. There were no such previous experiences to the samples. For 90 percent of the samples, who actively participated in the games, opined that their level of awareness on COVID-19 has increased after playing the games and about 10 percent reported that there was no change in their overall knowledge.

Table 4: Checklist of the older adults

	Category	No.	%
Ease of playing the game	Easy and it was ok	4	40%
	Very comfortable and easy	6	60%
Ease of understanding the instructions	Easy and it was ok	0	0%
	Very comfortable and easy	10	100%
Appeal of the game	Good	2	20%
	Very good	8	80%
Previous experiences	Yes	0	0%
	No	10	100%
Level of awareness on COVID-19 after playing the games	Increased	7	70%
	No change	3	30%

By analysing the checklist of the older adults, 60 percent of the samples responded as the games are very easy and 40 percent as easy. All the selected samples responded about the easiness of understanding the instructions as very easy. 80 percent of the selected samples responded about the appealing of the games as very good and 20 percent as good. In this samples also there was no such previous experiences playing of this type of games. For 70 percent of the samples the level of awareness on COVID-19 increased after playing the games and there was no change for 30 percent.

Ranking of the Games used for Creating the Awareness

The games were further ranked based on the responses shared by the samples.

Table 5: Ranking of the games and tasks

Games	6-9 years	10-12 Years	60-70 Years	71-80 Years	81+ Years
Corona Drawing	1	2	9	8	7
Quiz	8	1	2	2	2
Steps in hand washing (video)	2	6	8	9	9
Word puzzle	9	5	1	6	6
Aksharamala	7	3	6	3	3
Post covid practices	6	7	7	7	8
Peace maker vs peace breaker (jumbled words)	4	5	5	4	5
Find the difference	3	3	4	5	4
Lock down bingo	5	4	3	1	1

Age wise Rankings for each games and tasks were analyzed from the responses of the samples. It was interesting to note that , the task on drawing the picture of Corona was found to be the most excited one for children in the age group of 6-9 years, followed by steps in hand washing, find the difference, jumbled words and lockdown bingo. Fun Quiz related to COVID-19 was ranked no 1 by the 10-12 years of children, followed by drawing a corona virus, Aksharamala, Lockdown bingo and jumbled words. Word puzzle is the most rated game for the older adults in the age group of 60-7o years of age, followed by Quiz, Lockdown bingo, Find the difference, Jumbled words, whereas those between the age group of 71-80 and above 80 rated Lockdown bingo as the most interesting game. From this ranking of games and tasks it is evident that children preferred more activity oriented games and tasks and older adults preferred to have games that have less effort. Figure 3 shows the images of the games and activities done by the actively participated samples.

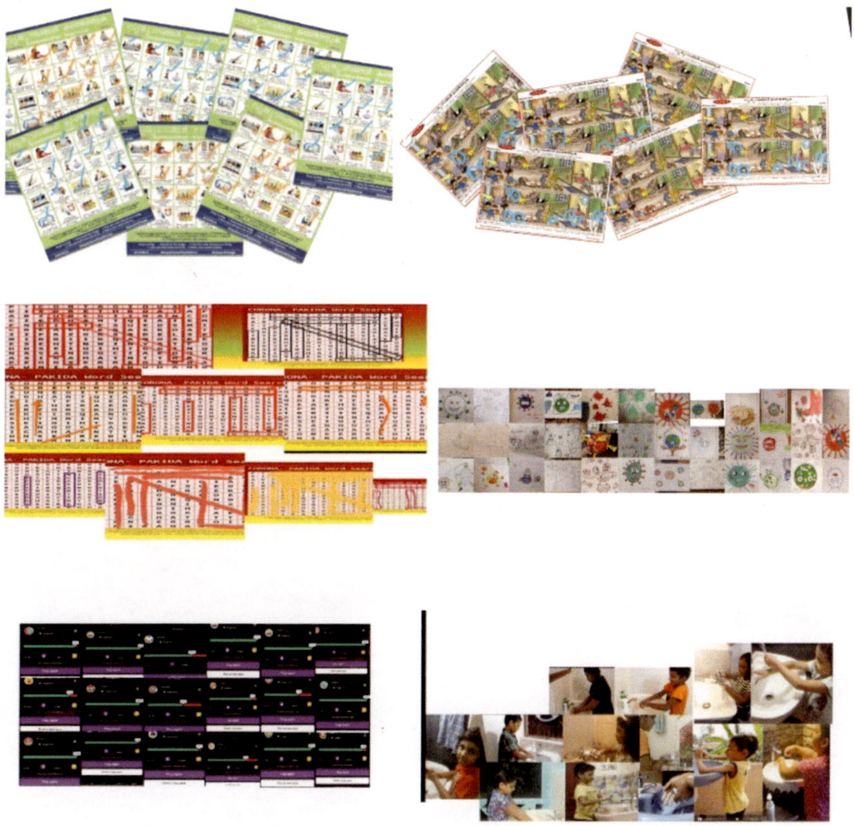

Fig. 3: Various games and tasks done by the samples.

Conclusion

Games are an interesting medium for bringing behaviour change among the masses. They are a useful way to understand, corelate and improve the level of knowledge towards a particular aspect. It was noticed that the colour, easiness of the games, instructions given on how to play the games, and design have an overall impact on the samples towards the games. The tasks given in between were found to be interesting and they helped to sustain the level of interest among the samples. Being, the first attempt, the study needs further exploration on understanding the effectiveness and mental wellbeing of the samples. A meta analysis could not be conducted due to the limited time frame and the prevailing conditions.

References

arogyakeralam.gov.in

https://covid19.who.int/

https://www.who.int/health-topics/coronavirus#tab=tab_1

https://www.ijsr.net/get_abstract.php?paper_id=02013823

https://www.who.int/emergencies/diseases/novel-coronavirus-2019/advice-for-public

https://www.sqlearn.com/coronavirus-disease-covid-19-awareness/

https://www.ncbi.nlm.nih.gov/pmc/articles/PMC7460945/

https://www.livemint.com/mint-lounge/features/how-to-talk-to-your-children-about-covid-19-11584687059393.html

https://www.unicef.org/india/media/3401/file/PSS-COVID19-Manual-ChildLine.pdf

https://www.ijhsr.org/IJHSR_Vol.10_Issue.8_Aug2020/8.pdf

https://www.voicesofyouth.org/blog/creating-covid-19-awareness-children-through-animation

https://www.webmd.com/lung/coronavirus

https://www.healthline.com/health/coronavirus-prevention#tips

https://www.healthline.com/health/coronavirus-covid-19

https://intermountainhealthcare.org/blogs/topics/live-well/2020/03/myths-about-coronavirus/

https://www.avert.org/coronavirus/covid-19-myths-and-facts

https://www.physio-pedia.com/COVID-19_and_Nutrition

http://www.emro.who.int/nutrition/nutrition-infocus/nutrition-advice-for-adults-during-the-covid-19-outbreak.html

https://www.ndtv.com/health/post-covid-19-care-6-things-you-need-to-do-after-recovering-from-coronavirus-2302135

19

Rural Education and Health in Tripura

Debajyoti Gupta

Department of Liberal Arts, Tripura University, Agartala, Tripura

Abstract

India has one of the largest education systems in the world with more than 15 lakhs schools and 80 lakhs school teachers, focused on education, rural schools also play an important role in the health of students and the community. From 1950 onwards the Government of India (GOI) has created policy instruments to promote Universal Elementary Education (UEE) in an attempt to eliminate all forms of discrimination based on caste, community and gender. Working on health-related initiatives makes sense for rural schools as physical and mental health status, physical activity, and good nutrition all impact how students learn. In Tripura, NGO Voluntary Health Association of Tripura, this organization has been working for the promotion of health and primary education, gender equity and promotion of care and support for the children in need. VHAT initiated their Community-Based Rehabilitation (CBR) Programme for the Persons with Disabilities (PWDs) in a few blocks of west and south districts of Tripura in 2007. School-based health centers can provide access to healthcare for students who may not otherwise receive care. Rural students may face barriers to accessing healthcare services including long distances to providers, lack of reliable transportation, work demands of students' parents, poverty, lack of health insurance, and the cost of healthcare. School-based health centers often operate as a partnership between the school and a community health center, hospital, or local health department to improve the health of students, as well as the community as a whole. Schools play a significant role in promoting the health and safety of students. Ideally, rural schools will strive to provide a healthy physical environment for students to learn, provide nutritious meals, and encourage physical activity. In addition to school health promotion initiatives and policies, some rural schools are able to employ a school nurse or house a school-based health center, both of which work to directly address the health needs of students.

Keywords: Opportunities, Counselling, Community, Students, Role.

Introduction

The present world is still dominated by the villages. In sprite of the craze towards city life majority of the people in this globe prefer to stay in their village's n the midst of nature. Hence the world population is predominantly rural. More than 2/3 the world's population still lives in villages. India is basically a land of villages. In spite of the increase trend towards urbanisation Indian societies become meaningful and fruitful only if the villages and there problems are studies carefully. Villages existed in India even during the Vedic period and also prior to that. Some even described these villages as self contained little republics.

Tripura is a state in North-East India which borders Bangladesh, Mizoram and Assam. It is surrounded by Bangladesh on its north, south and west: the length of its international border is 856 km. It shares a 53 km long border with Assam and a 109 km long border with Mizoram. The state is connected with the rest of India by only one road (NH-44) that runs through the hills to the border of Karimganj District in Assam and then winds through the states of Meghalaya, Assam and North Bengal to Calcutta. At the time of Tripura's merger with effect from October 15 1949 with the Indian Union, the major mode of farming was shifting cultivation or 'jhum', which produced little surplus. A small proportion of the plain lands of the State were under settled agriculture undertaken by Bengalis, and the main crop was rice. Most of the plain lands of the State were not under cultivation and were covered with cane-brakes and marshes. Thus at the time of formation of the State, the economy was predominantly agricultural and forest-based, with no industrial base, a low level of urbanization and limited infrastructure.

For administrative convenience and decentralisation of power Tripura which had once been a single district only is now divided into altogether the state has been divided into 8 districts, 23 subdivisions and 58 development blocks—with effect from 21 January 2012, after a Government of Tripura Decision, out of which the newly created districts are 4, subdivisions 6, development blocks 5. Besides, a special feature of the state is the existence of an Autonomous District Council (ADC) for tribal's based on 6th schedule of the Indian constitution. The ADC in Tripura encompasses 68.10% of the state's total geographical territory and is home to roughly one third of the state's population.

Dropping-out of school remains a concern, particularly among village households. I t has been found in the 2011 census Primary (I – V) 1.28 Upper Primary (VI – VIII) 2.26, Elementary (I – VIII) 1.62, Secondary (IX – X) 24.37. Children attend for the first three to four years of primary school and gain a smattering of knowledge, only to lapse into illiteracy later. Few who enter continue up to the tenth grade; of those who do, few manage to finish high school. Therefore, very few are eligible to attend institutions of higher education, where the high rate of attrition continues. Eldest members of tribes often are reluctant to send their children to school, needing them, they say, to work in the fields. In the field of higher and technical education, special provisions such as reservation of seats, relaxation in minimum qualifying cut-off percentages, remedial coaching and scholarships were being extended by the Department of Secondary and Higher Education.

Fig. 1: Map of Tripura District wise

Health and Education in the Rural Tripura

Many determinants of health such as nutrition, water supply and sanitation are not directly within the realm of health care services. However, other services with health effects are being addressed to a certain extent through a coordinated inter-sect oral approach, highlighting the specific role of the Health and Family Welfare Department with the state playing the role of collective mediator.

Table 1: Literacy in Tripura as per Census 2011

Category	Number of literates	Literacy rate
Overall	28,04,783	87.22%
Males 7+ yrs	15,01,369	91.53%
Females 7+ yrs	13,03,414	82.73%
Rural 7+ yrs	19,92,773	84.9%
Urban 7+ yrs	8,12,010	93.47%
Rural Males 7+ years	10,81,503	90.07%
Rural Females 7+ years	9,11,270	79.49%
Urban Males 7+ years	4,19,866	95.51%
Urban Females 7+ years	3,92,144	91.38%

Data Source – Census 2011

Apart from literacy rates, dropout rates specify the quality education in a state. The table below mentions the dropout rates at different levels of school in Tripura.

Table 2: Dropout Rates in Tripura

Level of Education	Drop-out Rate
Primary(I – V)	1.28
Upper Primary (VI – VIII)	Ment of 2.26
Elementary (I – VIII)	1.62
Secondary (IX – X)	24.37

The strategy uses the primary health care approach and public health principles for priority health problems. Human resource development and management with good governance and accountability are recognized as critical areas. Greater efforts are put into health care financing, financial management and accounting systems. Building partnership is a thrust area with greater scope and capacity building for community involvement and closer linkages with NGOs, civil society, the private sector and donors to meet shared social goals. Special attention is being paid to the poor and vulnerable sections of society. Despite financial constraints, the Government of Tripura continues its effort to systematically enhance its investment in health in terms of improving the different health parameters, nutritional status of women and child, immunization status, etc. It is also actively seeking alternative and additional sources of support to strengthen its health system. The State is confident that these efforts will make a positive contribution to the lives of its citizens.

Although it can also been seen that many health programs has been launched by the state government in the rural areas such as National Vector Borne Disease Control Programme, Revised National Tuberculosis Control Programme National Leprosy Eradication Control Programme, eye camp etc. However the schools in the rural areas with the help of NGOs also playing an important role for the health of the students as they have launched School Eye Screening Programme, Training School teachers and nurses, Aids control Board, blood donation camp.

Many community development centres and primary health centres have been set up in different districts in the state. The School Education Department has taken necessary steps to provide Iron & Folic Acid tablet (IFA tablet) to all the children of all schools covered under Mid-Day-Meal Programme in collaboration with Health Department, Govt. of Tripura.

The School Education Department, Govt. of Tripura has initiated to aware the school going children about the goodness of clean and hygiene and introduces the "HAND WASH" programme in each school unit along with Mid-Day-Meal programme.

In exercise of the powers conferred by sub-section (1) read with clause (b) of sub-section (2) of section 39 of the National Food Security Act, 2013 (20 of 2013), the Central Government has made Mid-day-Meal rules 2015 which have come into the effect from the date of publication in official Gazette i.e. 30.09.2015. These rules are as bellow:

Every child within the age group of six to fourteen years studying in classes I to VIII who enroll and attend the school, shall be provided meal having nutritional standards as specified in Schedule II of the Act, free of charge every day except on school holidays. The meal shall be served to children at school only.

NGOs working in Tripura

There are many NGOs which are working in Tripura, but among them Voluntary Health Association of Tripura, is the main NGO which work for the rural health and education however it has different others work to perform. Voluntary Health Association of Tripura(VHAT) was formed as a state level net work NOGs and CBOs, by some health and science activists in 1988 with the aim to promote the Health, Elementary Education & Environment status of the state through policy level advocacy and conceptual level intervention by capacity building. It got the registration under Societies Registration Act 1860 on 12th December,1988. Voluntary Health Association of India, New Delhi, was one of the main inspirer to form this organization as State federation of Voluntary Health Association of India for the state of Tripura. This NGO Promotion of preventive & curative health service for the poor. Promotion of health, environment, gender issue, rights of children, women & PWDs, through IEC and technical support. Promotion of Elementary Education of underprivileged children. Prevention of Child trafficking and violence against children & Women. Socio-economic empowerment of the women of the weaker sections. Networking with government bodies, policy makers, Corporate Bodies & NGOs for improvement of health, environment & socio-economic status. Rehabilitation of persons with disability through integrated approaches. To undertake any other activities related to health promotion.

Conclusion

The Government recognizes that investing in health results in invaluable gains in human development. It has its goal in the improvement of health and education of its people and equitable access to good quality health care and education. According to the 2011 census, Tripura was the fifth most literate state in the country with the literacy rate of 87.22 percent. This rate, according to the study conducted by the Indian statistical institution, touched 95.16 percent. However the high rate of literacy bear any fruit of success in the life of individual for the problem of huge unemployment. The economy of Tripura is basically aagririan and 74 percent of the people live in rural areas.

References

Desia A.R. (2009). Rural Sociology in India. Sage Publication.

Shankar Roa C.N. (2009). Sociology of Indian Society. S Chand & company limited.

Website

Govt of Tripura, Health and family welfare department,https://health.tripura.gov.in/

E Health in Tripura, 2016,https://darpg.gov.in/

20

Pre-maternal Services Availed by ICDS Beneficiary and Non-beneficiary Women of Agra District

Madhulika Gautam

Associate professor (Extension Education and Communication)
Department of Home Science, Dayalbagh Educational Institute
(Deemed University), Dayalbagh, Agra, Uttar Pradesh

Abstract

The possible reasons for high mortality and morbidity are the inadequacy of health services and ignorance regarding the need of obtaining medical care by pregnant or lactating women. ICDS has an important role to fill in this gap of knowledge and non-availing attitude of women. An assessment in the presently study of the frequency of availing pre maternal services and the coverage of beneficiaries can give important information regarding actual benefits to the vulnerable groups. The study was conducted on three hundred ICDS beneficiaries and non beneficiaries women of Agra District. The sample was selected by multi stage random sampling technique. The obtained $\chi 2$ value of 20.14 indicates that higher frequency of availed antenatal clinic services was associated with non beneficiary status of women. Significant association was observed between the beneficiary and non-beneficiary status of women with their status of TT vaccination during pregnancy ($\chi 2$=32.26). Significant association was obtained between the beneficiary and non-beneficiary status of women and consumption of iron folic acid tablet during pregnancy ($\chi 2$=11.17).

Keywords: ICDS Beneficiaries, ICDS Non Beneficiaries, Pre maternal services, Agra District

Introduction

Woman dies as a result of the complications arising due to pregnancy such as sepsis, hemorrhage or obstructed labour. The possible reasons for high mortality and morbidity are the inadequacy of health services and ignorance regarding the need of obtaining medical care by pregnant or lactating women. ICDS has an important role to fill in this gap of knowledge and non-availing attitude of women. Integrated Child Development Services (ICDS) is a prominent national program that provides package of services of children and women. Role of Anganwadi worker in ICDS scheme has to provide information to beneficiary mothers regarding reproductive health, balanced diet, pre and post maternal services , neo-natal care and family planning etc .Group discussions are often organized to impart education regarding health, breast feeding and immunization etc at AWCs. AWWs also provide information to beneficiary mothers regarding health camp, doctor's visit and immunization day by visiting the beneficiaries at home as well as at AWCs. Present research has focused on pre-maternal services. An assessment in the present study of the frequency of availing pre maternal services and the coverage of beneficiaries can give important information regarding actual benefits to the vulnerable groups.

Objective of the Study

To study the pre maternal services availed by ICDS beneficiary and non-beneficiary women

Material and Methods

Sample selection: The study was conducted on three hundred ICDS beneficiaries and non beneficiaries women of Agra District. The selected women were in the age range of 15 th 45 years. The sample was selected by multi stage random sampling technique.

Tool Used: A self made interview schedule was used.

Result and Discussions

The researcher wishes to point out that beneficiaries were the women who were obtaining the services directly through ICDS/ Aganwadi workers. The non beneficiary women were those women who were not registered in ICDS/ Anganwadi but were availing some health service that was of their choice.

Table 1: Frequency of antenatal clinic services availed by ICDS beneficiary and non-beneficiary women.

Availed antenatal clinic services	Beneficiaries		Non-Beneficiaries		Total	
	No.	%	No.	%	No.	%
Nil	59	19.67	48	16.00	107	35.67
One	41	13.66	17	5.67	58	19.33
Two – Three	50	16.67	85	28.33	135	45.00
Total	150	50	150	50	300	100.00

($\chi 2$=20.14, df=2, p=<0.01 significant)

Above table 1. shows the distribution of ICDS beneficiary and non-beneficiary women according to availed antenatal clinic services . Out of the 300 women,

19.67% beneficiary and 16.00% non-beneficiary women did not avail any kind of antenatal clinic services in their previous pregnancy. Again 13.66% Beneficiary and 5.67% non beneficiary respondents reported that they had availed antenatal clinic services only once in their previous pregnancy. On the other hand the 16.67% beneficiary and 28.33% non beneficiary women had availed antenatal clinic services two to three times in their previous pregnancy. Thus indicating that non beneficiaries were more regular with antenatal clinical services in comparison to their counterparts. Pathi & Das (2015) reported that 11.8% women did not receive any type of antenatal clinic services. Although Bhomik and Manna (2017) in West Bengal reported that most of the ICDS pregnant women had received health checkup regularly. The obtained $\chi2$ value of 20.14 indicates that higher frequency of availed antenatal clinical services was associated with non beneficiary status of women. Whereas, the reverse was true for the beneficiary group . Reasons given by the ICDS beneficiary women for not availing antenatal clinical services are as shown like women did not feel the need, transportation difficulty, expensiveness, non availability of female attendant at centre.

Table 2: Number of beneficiary and non-beneficiary women vaccinated for tetanus during pregnancy

	Beneficiaries		Non- beneficiaries		Total	
	No.	**%**	**No.**	**%**	**No.**	**%**
Vaccinated	44	14.66	93	31.00	137	45.66
Not vaccinated	106	35.34	57	19.00	163	54.34
Total	150	50	150	50	300	I00

($\chi2$=32.26 df=1, p=<0.01 significant)

Table 2. shows that 31.00% of non beneficiaries had been vaccinated for tetanus in comparison to their 14.66% counterparts. Significant association was observed between the **b**eneficiary and non-beneficiary status of women with their status of TT vaccination during their previous pregnancy ($\chi2$=32.26). On the other hand a study conducted by FORCES (2017) in Delhi stated that most of the pregnant beneficiaries (88.00%) were immunized. This result indicates strong need of an educative programes and consistent efforts by AWWs. However poor vaccination status of beneficiary women was also reported by Advanced Management Services (AMS) Consulting (2017) in Uttarakhand.

Above table shows that the Tetanus Toxoid (TT) vaccine is given during pregnancy to prevent tetanus in mother as well as her baby. Antibodies formed in the body, after the vaccination, are passed on to the baby and protect her for a few months after birth. The results of the present study show that only 45.66% subjects had received TT shots. Some reasons have given by the women for not taking TT shots during pregnancy like husband not want, delivered safe babies earlier without it, lack of knowledge, complications, vaccine not available.

Table 3: Regular consumption of iron folic acid tablets during pregnancy

Regular Consumption of iron folic acid	Beneficiaries		Non- beneficiaries		Total	
	No.	%	No.	%	No.	%
Consumed	108	36.00	80	26.66	188	62.66
Not Consumed	42	14.00	70	23.34	112	37.34
Total	150	50	150	50	300	100.00

($\chi^2=11.17$, df=1, $p=<0.01$ significant)

The table shows that 36.00%beneficiary women had taken iron folic acid tablets in their last pregnancy in comparison to only 26.66% non beneficiary who did so. Significant association was obtained between the beneficiary and non-beneficiary status of women and consumption of iron folic acid tablet during pregnancy ($\chi^2=11.17$).

Similarly, in a study conducted by Indian Institute of Social Development (IISD) (2019) in Orrisa it was stated that more than 50% of the pregnant women were regular in the consumption of iron folic acid tablets. On the other hand a study of ''Prevention of Low Birth Weight Babies in the State" conducted by AMS Consulting, (2018) in Uttra Khand reported that only one third of the mothers in ICDS villages had consumed iron folic acid tablets regularly. In fact Dash (2016) reported that most of the mothers had not received any IFA tablet.

It is suggested to take the iron folic acid tablets for at least 90 days after the first trimester of pregnancy. the respondents gave their reasons for not taking it like taste bad, IFA tablets not available, nausea and vomiting.

Table 4: Person conducting birth delivery

Delivery conducted by		Beneficiaries		Non-Beneficiaries		Total	
No.		%	No.	%	No.	%	
Trained	Doctor	65	21.66	81	27.00	146	48.66
	Dai	63	21.00	51	17.00	114	38.00
Untrained	Family member	22	7.34	18	6.00	40	13.34
Total		150	50	150	50	300	100.00

($\chi^2 = 3.42$, df=2, $p=>0.05$ insignificant)

Table 4 indicates the distribution of beneficiary and non-beneficiary women according to the person who conducted the delivery of their child. Majority of the women (86.66%) were taking the services of trained person (Doctor-48.66%, Dai-38.00%) for delivery of their child. Very few respondents (13.34%) were getting their child delivered by an untrained person. The table shows that in the case of both beneficiaries (7.34%) as well as non beneficiary women (6.00%) delivery by an untrained person was a less common practice. Most beneficiary women had their delivery done by trained person, that is by a doctor (21.66%) or dai (21.00%). Similarly in case of non beneficiary women most deliveries were conducted by a trained person, that is either by a doctor (27.00%) or by a trained dai (17.00%). Perhaps the reason for majority of women going for deliveries by a trained person could be contributed to general awareness, availability of such services nearing by, effect of media and AW generated education strategies. Taking the service of trained or untrained personal for delivery for child was insignificantly associated with the beneficiary and non beneficiary status of respondents ($\chi 2=3.42$).

Conclusion

In the present study result indicate that non beneficiary women were more regular with antenatal clinic service in comparison to beneficiary women. Again more non beneficiary women had been vaccinated for tetanus in comparison to their counterparts. Although more beneficiary women had found for taken iron folic acid tablets in their last pregnancy .

References

Advanced Management Services Consulting Pvt. Ltd. (2017). Infant Death Audit. Dehradun, Uttarakhand. Research on ICDS An Overview Volume 3, pp.11-13

Advanced Management Services Consulting Pvt. Ltd. (2018). Prevention of Low Birth Weight Babies in the State. Dehradun, Uttarakhand. Research on ICDS An Overview Volume 4, Received: September 22, 2018 from https://www.indianpediatrics.net/sep2018/707.pdf

Bhowmik, P. K. and Manna, S. (2017). Mortality and Morbidity Pattern in Children 0 to 6 Years Attending Anganwadi Extent to Primary Health Care through Awareness and Community Health Problem, Follow-Up Case of Grade III and IV of Malnutrition Children under ICDS, West Bengal Council for Child Welfare, Kolkata, West Bengal. Research on ICDS An Overview, Vol. 3, pp.43-45.

Dash, N. C. (2016). Assessment/ Evaluation of ICDS Programme in the State of Orissa, Centre for Rural Development, Institute of Media Studies, Bhubaneswar, Orissa. Research on ICDS An Overview, Vol. 3, pp-61-63.

FORCES (Forum for Creche and Child Care Services). (2017). A Report- ICDS in Delhi: A Reality Check. New Delhi. Research on ICDS An Overview Vol. 3, pp.56-58.

Indian Institute for Social Development (IISD). (2019). Nutritional, Health and PSE Status of Children covered under the ICDS Scheme. Bhubaneswar, Orissa. Research on ICDS An Overview Vol 3, pp.23-25.

Pathi, S. Das, B. C. (2015). Breast Feeding Practices in a Rural ICDS block of Khallikote, South Orisa. Indian Journal of Community Medicine, Vol.30, pp- 4.

21

Information and Communication Behaviour of Voluntary (NGO) Agri-Extension Workers in Coastal Odisha

***R.S.Panigrahi*[1] *and K. Ghadei*[2]**

Department of Extension Education, College of Agriculture, OUAT Bhubaneswar, Odisha
Department of Extension Education, Institute of Agricultural Sciences BHU, Varanasi, Uttar Pradesh

Abstract

The present study attempts to unearth the information flow among agro- extension workers of grassroot voluntary agencies (NGOs) in Odisha towards a viable cooperative extension service for the farming community of the state. The study was conducted with a sample size of 104 respondents equally drawn from the category of field, supervisory and administrative staff from 17 nos. of voluntary organizations in Balasore and Kendrapara district of Odisha. The volunteers were interviewed through an interview schedule and the responses were analysed following standard statistical measures. It was revealed that the responses of agro-extension workers of the voluntary organisations in both Kendrapara and Balasore district differ as the respondents of Balasoredistrict expressed more essentialities (56.8%) in developing capabilities of receiving information than Kendrapara district (50.9%) which expects developing capabilities to manage information independently. The study as a whole suggested that both the districts were not having the essentialities of developing capabilities to manage information independently which could be very well managed with receipt of the information and direct correspondence with the experts.

Introduction

Voluntary efforts are always focused on the search for an efficient extension mechanism towards a progressive rural development intervention. Extension is otherwise the organized exchange of information and purposive transfer of skills to fulfill the communication gap in transfer of technology process. The goal of extension varies between different extension organizations according to their system of administration and management. As such the voluntary organizations engaged in agriculture & rural development need an incessant backstopping of farm and allied information from the formal Government extension network, so as to complement & supplement the state sponsored extension efforts as an able stakeholder. Hence the study is an attempt to unearth the information flow in voluntary agencies for viable cooperative extension in the times ahead.

Materials and Methods

The study was conducted with 34 numbers of agro-based grass-root voluntary organizations of Kendrapara and Balasore districts of Odisha. One functionary each from the cadre of administrative, supervisory and field staffs were selected as respondents covering total sample size of 102 respondents. Random sampling technique was followed to select voluntary organizations and respondents. Responses were collected on a four point continuum with the use of mean score, pooled mean and rank order for analysis of the data. Results obtained on the information and communication behaviour of voluntary agro- extension workers towards field extension are tabulated and interpreted below.

Results and Discussion

Table 1: Information & Communication flow among Agro extension workers in Kendrapara district.

Sl. no.	Suggestion	HE	E	LE	NR	Mean score
1	Regular flow of information	5 (9.80)	26 (50.98)	10 (19.61)	10 (19.61)	2.51
2	Competency development on teaching methodology	0 (0.00)	42 (82.35)	9 (17.65)	0 (0.00)	2.82
3	Availability of required teaching aids	5 (9.80)	36 (70.59)	8 (15.69)	2 (3.92)	2.86
4	Developing capability to manage information independently	5 (9.80)	33 (64.71)	13 (25.49)	0 (0.00)	2.84
5	Linkage of staff with the source of information	10 (19.61)	31 (60.78)	10 (19.61)	0 (0.00)	3.00

(Figure in parenthesis indicate percentage)

Majority of the respondents stated for the essentialities of various aspects of information and communication management systems in the voluntary organisations of Kendrapara district as revealed from the table. Field staffs need competency on various teaching methods and its use for good understanding of the people. Various teaching aids particularly electronic and print media should be available in the voluntary organisations for better training and demonstration. The staffs should also go for intermediary exposure for developing their capability in successful use of the technologies. Therefore, they should be linked directly with the source of information for further clarification and additional information, if required.

Though majority of the respondents had stated for essentialities of all these aspects of information and communication, but the mean score value indicated differently being the value at lower side. It is therefore apprehended that the respondents might have not been exposed to the extension methodologies in transferring technologies for which they did not realize its importance.

Since; the diffusion and adoption of latest farm & home innovations are very much essential for its sustainable use by the target groups, the voluntary organisation needs to strengthen information and communication system for effective transfer of technologies.

Table 2: Information & Communication flow among Agro extension workers in Balasore district

Sl. no.	Suggestion	HE	E	LE	NR	Mean score
1.	Regular flow of information	22 (43.14)	29 (56.86)	0 (0.00)	0 (0.00)	3.43
2.	Competency development on teaching methodology	8 (15.69)	42 (84.31)	0 (0.00)	0 (0.00)	3.16
3.	Availability of required teaching aids	19 (37.26)	29 (56.86)	3 (5.88)	0 (0.00)	3.20
4.	Developing capability to manage information independently	0 (0.00)	21 (41.18)	30 (58.82)	0 (0.00)	2.41
5.	Linkage of staff with the source of information	5 (9.80)	32 (62.75)	10 (19.16)	4 (7.84)	2.75

(Figure in parenthesis indicate percentage)

Majority of the respondents of the voluntary organisations in Balasore district had stated the essentialities of regular flow of information (56.86%), competency development on teaching methodology (84.31%), availability of required teaching aids (56.86%) and linkage of staffs with the source of information (62.75%). Majority of (58.82%) the respondents had stated the non-essentialities of developing capabilities to manage information independently which indicate that the staffs develop skills to manage due to exposure on teaching methodology.

Comparative analysis of the suggestion of the respondents of the voluntary organisations in both Kendrapara and Balasore district were taken up to examine the differential opinion and arrive at conclusion. The analysis made with the mean score value have been presented in table 3 below.

Table 3: Comparative analysis of information flow among Agro extension workers of both districts

Sl. no.	Suggestion	Kendrapara district		Balasore district		Pooled mean score	Rank
		Mean score	Rank	Mean score	Rank		
1.	Regular flow of information	2.43	I	3.43	I	2.93	I
2.	Competency development in teaching methodology	2.16	V	3.16	III	2.66	III
3.	Availability of required teaching aids	2.31	IV	3.20	II	2.78	II
4.	Developing capability to manage information independently	2.41	II	2.41	V	2.41	V
5.	Linkage staff with the source of information	2.41	III	2.75	IV	2.58	IV

(Maximum obtainable score-4)

The responses of agro-extension workers of the voluntary organisations in both Kendrapara and Balasore district differ in their opinion and the respondents of Balasoredistrict expressed more essentialities in developing capabilities than Kendrapara district which expects developing capabilities to manage information independently. The study as a whole revealed that the respondents of both the districts were not in need of developing capabilities to manage information independently which could be very well managed with receipt of the information and direct correspondence with the experts. The respondents of Kendrapara district did not perceive the utility of competency development in use of teaching aids to transfer the messages to the farmers skillfully for which they did not express its essentialities.

Conclusion

The findings therefore suggested that the voluntary organisations need to analyse the importance in use of teaching aids, competency development in methodologies for skilful transfer of information to the farming community for use along with exposing staffs to the latest information regularly and strengthen their organizational efficiency on information and communication system.

References

Dhillon, D.S. and Sangha, G.S. (1990) Voluntary agencies : Identification and role in rural development, Kurukshetra. June, 1990, pp.10-11.

Maheshwari, S. (1987) Voluntary action in rural development in India, Indian Journal of Public Administration, 33(3): 559-568.

Roy, S. (1977) The role of voluntary agencies in development, In rural development, Role of voluntary agencies and business houses, Bombay, Indian Merchants Chamber p.559-568.

Shastri, Rajesh Kumar (2008) A strategic action plan for managing non government, with special reference of India definition of NGOs, International NGO Journal, 3(3): 74-76.

22

Community Voices in Rural Development: The Role of NGO Based Community Radio

Pragya Goswamy and S.K. Kashyap

Department of Agricultural Communication, College of Agriculture, G.B. Pant University of Agriculture & Technology, Pantnagar, Uttarakhand

Abstract

Community is referred as a group of people residing in a common location, who form relationships over time by interacting regularly around shared experiences and common interest. In India, community has been identified as the core of development. The concept of community radio has emerged with a thrust of participatory development with community as the prime stakeholder. Multiple efforts and approaches of agricultural extension till time has proved that unless the community is involved in the realm of development genuinely, right from the need assessment stage, the community doesn't emerge as the prime stakeholder and therefore a sustainable and consistent developmental process does not emerge. Equally, the role of communication is critical in developmental process. The grassroot communication involing community voices in development models has a prime and critical role in overall development. Community radio has emerged as an efficient tool to instigate community involvement in grassroot development process in which the community radio stations owned by Non-government organizations have played a distinct and crucial role at microlevel. In 2006, Government of India notified new Community Radio Guidelines, which permitted Non-governmental Organizations (NGOs) to own and operate community radio stations. As on 18/11/2020 (mib.gov.in), out of total 310 operational community radio stations in India approx. 50% are owned by NGOs.

The study revealed that the community radio owned by Non-governmental organization were more aligned to the community needs and aspirations. It was found that the unheard voices at grassroots got a channel through these radios which mostly depended on community based programs which raised critical issues o livelihood generation, youth aspiration, women empowerment, primary education, health and hygiene and environmental issues. On the otber hand it was also found that the NGO owned community radios regularly confronted the issues of financial sustainability and a major thrust of these radio was focused towards the struggle for financial assistance for survival. Looking into the impact and contribution of NGO based community radios in contrast to the government support, the impact was found to be substantial and significant whereas the government support was found to be negligible. It is inferred that the government should make structured investment into programme production and community participation processes of community radios across the country to strengthen this very effective grassroot media which will payback manyfold as compared to the investment ever made to grassroot development.

Keywords: NGO based Community Radio, grassroot communication, rural development, community participation, financial sustainability.

Introduction

The concept of communication for development emerged with the Modernization theory, which developed its roots after World War II. Wilbur Schramm and Daniel Lerner, the two major proponents of this theory opined that the major obstacle to development was the psychology of the undeveloped citizens. The communication models under modernization paradigm observed communication process as the transfer of message from sender to receiver (Lasswell, 1948). It was found that communication theories like two step-flow, diffusion theory, or the early extension approaches were quite harmonious with the top-down modernization theory. After the Modernization paradigm, an another theory of communication for development emerged named as Dependency theory. Dependency theorist like Jean Servaes (1996) criticized the broadcasting and other mass-media systems as they tended to support modernization and were anti-development; also they were inclined to promote the political agenda. Solution to underdevelopment of third world countries require major changes in media structure that were dominated by commercial principles & foreign interests into media structures which are self-reliant and community oriented (Servaes, 2002). In 1970s scholars such as Paulo Freire and Nora C. Quebral started pleading in favour of bottom-up development theories which led to evolution of a new development paradigm called the multiplicity paradigm. This paradigm conceptualised the basic idea that development is integral, multidimensional and dialectic process which can differ from one society to another (Okigbo and Eribo, 2004). Multiplicity paradigm led to emergence of a new form of communication called participatory communication. Though the perspectives of communication for development has changed over the years but multiple researches revealed that communication played a critical role in the process of development. FAO's Communication for Development Group has been one of the leading practitioners of applied communication for improvising agriculture and related sectors in the developing world since its establishment in1969. The corresponding shift towards approaches of participatory development (Chambers, 1994; Brown *et al.*, 2002) signifies that a greater apprehension of community perspectives is required to identify the local resources that can be built on to cater local needs.

Community Radio: A Participatory Communication Tool

According to UNESCO, the community can be territorial or geographical or a group of people with common interests, who are not necessarily living in one defined territory (Tabing, 2002). Similarly, AMARC defined it as a "collective or group of people sharing common characters or interests" (Mtimde*et al.*, 1998). The important point in the various definitions is that, the community is identified as a group, not simply a mass of people that the broadcasting signal happens to cover. A group of people who form relationships over time by interacting regularly around shared experiences, which are of interest to all of them for varying individual reasons is a community.

'Participation' is often highlighted as a key characteristic, distinguishing community radio from most other kinds of media (Mtimde*et al.*, 1998). Bruce Girard (2001), proposed that the "most distinguishing characteristic of community radio, is its commitment to community participation at all levels". AMARC reffered community participation to continuous interaction between the broadcast station and the community about their needs, concerns and interests (Plaugher &Solervicens, 2007). Community radio promotes local discussion within the community to assess their environment, the causes for their problems and explores possible actions can be took to address the situations. According to Sharma (2013), Community Radio reaches out to various people simultaneously and in diverse locations. The end user just need to have FM access to listen to the broadacast. Radio can play an important role in holding local authorities to account by playing the watchdog's role for the community. In its original design, community radio actively encourages, support and seeks out the participation of members of the community in all sorts of different programme and roles in the radio. Listeners can get inspired, listening to their peers, and learn about best practices of others. Over the time community radio station has emerged as the voice of community being people centered, development oriented and inclusive. Another difference with the community radio is that there is a democratisation of communication wherein community participation is of paramount importance (Elliot, 2010). People feel ownership of community radio as one can create, access, utilize, share information and knowledge. It helps people to achieve their full potential in promoting and improving their quality of life. As a 2007 UNESCO report put it, "Just hearing themselves on radio is an empowering experience for people, who realize that this, truly, can be their voice and help them get a better deal from the government and other castes. It is the first time Dalits, tribals and women have found an opportunity to speak out."

Non-Government Organization based Community Radio

The grassroot communication involving community voices in development models has a prime and instrumental role in rural development. Community radio has emerged as an efficient tool to instigate community participation in grassroot development process in which the community radio stations owned by Non-government organizations have played a distinct and crucial role at microlevel. On February 1995, the Supreme Court of India declared in its historic judgment that "airwaves are public property" and inspired the campaign to legitimize community radio in India. The famous "Pastapur Initiative" issued in a UNESCO sponsored workshop, hosted by an NGO, Deccan Development Society in July 2000 urged the government to decentralize the broadcasting service not only to private players but also to communities. In the year 2003, government released the first community radio guidelines, but restricted the eligibility to only educational institutions. However, this could not limit the intense urge of the grassroot people to reach out to their communities and voice their concerns and opinion. Pioneer Community radio supported by NGOs like Sangham radio and Namma Dhwani started producing and recording radio programmes but were either narrowcasted or telecasted through TV's without video. In 2006, Government of India notified new Community Radio Guidelines, which permitted Non-governmental Organizations (NGOs) to own and operate community radio stations. At present NGOs with an experience of three years in community development work are provided with the license to broadcast within a radius of 12-30 km. As on 18/11/2020 (mib.gov.in), out of total 310 operational community radio stations in India approx. 50% are owned by NGOs. The present study highlights the important role played by Non-government organizations in raising community voices through community radio stations. The study reviews the cases of two successful and pioneer NGO based community radio stations in terms of their focus areas, target audience, key programmes, strategy of raising community voices, sustainability and overall impact.

CASE of Two Pioneer and Demonstrative Community Radio Stations of India

1. Deccan Development Society, Medak, Telangana (Radio Sangham)

Deccan Development Society (DDS) is a non-government organization working with Sanghams (village level groups) of rural women, most of which are Dalits. The Society envisions empowering the marginalized groups of women to emerge as vibrant organs of the local governance. It aims at facilitating rural communities to attain autonomy in access and contol and with that inspiration the groups have worked to achieve the autonomy over food production, seeds, natural resources, autonomous market and autonomous media. Transcending their barrier of illiteracy the women members of Deccan Development Society (DDS) strived hard to learn media usage to raise dialogue within their groups and inform the world about the dent they are making in the society (http://www.ddsindia.com/). On October 15, 2008, Sangham Radio, the first NGO based community radio station of India was launched at Machnoor village in Medak district. Sangham Radio emerged as a platform to raise the needs, aspirations and experiences of the community around.

DDS has worked with the people of Medak district, most of which are farmers, for more than a decade and identified the potential of community radio to encourage two-way communication between the people of one the most undeveloped regions of the nation with the rest of the country (Rikzana and Neelamalar, 2015).

Hyderabad Narsamma and Nalugindla Narasamma, the two Dalit women who look after the community radio broadcast are educated upto 10th Class with a team of reporters including both men and women mostly illiterate or semi-literate. The entire staff is paid by DDS. In day time, they work as farmers and as the night approaches, they turn into broadcasters. The radio covers a radius of 25 km broadcasting in about 100 villages with about 50,000 listeners (Nirmala, 2015). While speaking about the establishment of the radio station, the Director of DDS, PV Satheesh said, "We were identified as a partner for UNESCO's 'Women Speak to Women' project. As part of this, we established the radio station with the programming content aimed at serving the information, education, and cultural needs of the region." The women are using community radio to save their dying language and cultures, spread their message of sisterhood, ecological agriculture, women's control over seeds and a host of different issues (Avadhani, 2017).

Sangham radio broadcast for 2 hours daily from 7 to 9 PM in Telugu language, the programs are related to social messages, folk songs, farming and health tips, and also has information about fairs and festivals in their area. They mainly use short skits and dramas as programming methodology share educational messages to women and other people. "Villagers feel happy when they hear their voice on the radio in the form of songs or discussions. They will also share this joyfully with their neighbours. The song the community radio plays depends on the season, whether it time to plough, plant seeds, irrigate or harvest. The songs recall our tiring work at the field in the morning," said BegariSammamma of the village of Bidakanne in an interview with the Hindu newspaper (Avadhani, 2017). Over the years, Sangham radio has gathered about 2,500 songs sung by community members. In this way, Sangham radio in its journey of about 12 years, has created an aware community, given voice to the marginalized women of the society and promoted local culture, language, indigenous knowledge, innovations and good governance in the area.

Sustainability:Sangham Radio was initially established by DDS with financial support of the donor agency, UNESCO. UNESCO sponsored DDS through its projects to purchase the transmitter and other equipments essential for establishing community radio. The women of various sanghams (self-help groups) unders DDS also contributed Rs. 5 each to bring up their own radio station (Negash, 2016). The radio sustained somehow for ten years, but in the year 2018, a crowdfunding plea was raised as campaign to help the radio station overcome its financial and technical crisis. The technical problem was reported due to its impaired transmitter, which is required to reach an area of 30km but now barely covers a 3 km radius. "Very soon even this may fail and Sangham Radio may be silenced forever," said the statement, which hopes to raise a fund of INR 10 lakh, the minimum needed for continued survival . The government has not paid for the advertisements aired and its dues to Sangham Radio over the last three years has totaled to INR 3.25 lakhs," declared the appeal (Avadhani, 2018). Thus, though the community radio has been able to put the women of the community at the frontiers of communication, but the financial sustainability issue often threatens the existence of the radio.

2. The Mysore Resettlement and Development Agency, Kolar, Karnataka (Namma Dhwani)

MYRADA NGO was established in 1968. It is a NGO working in backward areas of Tamil Nadu, Andhra Pradesh and Karnataka. It aims at supporting the marginalized segment to develop their own livelihood strategies, access resources and manage their own institutions. Budikote is a poor village in southeastern part of Karnataka, with population involved in agriculture and people were organized into self-help groups. It also has media-penetration and cable connection. Although, the programmes of All India Radio were transmitted to Budikote in three languages, but the community felt that they were not suited to their needs. In 1999 a baseline survey was conducted to assess information needs and preferred media revealed that people preferred an audio channel because of low comprehension, illiteracy and poverty. They revealed to have locally relevant information about agriculture, health, governance etc. Keeping in view the needs and preference of the local community and their urge to have a radio of their own in their own dialect and with voices of their own people, Budikote community with two NGOs MYRADA and VOICES, started creating locally relevant audio programme and narrowcasted them in the meeting of Self-help groups. The success of the narrowcasting encouraged the establishment of an audio production center "Namma Dhwani" in Budikote with the help of UNESCO in the year 2001. The audio programmes were scripted, produced and recorded by the volunteers of various self-help groups, who were initially trained in interviewing, recording, scripting, editing, and mixing skills. To reach every household of the nearby villages, in 2003, a direct to home cable connection was established in collaboration with the local cable operator (Kumar, R.K., 2008).

With the support of NGOs and UNESCO funds, Namma Dhwani also bought few computers for its center through which it acted as a resource center for villagers providing computer trainings and other market related information through internet. Namma Dhwani established itself as a demonstrative model of empowering community voices even before the grant of permission to NGOs to get license of community radio. After 2006, Namma Dhwani got its own community radio and was ready with its already prepared 800 hours of audio programmes which could now broadcast through radio. Namma Dhwani is a fully operational community multimedia center with radio, video and internet facility (Anthony, 2004). All the programmes are made by volunteers of Namma Dhwani, who are mostly the self-help group members and with participation of the community. It has made impact by creating awareness among the women regarding health, sanitation, family planning, education etc. in particular and all the members of community in general. It was also found capable in generating leadership behaviour among the women of community (Nirmala, 2015).

Few cases highlight the impact Namm Dhwani has created in the lives of the local people. For about eight days budikote village was deprived of water supply because of some issue with pipes. The community women gathered in front of the building, discussing their problems. One of volunteers of Namma Dhwani, Nagaraj, 20, recorded the opinions of the women ad hoc and played back the audio recording infront of the Panchayat Secretary, who promised to look into the matter and got it fixed within the next two days. Community radio has proved to be helpful in solving various other problem of the community related to electricity, streetlights, drainage etc.

Moreover, it has been of great help in solving the problems of individuals. Narayanswamy was earning livelihood by selling milk, but after the death of her only cow, her sole hope was on the insurance money. She went to the insurance office a dozen times to claim for the insurance money, but the agent was trying to cheat her. Then, she talked about her problem on the community radio station. The next day, the agent gave her the entire money. Narayanswamy while reflecting on the importance of radio said, "This radio station is ours because it speaks about us, in our language and in our accent. When I turn it on, I hear the voices of people I know." (Kantharaj, 2014). Thus, Namma Dhwani represents the community's voice airing content focused on social evils such as dowry harassment, environment protection, local news, general knowledge, folk music, agriculture etc.

Sustainability:Namma Dhwani has broadened its scope by acting as a community resource center. In 2001, UNESCO provided funds to set up an audio recording center in the space allotted by MYRADA, and a Namma Dhwani Management Committee (NDMC) was formed to look after the operations of community radio. At its initial phase, each Self Help Group involved made a onetime contribution of Rs.1000.00 to cover the cost of running the studio. Namma Dhwani generates funds for its operations by providing additional services and training to rural youth and women through its setup of computers. These computers have created a significant impression on the youth of Budikote. In association with a computer institute in Kolar, Namma Dhwani also offered certificate courses on Microsoft Office package to young people from Budikote.

Opportunities and Challenges for NGO based Community Radio

As depicted by the cases of Sangham Radio and Namma Dhwani, it is evident that Non-Government Organizations have lead the community radio movement in the country and have been successful in mobilizing communities to establish and run their own radio station for their own development. Many other successful cases like Manadakini ki Aawaz (Mandakini ki Aawaz Sewa Samiti, Uttarakhand), Radio Tilonia (Social Work and Research Center, Rajasthan), Radio Alfaz-e-Mewat (S.M. Sehghal Foundation, Haryana), Radio Bundelkhand (The Society for Development Alternatives, Madhya-Pradesh), Radio Kotagiri (Key-stone foundation, Tamil Nadu) etc. have also highlighted that community radio leads to social, economic and cultural empowerment of the communities. Its programmes ensures a wide diversity of voices and views from marginalized groups, such as women and youth, and it promotes and protects the interests, culture, and linguistic diversity of ethnic minorities in the community. However, financial constraints always pose diverse challenges to the functioning of community radio. It primarily relies on varied sources for financial support, which many include, grants, membership fees, donations, advertising or sponsorship.

According to the new policy guidelines of the Ministry of Information & Broadcasting applicants who can arrange for foreign funding for setting up of community radio stations can now obtain it through FCRA Clearance under Foreign Contribution Regulation Act, 1976. Apart from this the policy guidelines of the Government of India in regard to community radio clearly state: "Transmission of sponsored programmes shall not be permitted except programmes sponsored by Central & State Governments and other organisations to broadcast public interest information. In addition, limited advertising and announcements relating to local events, local businesses and services and employment opportunities shall be allowed. The maximum duration of such limited advertising will be restricted to 5 (Five) minutes per hour of broadcast." The liberty of broadcasting of advertisements for 5 minute/ hour can also prove as mutually beneficial for both, community radio stations and local business, entrepreneurs, etc. Many researches, also suggested that community radio being non-profit and owned by grassroot community, the license fees of getting a community radio is very high which can be waived off, especially for the peoples' radio. Looking into the significant role of the NGO operated community radio, it is desirable to have a large number of radio stations in the country, which is at present very less in comparison to the expectation. Structured investment by the government into programme production and community participation process of community radio across the country has been refereed as one of the most important solutions for overcoming financial constraint of these radio stations to make them sustainable.

Conclusion

Communication has a critical role in the process of development. However, communication needs to be participatory, involving people's voices and dialogical process. With participation as its key component and the power to broadcast the concerns, voices, opinions, needs and experiences of poor and marginalized people, community radio has emerged as one of the most essential tool of communication for development. Non-government organization like MYRADA and Deccan

Development Society have been pioneers in the community radio movement of the country. NGOs with their grassroot connect, reach among communities and focus on developmental aspects; were found to be successful in mobilizing and supporting the rural people to establish and run their community radio station. The study revealed that the community radio owned by Non-governmental organization was more aligned to the community needs and aspirations. It was found that the unheard voices at grassroots got a channel through these radios which mostly depended on community based programs which raised critical issues o livelihood generation, youth aspiration, women empowerment, primary education, health and hygiene and environmental issues. On the other hand, it was also found that the NGO owned community radios regularly confronted the issues of financial sustainability and a major thrust of the radio was focused towards the struggle for financial assistance for survival. Looking into the impact and contribution of NGO based community radios in contrast to the government support, the impact was found to be substantial and significant whereas the government support was found to be negligible. It is inferred that the government should make structured investment into programme production and community participation processes of community radios across the country to strengthen this very effective grassroot media, which will payback manyfold as compared to the investment ever made to grassroot development.

References

Anthony, J. (2004). Namma Dhwani: our voices. LEISA India: 26-28. Retreived from https://edepot.wur.nl/89956

Avadhani, R. (2017, April 08). India's first community radio still makes the right connect. The Hindu. Retrieved from https://www.thehindu.com/news/national/

Avadhani, R. (2018, August 20). Crowdfunding plea to save Dalit women's radio. The Hindu. Retrieved from https://www.thehindu.com/news/national/

Brown, D., Howes, M., Hussein, K., Longley, C. and Swindell, K. (2002). Participatory methodologies and practices: Assessing PRA use in the Gambia. Agricultural Research and Extension Network Paper No. 124, London ODI.

Chambers R. (1994). Paradigm shifts and the practice of participatory research and development. IDS working paper no. 2. Brighton: IDS.

Girard, B. (2001). 'The challenges of ICTs and rural radio'. Paper presented at the First International Workshop on Farm Radio Broadcasting, 19–22 February 2001, FAO, Rome.

Kantharaj (2014, August 11). Echoing the voice of the people. The Deccan Herald. Retrieved from https://www.deccanherald.com/content/424836/echoing-voice-people.html

Kumar, R.K. (2008). Impact of Community Radio Station "Namma Dhwani" on people's socio-economic and cultural development in Budikote, Karnataka. Masters' Thesis, IARI, New Delhi. Retrieved from https://krishikosh.egranth.ac.in/displaybitstream?handle=1/96573

Laswell, H.D. (1948). The structure and function of communication in society. In L. Bryson (Ed.), The communication of ideas. New York: Harper and Brothers.

Mtimde, L. (1998). What is Community Radio? Panos. Southern Africa: AMARC Africa

Negash, S. 2016. Role of Community Radio in Development: A Case Study on Sangham Radio, India. Asian Mirror - International Journal of Research. 3(1): 5-13

Nirmala, Y. (2015) The role of community radio in empowering women in India. Media Asia, 42:1-2, 41-46, DOI: 10.1080/01296612.2015.1072335

Okigbo, C.C. &Eribo, F. (2004). Development and Communication in Africa. New York: Rowman &Littelfield, Inc. Publication

Plaugher, Z and Solervicens, M. ed (2007). Community Radio Social Impact Assessment: Removing Barriers, Increasing Effectiveness; AMARC; Montreal. (http://www.amarc.org/documents/articles/evaluation_2007.pdf)

Rikzana, A. A., &Neelamalar, M. (2015). The Role of Community Radio in Women ' s Empowerment : Case Studies from Developing Countries. Advances in Computer Science and Information Technology (ACSIT), 2(13), 22–26. https://www.krishisanskriti.org/

Servaes, J. (1996). Participatory Communication Research from a Freirean Perspective in Africa Review 10/ 1. 73-83. UNGA Resolution 41/128.

Servaes, J. (2002). Communication for Development: One World, Multiple Cultures. New Jarsey, Hamptom press Inc.

Sharma, A. (2013). Developing Participatory Community Radio Module for empowerment of rural women: an action research in Tarai region of Uttarakhand. Unpublished PhD. Thesis. Govind Ballabh Pant University of Agriculture & Technology, Pantnagar.

Tabing, L. (2002). How to do Community Radio: A Primer for Community Radio Operators. New Delhi: UNESCO.

23

Assessment of NGOs Activities for Disseminating SRI Technique Among Paddy Growers

Pooja Jena[1], Manoher Saryam[2] and Debaprasad Patra[3]

[1]*Department of Extension Education, College of Agriculture, JNKVV Jabalpur (M.P.)*
[2]*Department of Extension Education, Institute of Agricultural Sciences Banaras Hindu University, Varanasi, (U.P.)*
[3]*AFPRO, Korba, (C.G.)*

Abstract

The system of rice intensification (SRI) began in Madagascar in the mid-1980s and has spread today to 40 nations on the planet. SRI expands paddy yields including while low requirement of seeds, water and other agricultural inputs. SRI is a planting strategy dependent on the standards of utilizing single, healthy transplants at wide dispersing, the utilization of manure and fertilizer, mechanical weed control and discontinuous water system. To fulfil of the demand of increasing population and keeping up self-sustainability, the present situation should be increased the production 120 million tons by 2020 with a normal efficiency of 4.03 t ha-1. The expansion underway must be accomplished in the setting of declining and breaking down asset based, for example, arrive, water, work and different information sources and without unfavorably influencing the nature of condition. All SRI paddy growers were participants, the vast majority of them accomplishing exceptionally better yields and income thought about than their present method by providing NGOs (AFPRO, Korba) activities. SRI paddy growers have instigated a sensational move in the observation and comprehension of how to achieve sustainable and profitable production, agricultural agents and experts to start a progression of advancements motivated by the SRI method.

Keywords: SRI, CSR, BALCO, NGOs, yield enhancement and profit maximization etc.

Study Area: Project villages in Korba district of Chhattisgarh state under CSR initiative of Bharat Aluminium Company LTD (BALCO).

Introduction

The success of green revolution strategies for raising agricultural productivity and production has depended on the development of crop varieties that are more responsive to increased applications of water, manure, fertilizer, and agrochemical means of protection against pests. For farmers to benefit from this process whereby investing in more inputs can lead to production of more outputs, they had to begin by buying and using improved seeds. The green revolution was clearly seed dependent. The system of rice intensification (SRI) and other agro ecological approaches follow a different strategy, on the other hand. They work with whatever crop varieties the farmer already has or prefers to use (recognizing that it is always most productive to start with whatever genotype is best suited to the farmers' circumstances and goals). Agro ecological methods are more likely to reduce the use of external input than to increase this as they mobilize and capitalize upon potentials that already within plants and upon the interactions between plants and their biotic and abiotic environments, i.e., mutual influences among plants, other organisms, climate, soil and other inanimate factors (Altieri, 1995; Gliessman, 1998; Uphoff, 2002). This is a very different envisioning of the agricultural domain.

Ever since the validity of SRI methods was first demonstrated outside of Madagascar, its country of origin, in China some 10 years ago, we have reiterated that SRI is not a technology, and further, that it is a work in progress, continuing to evolve, diversify and have greater impacts. SRI should be understood initially as a set of ideas and insights that derive from the life's work of Fr. Henri de Laulanié (1993).

Chhattisgarh is one among the few states of India, which has vast natural resources. Proper and planned exploitation of the untapped resources can change the present scenario of this state, which is also known as ***"Poor People's Rich Land".*** Here, more than seventy percent of its farmers belong to the small and marginal farmer category.

The various effects that climate change is likely to have on ecological, social and economic system, many are in some way or other related to change in water cycle. Increasing temperature will affect hydrological systems; alter rainfall, magnitude and timing of run-off, moisture levels in the air and soils and groundwater levels. These changes are being felt by the tribal & other farmers of Chhattisgarh State whose livelihood activities are directly dependent on rainfall and availability of surface and ground water resources. The Climate of Chhattisgarh is mainly tropical, humid and sub-humid.

Agriculture is the main occupation of the people and the foundation of the economy. Even in the districts in the north and south, where forests play a significant role, agriculture is very important. The norm is single cropped and rain-fed agriculture, with paddy as the main kharif crop, in about 80 percent of the net sown area.

Due to vagary of monsoon, ensuring protective/secured irrigation for this single rain-fed crop is quite difficult proposition for the farmer's perspective due to dry spell of monsoon season. Initiation of Rabi cropping is also a challenging situation for the farmers due to water availability. Due to this reason, it is directly affecting the livelihoods of the rural communities in the farm sector interventions.

In the present context, it can be said that proper and planned development & management of water resources, improved agricultural practices, improved skill & knowledge at local level can enhanced the overall condition of the rural farmers of the State.

In this regard, Land and water Management Project (run by a NGO named AFPRO) was initiated in selected villages of Korba district as a CSR initiative of Bharat Aluminium Company LTD (BALCO). Varieties of activities are planned for creation/construction of water conservation structures and accordingly promotion of improved agricultural and allied practices during kharif & Rabi season.

SRI technique is one of the advanced techniques for enhanced yield with low input cost in comparison with normal broadcasting method. As the quantity of water required is much lesser than the traditional method. The amount of seeds used for this technique also quite less in this method.

Research Methodology

It is a very much difficult proposition to introduce a new technique of paddy cultivation instead of normal broadcasting method as the farmers were completely acquainted with the primitive techniques and also had less tendency towards adoption of new one. There were no such farmers in the project villages who had ever applied this technique in the fields so lot of capacity building and information sharing was required in the initial stage.

Through Regular meetings with audio- visual aids and discussions the basic need for SRI method application was assessed. Basically, it was the capacity building and some input support to minimize the loss if occurred.

- Interested farmers were selected by the respective VDCs of the village.
- Technical demonstration has been done regarding the new technique. Its methods of application and field requirements with the help of agricultural expert from implementing agency and also from the Government agriculture departments through videos and PPTs.
- In the initial years, manual method was applied for marking and planting and weeding which subsequently evolved to application of equipments like Markers, Power Weeder etc.

Details of Activities

SRI technique was introduced and implemented ***first time*** in the project area in the year 2013 with only 9 interested farmers of Dondro village which gradually increased with passage of years and reached to 78 in 2017 and 110 beneficiaries in the year 2018.

Table 1: The Year Wise Farmers and Acres Details Covered Under Sri Through the Support Is as Follows

Sl. No.	Year	No of farmers covered under SRI cultivation	No of Acres covered under SRI cultivation	Yield in Quintal	Income Generated
1	2013	9	6.5	84.5	114075
2	2014	8	9	126	170100
3	2015	23	25	400	560000
4	2016	36	43	768	1075200
5	2017	78	107	2279	3304550
6	2018	110	136	3128	4535600
7	2019	220	235	5875	10868750
8	2020	360	435	12180	21728250
	Total			7057.5	42356525

Table 2: The Details are as Follows -

Sl. No.	Year	No of farmers covered under SRI cultivation	No of Acres covered under SRI cultivation	Yield in Quintal	Income generated - SRI (in Rs.)
1	2017	19	16	320	464000
2	2018	22	18	414	600300
	Total			734	1064300

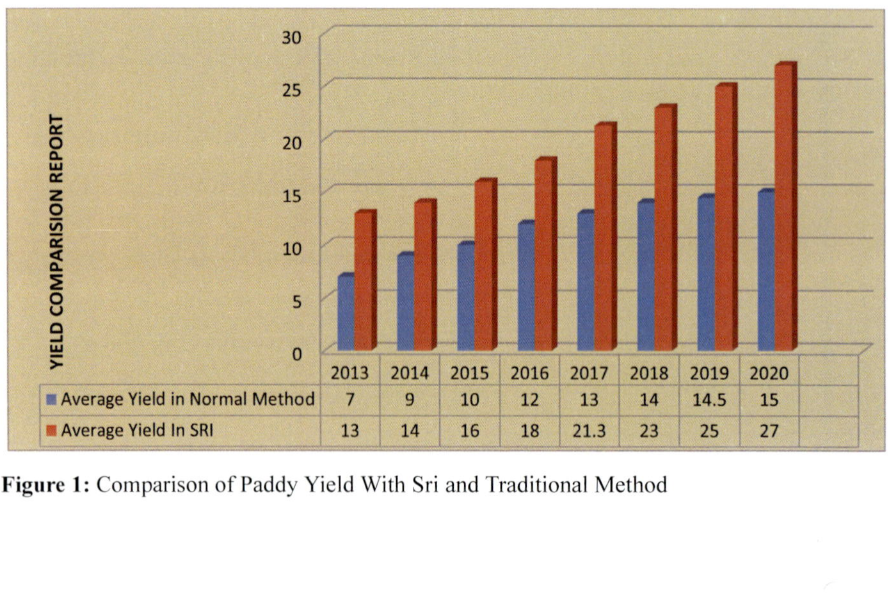

Figure 1: Comparison of Paddy Yield With Sri and Traditional Method

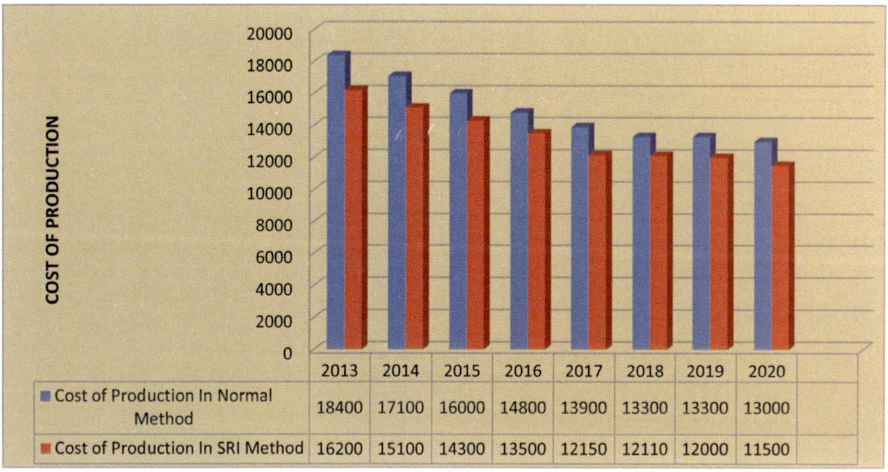

Figure 2: Per Acres Cost Analysis for Sri and Normal Method -

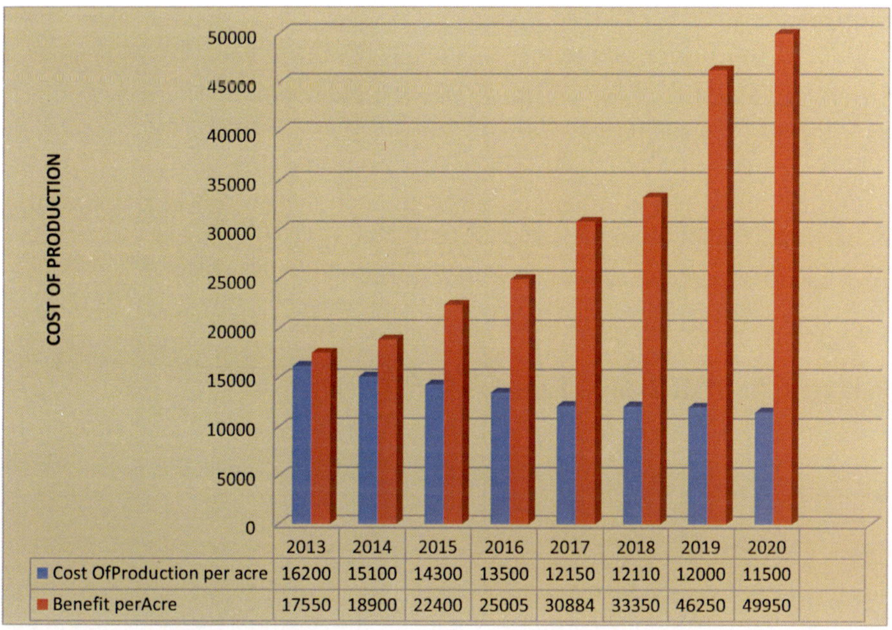

Figure 3: Cost Benefit Analysis

Selection of Beneficiaries

Considering the high labour cost and high water consumption for the broadcasting (normal) method of sowing of paddy in the field, farmers took interest in SRI method with having low input and high production. The interested farmers in the different villages came forward to adopt the SRI technique. Beneficiaries have been selected by the VDC committee, considering the interest of the farmers & availability of water bodies for supplementary irrigational facilities during lean period of monsoon season. .

Support Provided Through Project

From the initiation of the project in 2013, supports were provided in terms of different input materials like seed, fertilizer and pesticides or insecticides time to time as per the need and mandate in a reducing manner to increase the effective community Participation.

In year 2017, fertilizers were provided to the farmers in the tune of Rs. 960/Acre as hand holding support through the project followed by Rs. 760/Acre in year 2018 in a reducing manner to increase community participation and to reduce remaining burden of input on the farmer support was mobilized from the Agriculture department in the tune of 1000/ Acre in 2017 which further increased to 2500/ acre in 2018 and the remaining input cost was provided by the farmer itself in the tune of Rs. 10190/ Acre in 2017 which further reduced to Rs. 8960/ acre in 2018.

Impact of the Intervention

Before the initiation of the project it was a general conceive of a farmer that plantation of more saplings results in more production, so to promote the plantation of lesser saplings in a line wise manner was a very tough task.

During the initial phase in the year 2013, through the support of Balco total 9 farmers applied SRI technique in 6.5 acres of land which may not be eye catching but it provided a major thrust towards the goal and with passage of years nos. gradually increased and reached a mark of 120 acres benefitting 110 farmers in the year 2018.

Inducted/ Self Motivated Farmers for SRI

Apart from HDFC support, by realising the benefits from SRI, total 60 farmers from 6 villages got self motivated and applied SRI in their respective fields, which is a big success and brought massive change in the area.

Earlier Situation

SRI technique is not adopted by the farmers at all prior to the initiation of the project of Balco. It is the first time initiation of SRI by the farmers in subsequent years after getting exposed through this technique.

Earlier with the traditional method of cropping, after the arrival of monsoon, they were starting the preparation of their land with the help of bullock cart or tractors and after that they were sowing through the throwing technique of paddy cultivation. In this throwing method, near about 50-60 kg of seeds/acre and also consumption of fertilizer & manpower/labour work are on the higher side.

Through normal broadcasting method the farmers were usually getting production is found 7-8 quintal per acre.

After Implementation Process Action

Initially, it was found difficult to change the mindset of the farmers for SRI as the technique was new for them, but after discussion, orientation, demonstration & sharing of success stories in Chhattisgarh, farmers came forward for SRI. Only the labour work to plant the nursery bed in a line is a bit tough task but the outcome is tremendous in comparison to the traditional method.

In SRI method of paddy cultivation, the seed requirement falls down up to 3-6 Kg Per Acre kg/acre. The requirement of water, fertilizers and labour charges are reduced. Each stalk of the plant with full growth contains 100-130 nos. of rice grains on it.

Also the requirement of quantity of paddy seeds, fertilizer and pesticides is less in SRI. The water consumption is quite less which in turn saves the ground water. In SRI technique, each stalk of the plant produced around 300 nos. of rice grains in full growth, which increases the production per acre. The average production is gradually increased through the years from 14-22 Q/Acre So there is average enhancement observed in production is in the tune of 9 quintals per acre of land.

Conclusion

- Coverage area under area increased subsequently from 6.5 acres in 2013 to 136 acres in 2018 so as the beneficiary from 9 farmers in 2013 to 117 farmers in 2018.
- Net Profit enhanced in the tune of Rs.1350/acre in 2013 to Rs.24140/acre in 2018.
- Farmers are very much influenced with the technique and are very keen in increasing the coverage area to maximize profit.
- Farmers are very much interested in application of modern agroequipments for further reduction of input cost

Suggestion

Timely use of implements on field as well as strictly following the crop calendar will enhance the productivity.

References

Altieri, M. (1995). Agroecology: The Science of Sustainable Agriculture. Westview Press, Boulder, CO.

Gliessman, S. (1998). Agroecology: Ecological Processes in Sustainable Agriculture. Ann Arbor Press, Chelsea, MI.

Laulanié, H. de. (1993). Le système de riziculture intensive malgache. Tropicultura11 : 110-114.

Uphoff, N., (2002). Agro ecological Innovations: Increasing Food Production with Participatory Development. Earthscan, London.

Abbreviation

SRI- System of Rice Intensification.

BHU- Banaras Hindu University

CSR-Corporate Social Responsibility

JNKVV –Jawaharlal Nehru Krishi Viswa Vidyalaya

BALCO-Bharat Aluminium Company (Funding agency)

NGO – Non- Government Organisation

AFPRO - Action for Food and Production (Implementing agency)

VDC – Village Development Committee

C:B – Cost Benefit ratio

24

Legal Perspectives of NGOs in India

D. Swetha[1], T. Sravan Kumar[2] and Neha Upreti[3]

[1&2](Agricultural Extension), Agricultural College, Jagtial, PJTSAU, Telangana and
[3]Extensionist, Uttarakhand

Abstract

Non Governmental Organizations (NGOs) are generally non-profit groups not affiliated with any government operating at local, national and international. These can be registered as societies or trust or a private limited non-profit company commonly referred to as Section 8 (Indian companies Act, 1956). An NGO can be registered under any of the three acts i.e., Societies Registration Act (1860), Indian Trust Act (1882) and Companies Act (2013). In India there are around 3.2 million registered NGOs. They have a greater influence on the environmental policies, women issues and human rights. NGO is used as a brolly to cover all legal entities that seek charitable and philanthropic funds towards betterment of society without the motive to derive profit. NGOs often play a critical role in advocating for changes in law policy, procedure and administrative rules. NGOs bring the stories of the individuals they serve to the process which form an important part of the evidence needed to convince the policy makers of the needed changes. The rules for the NGO participation in the legislative process differ in every country. In some cases NGOs may participate in the drafting process but not in the introduction of the legislation. NGOs may have difficulty in accessing the information on the draft laws, but can sometimes use national laws on freedom of access to information to assist the public relations offices of the parliament. Indians have more per capita NGOs than hospital beds. The Central Statistical Organization of India states that there are around four NGOs for every 1000 people in urban areas and 2 or 3 NGOs for every 1000 rural populations. This paper mainly focused on the legal exposure and status of NGOs in India.

Keywords: Law policy, Legal exposure, NGO, Non-profit Company, Policy makers

Introduction

Non Governmental Organizations (NGOs) are the organizations engaged in particular plethora of activities aimed at protecting and promoting the varied interests of the people. As the name suggests, NGOs are the state independent voluntary groups working towards various social causes – 'Charitable Purposes'. Section 2 (15) of the Income Tax Act 1961 explains charitable purpose as those activities aimed at the relief of the poor, education, yoga, medical relief, preservation of environment including the watersheds, wild life forests and preservation of monuments or places or objects of artistic or historic interest and the advancement of any other object of general public utility. These 'Not–for–Profit' organizations differ from other business as the main aim of such ventures is charity not generating profits. These NGOs are prohibited from distributing the income generated from their working among its members. There is no single consolidated act dealing with the formation and registration of NGOs. This is because an NGO is an umbrella term for all the charitable ventures. An NGO, in itself, can have various structures. These can be registered as societies, trusts and as private limited. Non-profit companies commonly referred to as section 8 companies therefore an NGO can be registered under any of the three acts i.e., Company Act (2013), Societies Registration Act (1860) and Indian Trust Act (1862).

No organization can work in isolation. Therefore it is necessary for an NGO to develop its network with government agencies, other NGOs median institutions such as schools, hospitals etc. This would not only generate more funds and participation in the NGO but would also be important for spreading awareness and the benefits of the activities undertaken by the NGO. Thus, helping the NGO to achieve its objective in a more efficient and seamless manner. After understanding the steps for the formation of an NGO, it is imperative to look at the various procedures involved and the benefits provided by the three structures of the NGO i.e., NGO as a section 8, company, society or a trust. This would help choose the ideal form for the purpose of which the NGO has to be established. Therefore these are certain factors upon which this decision rests. These factors include time, constraints, aims and objectives of the NGO required funds and funds raising ability, location of the NGO, amount of flexibility deserved etc.

Section 8 Companies Act, 2013

To register as a Section 8 Company, the provision regarding the incorporation of a company as contained in the Company Act 2013 would be applicable. Thus, Section 8 Companies would be governed by the same act throughout the country. In this sense they are very similar to other limited companies. Only few differences exist such as a Section 8 Company does not have to add the word 'Limited' at the end. The income generated through its working cannot be paid as dividend to the shareholders. Then incorporation and the registration of the Section 8 Companies are also similar to that of normal companies. Thus Memorandum of Association (MoA) and Articles of Association (AoA) form the cardinal documents of the Section 8 Companies too. Apart from the normal procedure, an application 'Form INC-12' has to be submitted to the Registrar to grant the permission to drop the word 'Limited'. A minimum of three trustees are required for the incorporation of such a Company. However, no upper limit has been provided for it is important to note that such a company can be found only if the intention of such a company is for promoting commerce, art, science, sports, education, social welfare, religion, charity, research, protection of environment or any such other object.

It usually takes around three to six months for completing the formalities and legal requirements. Even though it takes comparatively longer time to establish, it is preferred by many because it can generate lot of funds on one hand (by issuing new shares) and also lead to democratization of the decision making power as every member gets voting right proportionate to their shares. Thus it ensures higher level of participation and involvement of the maximum people.

NGO in India

NGO or Voluntary organization is not a new phenomena and the concept of voluntary action is very ancient (Aliband, 1983). During ancient and medieval times, voluntarism operated freely and exclusively in the fields of education, medicine, cultural promotion and even acted as succours in crises like drought, flood, epidemics and foreign invasions (Inamdar, 1987).

After independence, leadership in India was provided by social workers who had worked under the leadership of Gandhi. As a matter of fact they were the one who started the movement of voluntary action both in urban and rural areas in the field of health, education, social welfare, adult education, rural development etc. (Chowdary, 1987). The government undertook welfare schemes under various plans and policies besides encouraging voluntary organizations to undertake social welfare programmes under the grant-in-aid programme and set up autonomous bodies like Central Social Welfare Board, Indian Council of Social Welfare etc.

Some of the institutions started by Gandhi and by the wives of higher officials and officers with the support of British Government and those started by the Indian Philanthropists, Rama Krishna Mission etc., continued to function. Although national organizations like Indian Red Cross Society, Young Men's Christian Association (YMCA), Young Women's Christian Association (YWCA), Harijann Sevak Sangh etc. were functioning it was around this time that several all India level voluntary organizations such as Kasturba Gandhi National Memorial Trust, Indian Council of Child Welfare, Youth Hostel Association, Association of Social Health etc. were set up in 1999. A parallel development was the creation of resource agencies which work directly with the poor and also provide support services to other NGOs in the form of training, evaluation and documentation.

In the year 1983, a new organization called Council for Advancement of Rural Technologies (CART) was set up to improve the conditions in rural areas. In 1986, CART was merged with People Action for Development in India (PADI) to form Council for Advancement of Peoples Action and Rural Technology (CAPART) and its main thrust was in the areas of employment, income generation, creation of community assets and fulfilment of basic needs like housing and drinking water (Riddel and Robinson, 1995).

In 1990s, empowerment approach gained momentum, NGOs began to perform advocacy and lobbying in order to meet the challenges and threats of macro forces towards the rural poor and marginalized. As NGOs have grown in size and influence, their activities have brought them into close contact with the Government in rural reconstruction works. They received explicit recognition from the Government in the latter half of 1980s. The Seventh Five Year Plan (1986-1990) emphasized the involvement of voluntary agencies in various proposed projects. The Eight Plan further gave credit to NGOs and encouraged them to participate in the innovative projects like Agro Climatic Regional Planning (ACRP), The Watershed Project under DPAP etc. The Ninth Plan envisaged involving NGOs right from planning process. In 1999, the SGSY Scheme had been launched merging various rural development schemes like IRDP, JRY etc. Group lending remains the major thrust. As NGOs have done pioneering work in SHG promotion, they are being extensively involved in influencing the SHGs (Riddel and Robinson, 1995).

Problems of NGOs

In India, the proliferation of NGOs is extremely palpable. As per the estimate, there are a total of around 31 lakh NGOs in 26 states. It indicates that there has been tremendous growth of NGOs in the last two decades. In spite of their tremendous growth, the NGOs face several problems like lack of funds, misuse of funds, inadequate trained personnel's, lack of public participation, monopolization of leadership, lack of social work among youth, centralization in urban area, lack of coordination, area of interest etc.

Lack of Funds

Most of the NGOs in India are suffering from paucity of funds. Government does not give cent per cent grants in aid or make delay in sanction of grants for numerous programmes. NGOs have to make matching contributions which they are sometimes unable to manage and are therefore unable to avail themselves of the grants. Now a day's charity is not so strong in the minds and hearts of the people as it was in the ancient society. This was another reason for languish of funds for NGOs.

Misuse of Funds

It is the matter of fact that some unscrupulous elements have made fortunes by floating NGOs for their personnel gains and managing the grants from the government. It is a common experience that there have been serious charges of misuse and misappropriation of funds received as grant in aid from the government, foreign donors and raised through their own resources. There NGOs may reflect its image to other NGOs who are working with dedication and commitment.

Inadequate Trained Personnel's

The personnel's working in the NGOs may be of personnel working in such organizations is a sense of dedication and commitment and interest in the social services. NGOs earlier were assumed to be served by the unpaid social workers. Imbibed with the spirit of service and did not require any special education or training. But in the present trend the people with professional education are not interested to work with NGOs. Their visions have been changed and are interested to work in urban areas only. Therefore, it is very difficult to get the trained persons who are either willing or trained to work in the rural society where most of the NGOs work. Moreover, the professionally trained persons have high expectations in terms of salaries, status, opportunities for their growth in the career of their choice (Gangrade, 1987).

Lack of Public Participation

NGOs are meant to provide opportunities to the citizens for democratic participation but they have not been able to fulfil this obligation due to the method and manner in which they function and failed to attract rural people, interested in ground level work and develop channels for people's enthusiastic participation. Some of the factors responsible for such a state of affairs are general backwardness of the people, absence of adequate number of dedicated persons, over emphasis on targets and time bound programmes, political interferences and vested interests, easy availability of felt needs and safeguards for the community. Distrust of agencies and workers who do not have a base in the community and are unable to win its support and lack of decentralization which could give a feeling of being partners in development (Gangrade, 1987).

Monopolization of Leadership

The growing tendency towards monopolization and interlocking of leadership at the top level of voluntary action groups and organizations is reflected by the person being the president in one organization, secretary in the other, treasurer in the third and a member of the executive in the fourth. The interlocking of leadership can be advantageous in formulating coordinated policies, programmes and activities facilitating exchange of technical knowhow and experience and mobilizing the people for common goal. But the disadvantage of such leadership is that fresh blood is not allowed to flow in to the organization and leadership (Kevin, 1994 and Lawani, 1999).

Lack of Social Work Among Youth

The basic characteristic of NGO is volunteerism. In early days, youth are making their career in volunteerism but that enthusiasm seems to have faded these days. The extent of volunteerism is declining day by day and turning it into professionalism. Even the young graduates from social work are interested in making their career in professionalism. This leads to lack of efficient volunteers in NGOs (Aliband, 1983).

Modernization

Because of modernization, professionalization and introduction of management techniques the traditional NGOs need certain minimum infrastructure and administrative expenses. Unfortunately grants in aid rules do not allow for such administrative expenditure except contingencies. This leads to evaporization of traditional NGOs in the country which used to lead by great leaders like Raja Ram Mohan Roy (1772-1833), Swami Vivekananda (1863-1902) (Almond and Sidney, 1963).

Centralization in Urban Areas

NGOs are more developed in the urban areas as compared to rural areas. The backwardness and ignorance of the rural people and lack of enthusiasm among social workers to among them in the absence of availability of minimum comforts are the two important reasons for the backwardness of the NGOs in the rural areas.

Lack of Coordination

The absence of coordination between NGOs existing at local, state and national level has laid to the common problems such as overlapping, duplications, non coordination etc. The absence of such a common forum also incapacitates NGOs to offer united stand against the government (Bunker, 1994).

Area of interest

Sometimes, NGO will prefer the donor area of interest in welfare activities going beyond his own interest or working in more than one programme. This also leads to deterioration in quality of service in particular programme (Gangrade, 1987 and Vanitha, 1993).

Conclusion

NGOs are the ones who really intend to care the uncared sections and the people at the bottom of the social stratum. They mobilize the public opinion and contribute a lot to the society. With governments in the developing country like India are facing ever increasing difficulty in meeting the basic needs of the people. NGOs provide a complementary approach by promoting the mobilization of local resources for the development of the targeted people. They also serve as a channel for development of resources from external sources. The government and other stakeholders should support these organizations and help them to solve their problems at the gross root level. The services of the NGOs are undoubtedly commendable in uplifting the livelihoods of the rural poor.

References

Aliband, T. (1983). Catalysts of Development: Voluntary Agencies in India. Kumarian Press, Connecticut., US.

Almond, G. A and Sidney, V. (1963). The Civic Culture. Princeton University Press, Princeton, New Jersey, US.

Bunker, R. (1994). Spirit of Voluntarism: The New Parasites. Indian Express, Pg. 8.

Chowdary, D. P. (1987). Critical Appraisal of Voluntary Effect in Social Welfare and Development since Independence. The Indian Journal of Public Administration, 33 (3), 492-500.

Gangrade, K. D. (1987). Development of Voluntary Action: Encyclopaedia of Social Work, Ministry of Welfare, Government of India, New Delhi. Pg. 227.

Inamdar, N. R. (1987). Role of Voluntarism in Development. The Indian Journal of Public Administration, 33 (3), 420-432.

Kevin, B. (1994). NGOs: A Resorting to Power. News Week, August, XXIV (5): 14-15.

Lawani, B. T. (1999). NGOs in Development: A Case Study of Solapur District. Rawat Publications, Jaipur.

Riddel, R. C and Robinson, M. (1995). Non Governmental Organizations and Rural Poverty Alleviation. Oxford, Clarendon Press., UK.

Section 2 (15) of the Income Tax Act 1961.

Vanitha, V. (1993). NGOs and Women Development in Rural South India: A Comparative Analysis. Vistar Publications, New Delhi.

25

Role of Media in Extension Education in the Domain of Rural Development in India in Post-Liberalization Period

Pradip Kumar Parida

Public Administration, FSS, BHU, Uttar Pradesh

Abstract

This paper is a humble attempt to critically examine the strategic issues involved in the context of role of media in extension education in the domain of rural development in the present scenario of our country. As always it is expected from the fourth pillar of democracy to be vigilant and provide proper information to all sections of our society, including the rural hinterlands. However some of the empirical analysis proves it otherwise. On number of occasion, it is found that media is not proactive to the news concerned about the 'problems of governance' in rural areas of our country. In that context, it is pertinent to examine the issues in detail, why rural development is not important for media? It is evident from the field that media is important but it will play its proactive role in rural development when there is motivation among media people and society taking pride in highlighting these issues. In that background, this paper is a humble attempt to examine the role of media in extension education of rural masses of India.

Keywords: Media, News, Rural governance, Extension Education, Rural Development.

Methodology

In the present study, the researcher has tried to present a broad picture of the role of media with his empirical observations on news in some selected print media and TV channels. The content analysis of news along with comparative method between print and electronic, news related to problems in rural governance and other news were taken into consideration.

Introduction

As Mahatma Gandhi, the Father of Nation, rightly pointed out that "*The true India is to be found not in its few cities, but in its seven hundred thousand villages. If the villages perish, India will perish too*". (GoI website) Primarily India is an agrarian society, where more than 70% people lives in countryside. As majority of our population reside in rural hinterland and their livelihood depends on agricultural and economic activities related to it. People from rural areas migrating to mega cities or urban areas or industrial townships to seek jobs in service sectors. The seasonal migration of laborers to big cities has become a regular feature in our economy. These uprooted people feel lucky to get something to earn for near and dear ones. 58 per cent of these rural households are employed in farming sector i.e. cultivation, livestock rearing and other agricultural activity, which in turn contributes less than 60 per cent to their average monthly income. These were the findings of the 'Situation Assessment Survey of Agriculture Households' for the 2012-2013 crop years, from July to June. A brief overview of Indian economy reveals that agriculture contributes 16.5 per cent of the GDP contribution, compared to Industry's 29.8 per cent and service sector's 54.4, as per the census of 2015. It is also worth to recall that despite the big claim of booming Indian economy, the share of informal employment in total employment has remained above 90 per cent throughout the period 2004-05 to 2011-12, the Economic Survey 2014-15 revealed.

Rural Development in India

Rural development means growth and progress of people living in rural areas by increasing their income, assets, livelihood generation activities and providing them the basic amenities of civic life, i.e. food, cloth, shelter- housing, primary education, primary health care facilities, providing the minimum platform, where they can live their life with dignity, as per the provisions of the constitution of India. In the post-liberalization period, there has not been much emphasis on holistic development of rural and urban economy. The agrarian economy as well as rural parts of the country is in great distress. With agriculture becoming more and more capital intensive, farmers are becoming more and more vulnerable. Rural areas are yet to take off as the region lacks basic amenities of civic life, poor infrastructure, low level of investments in rural economy, increasing cost of production, weather playing havoc on crops, usurpation of cultivable land by real estate sharks, deterioration of governance system day by day. For the majority of rural population in India, the MNREGA scheme is like breathing space. Though there are numerous rural development programmes in India, the intended beneficiaries are yet to achieve the results, primarily due to lack of proper information, awareness. Here the role of extension education with the help of media - be it print, electronic, folk, even social media are significant.

Role of Media in Development of the Country

Media is the fourth pillar of democracy, one of the most important medium to check the role and responsibilities of other organs like- Executive, Legislature & Judiciary. It serves as a platform to share news and views of all stakeholders in a democratic setup. A vigilant and active media connects and integrates a nation by linking people with government. It is also one of the important institutions of governance. In modern democracies, media disseminates information about policies and plans to public and also conveys concerns of the public and issues of governance to the government. When there is no prompt and formal mechanism to know and address the public grievances, media outlets are first line of communications between the government and public. It is a platform of sharing information, airing public grievances and taking note of shortcomings. A liberal democratic governance structure always promotes media activism and proper articulation of the problems of masses in a positive spirit. However, despite considerable amount of presence of media in our country, its coverage of problems related to rural development and depiction of true picture of rural India is not credible & proper.

In India most of news and entertainment media is in private hands, though the presence of official television and radio networks add values to it. However, the public depends upon these sources of information. Reports and analysis of many major print and television networks on rural India lack proper research and analysis of the real issues related to crisis in governance. Serious reader/ viewer one can easily find out superficiality in reporting. With more publications coming out of small towns and big media houses launching local/ regional/district editions, readers/ viewers have many newspapers/ TV channels, but hardly can they make any sense out of them. Now a day, TV channels are hiring more stringers, but they select only those reports that can attract sensitive news & more Television Rating Points. They are more interested in politicking and trivial matters, than focusing on real issues. On the other hand, official public service broadcasting of radio and television on many occasions gives much prominence to ruling regimes and tries to keep them in good humor. Though the problems of people gets coverage, only during natural calamity, such floods, cyclone, earthquake, major accidents only. However, the mandate of a public service broadcasting institutions is not to become propaganda machine of a ruling party like a sycophant, but to report truthfully the ground realities and produce public news and views. Rural reporting demands that flaw of governance & appalling human conditions must be reported truthfully and objectively.

Role of Media in Rural Development

Media gives information to public in general, on rudimentary issues and on occasion after processing the raw information. The quality of media information, educational and cultural environment of the time also play an important role in progress of human beings. Media has power to provide information, giving latest happenings around the world and interpreting and analyzing happenings and outcomes as well as anticipating scenarios. Now media affects entire edifice of human lives. Quality of news reports and analysis is also important. There are numerous rural development programmes but we see there is much shortcomings on part of official machinery as it fails to reach intended targets. In recent time, there are plethora of central sponsored and state sponsored programmes addressing the rural crisis, governance issues and rural development schemes in this country. This includes Mahatma Gandhi National Rural Employment Guarantee Act (MNREGA-the largest one among the CSS), Sarva Siksha Abhiyan (SSA), National Rural Health Mission (NRHM), Rashtriya Swasthya Bima Yojna (RSBY), Prdahn Mantri Awas Yojana (PMAY) - the modified version of Indira Awas Yojna (IAY), Integrated Child Development Schemes (ICDS), Pradhan Mantri Swasthay Yojana (PMSY) etc. Thus, the rural development domain was propelled by a slew of Centrally Sponsored Schemes (CSS), financed wholly or largely by the federal government. Unlike the previous rural development schemes, MNREGA was backed by an Act. It is important to mention here that the rural development has been the focused area in the domain of planned development in the post- independence period. It was almost co-terminus with the five year plans. Starting from Community Development Programme (CDP) that began from 1952, both its form and contents have undergone changes over decades. It led to creation of several implementing, monitoring and evaluating agencies: the CD Blocks, PRIs, Voluntary Organizations (VOs), NGOs and now Civil Society Organizations (CSOs) along with a host of associated concepts. The buzz words are: Participation, Empowerment, inclusive growth, social justice, Social Inclusion, decentralized planning, social entrepreneurship, Social Audit, etc.

Newspapers and Magazines

Despite the presence of large number of newspapers & magazines- both vernacular and mainstream languages, there is dearth coverage on rural affairs and rural development. The geographical presence of newspapers and magazines reflects a disturbing trend in India. According to RNI's Press in India-2013-2014 Report, publications are registered in English and 22 main languages listed in the Eighth Schedule to the Constitution as well as publications also registered in 149 other languages/dialects. The total numbers of publications registered in four Metropolitan Cities were 27,333 accounting to 27.43 per cent of the all India total. State capitals were publishing 19,945 registered publications or 20.01 per cent of total. Big cities published a total of 36,942 publications constituting 37.07 per cent. And, merely 11,054 publications constituting 11.09 per cent of total registered periodicals were coming out from small towns.

Periodicals

No doubt there are a few publications like *Grassroot* (of the Indian Press Institute), which focus on rural issues, but pity is that even very few outside the mainstream realm know its existence. Such publications are not in popular imagination and it is also difficult to find them in market. Such publications are not multilingual and also not have pan Indian footprint. Besides their limited availability, very few journalists are willing to cover such issues of Indian backyard. *The Hindu*'s former Rural Affairs Editor, and perhaps the country first such editor, P Sainath in an interview to *The Sunday Indian*, now defunct, recalled that "*At the time of the Lakme Fashion Week in Mumbai in 2006, one of the worst years ever for the farmers' suicides, there were six journalists covering Vidarbha, all outsiders, while 512 accredited journalists covered the Fashion Week for the full seven days and another 100 did it on daily passes. So what copycat effect are we talking about?*". It is also a pity that not just English and Hindi publications lack rural coverage, even regional languages have neglected their rural readers.

Television

As of now there are two major television channels *Green TV* and *DD Kisan* devoted to the cause of rural India and farming sector. *Green TV* is the first private television channel dedicated rural population. It was launched in October 2014. In one of the writings in The Hindu, journalist S. Ravi mentioned that, ' the channel is not dedicated solely to agriculture but rural India and supposed to cover everything which affects rural people in general. Honorable Prime Minster of India inaugurated DD Kisan, DD's dedicated channel addressing agriculture and farmers issues on 26th May 2015. The channel has tied up with all agriculture Universities, ICAR, IARI, KVK, Ministry of agriculture, Government of India, department of agriculture of all state governments. All India Radio and Doordarshan are right in taking pride for presenting agricultural and educational programmes. However, it is gross injustice to rural development for a country like India with so many geographical and linguistic variations. Earlier, the 'Satellite Instructional Television Experiment' (SITE) was launched in 1975-76. Nalin Mehta noted that with this launch, India became the first country in world to use a direct broadcast satellite to reach remote villages directly with educational information. He also goes on to say, "It is important that the both the organization helped in forging a sense of oneness and belonging among the nation's disparate linguistic groups". DD has presented regular programmes on farming and allied activities, but it has much to do for rural India, even after launch of first public network on agriculture. But on whole, the state media coverage of rural issues could not be termed much satisfactory as it has been following top-down approach and not participatory kind of things. Another important aspect is that private media is not coming into the sector for public welfare but to maximizes profits in term of TRPS or incomes.

Radio

All India Radio service comprises 420 radio stations located across the country. AIR presents programming in 23 languages and 146 dialects. It claims that its transmission reaches to nearly 92 per cent of the country's area and 99.19 per cent of the total population. Addition to this, All India Radio has 391 FM broadcasting stations as on 28th February 2017. Besides these, there were 290 private FM channels in the country, with Maharashtra topping the list with 47 and Uttar Pradesh and Tamil Nadu having 28 stations each. Much credit has been given to AIR for its coverage to rural affairs and taking the cause of rural development and espousing the cause of modern agriculture. With the emergence of community radio, one can hope that it will help in disseminating information at local level and help in building bondage amongst locals. It is also hoped that community will transmit immediate information and authorities will take care of immediate escalating crisis. It is very unfortunate to know that community radio stations and private FM channels are not allowed to air their news or comment on current events, though some anchors also find innovative ways to comment on existing scenario.

Depiction by Media on Rural Issues

Media as a means of communication has components like newspapers, leaflet, pamphlets, advertisements, magazines, journals, books, radio, television, films, music, computer aided presentations, mobile phones networks facilitated by internet and ever increasing social media and traditional media like skit, docudrama, nukkad natak etc. It seems to be historical legacy that media has not adequately presented the views of farmers, peasants and others living outside the state capitals and metropolises, to the public or on glowing television screens. As a matter of fact, the media schools and media outlets not feel comfortable in covering the rural areas. Historically, there have been very few incisive reports on rural India in news media. Prof. Pradeep Mathur poignantly points out that "*Media and the information systems in the developing countries are being governed by dominant concepts of the West. The developing countries can meet the challenge only through developing alternative concepts of media programming, which have roots in their own social and cultural settings. This demands a great amount of will to experiment on the part of entire gamut of media industry*".

India has thousands of publications, hundreds of television channels and a large network of radio stations and huge presence of social media, but there have been very few occasions on rural India coverage. News media are hardly report rural backyard of the nation regularly. A cursory look at any leading daily or general news magazine or in any TV channel or radio broadcast, one will fail to notice happenings around rural India. There has been diversification of news media in post liberalized era, but there is very few to show the seriousness in reports and analysis while reporting about rural India. Business dailies see how the rural folks have increasing purchasing power and advertising agencies calculate how it is easy to reach the masses. Very few mainstream English as well Hindi dailies religiously devote a page or carry reports on farming sector and happenings in rural hinterlands.

Critical Concern

Indian media is not objectively reporting the rural affairs. It gives much importance to trivial issues just for earning TRP rate and accordingly revenue for advertisement. Political news coverage dominates in news and plight of poor does not have much value, as society gives much importance to such news. Though there are number of national dailies launching local editions at the district/ city/ town levels, their coverage is not adequate. This has been going on as there are not dedicated or full time correspondents reporting on the issue. Many newspapers/ agencies have no trained personal coverage on crisis on rural governance and rural development. There is lack of coordination between media networks and press and persons involved in rural development activities. Government officials rarely connect with local media. Hardly there is any critical insight to assess all these. Hence, media reports what is being informed to them by the PRO of the department or government agencies.

Suggestions

Media and journalists at large should change its attitudes towards rural society, its problems of governance, issues affecting development & economy. There is need for attitudinal change. Media institutes can revise their courses by giving prominence to issues of rural India to train media personnel who can communicate with people speaking local language addressing their problems and local issues. They should cover news and happenings of rural India, especially farmers problems, poverty, unemployment, distress migration, lack of welfare schemes reaching at grass root level, massive corruption in trickle down of benefits to poor, administrative apathy etc.

Conclusion

A developing countries media has power to influence decision making process at various levels. If India has to emerge as a developed country, its media should first take care of its rural India. It should properly reflect all happenings in real perspective. Media in general and journalists in particular, must change their attitudes & perspectives towards rural issues to bring change in the lives of people living in the remotest part of the country. Media credibility is at stake as it fails to report on real happenings in rural areas. Media can transform the vulnerable section of society by addressing the problems related to agriculture, animal husbandry, forestry, fisheries, livelihood generation activities, small scale industries based on local resources, various scheme and programmes & welfare measures by the state and central government, be it CSS or state sponsored. In such scenario, media must play a pivotal role in improving the condition of rural masses by informing them properly & education them. It should not be a parrot in a cage in the hands of the government. It should act like the fourth pillar of democracy in true sense.

References

Drahos, P.; Braithwaite, J. (2002). Information Feudalism: who owns the knowledge economy. London: Earthscan. ISBN 1-85383-917-5.

http://www.pib.nic.in/newsite/erelcontent.aspx?relid=28193)

Maheswari, Shriram (1987) . 'Voluntary Action in Rural Development in India', in the Indian Journal of Public Administration, Jul- September: 560-561.

Nalin Mehta (2008): India on Television: How Satellite channels have changed the way we think and act, Harper Collins Publishers, New Delhi.

O'Neill, J. (1998). The Market: Ethics, Knowledge and Politics. London: Routledge. ISBN 0-415-09827-0.

Paul Samuel (1982): Managing Development Programmes: The Lessons of Succ*ess. Boulder, Colorado, West* View Press.

Robin Jeffery (2000): India's Newspaper Revolution, Oxford University Press, New Delhi.

Sevanti Ninan (1995): Through the Magic Window: Television and Change in India, Penguin, New Delhi.

Sevanti Ninan (2007): Headlines from the Heartland, Sage Publications, New Delhi.

Singh, Kartar (1999): Rural Development- Principles, Policies *a*nd Management, New Delhi, Sage.

Vinita Kohli-Khandekar (2006): The Indian Media Business, Response Boo*k*s, New Delhi.

26

Zinc Deficiency: Its Relation to Malnutrition and Strategies for Its Prevention

Ankur Aggarwal, Tarun Verma and Manish Kumar Singh

Department of Dairy Science and Food Technology, Institute of Agricultural Sciences, Banaras Hindu University, U.P, India

Abstract

In the body, zinc plays a variety of roles, and its absence has a significant impact on public health and economic development. To address zinc insufficiency, several treatments are employed, including supplementation, food fortification, bio-fortification, and dietary diversification. Among these options, food fortification stands out as the most cost-effective and successful strategy for combating zinc deficiency. Food fortification is widely regarded as an effective approach to zinc deficiency. Since milk is a poor source of zinc (2–6 mg/L), fortifying it with zinc presents a viable solution to the nation's zinc deficit. GRAS-classified substances such as zinc sulfate, zinc gluconate, zinc acetate, zinc stearate, and zinc chloride can safely be utilized for zinc fortification. In this chapter, we provide an extensive review of the biochemical and physiological functions of zinc, its metabolism, including factors influencing its bio-availability (both inhibitors and enhancers), the human requirement for zinc, factors affecting zinc absorption, high-risk groups, consequences and causes of zinc deficiency, evaluation of zinc status, prevention strategies for zinc deficiency, and the concept of zinc-fortified food

Keywords: *Zinc absorption, Zinc fortified foods, Zinc deficiency, Zinc intervention, Zinc nutrition, Zinc requirement*

1. Introduction

Micronutrients are essential nutrients required in small quantities but play a crucial role in human health and productivity. Key micronutrients include iodine, iron, zinc, and vitamins A, B6, and B12 (Godswill et al., 2020). Zinc deficiency, a common micronutrient shortfall, is prevalent not only in developed countries but particularly in impoverished nations. It can affect individuals of all ages, with young children and breastfeeding women being the most vulnerable. Addressing zinc insufficiency through zinc supplementation or diets rich in animal protein sources has shown promising results in promoting growth and reversing hypogonadism. Subsequent research has revealed that inadequate zinc consumption in the diet is the primary cause of this hypogonadism syndrome. Since the identification of zinc deficiency as a health concern in 1961, there has been a noticeable increase in interest in the biochemical and clinical aspects of zinc nutrition (King and Cousins, 2006). In this paper, we provide a comprehensive review of the physiological functions of zinc, zinc bioavailability (including inhibitors and enhancers), the human requirement for zinc, factors affecting zinc absorption, consequences and causes of zinc deficiency, evaluation of zinc status, and prevention strategies for zinc deficiency, focusing on the use of zinc-fortified foods.

2. Zinc and its Deficiency

Zinc is required for a variety of biological functions and essential for the structural & catalytic function of numerous enzymes that control the body's metabolic process i.e. such as diabetes, cancer (such as esophageal, hepatocellular, breast, and colon cancer), neurological disorders like Alzheimer's disease, and intestinal diseases including inflammatory bowel disease and irritable bowel syndrome, are affected by zinc deficiency (Ackland and Michalczyk, 2016). According to the WHO (2012), severe zinc deficiency poses a hazard to one-third of the world's population and causes 80,000 fatalities annually including 16% of respiratory system problems, 18% of malarial disorders, 10% of diarrhoea, and other conditions such skin lesions, hair loss, and nail spooning (White & Broadley 2005). Orissa (51.3%), Uttar Pradesh (48.1%), Gujarat (44.2%), Madhya Pradesh (38.9%), and Karnataka (36.2%) had the highest zinc deficiency rates (Kapil and Jain, 2011).

3. Zinc Bioavailability

Bio-accessibility refers to the quantity of an eaten nutrient that is potentially available for absorption after digestion and release from the food matrix. The interference of zinc with the availability of other food components is a major issue with bioavailability. In vivo (animal studies) and in vitro (stimulated stomach pH settings) bioavailability has been investigated (Ollig et al.,2016). Solubility, dialyzability, or a gastrointestinal model for bioaccessibility, and the Caco-2 models for bioavailability are the four main in vitro approaches for assessing bio-accessibility and/or bioavailability (Cilla et al., 2018).

The Caco-2 cell model is commonly used to determine zinc bioavailability from foods and diets. Hunt et al.,1989 found that neither the quantity of zinc associated with low molecular weight fractions (i.e. dialyzable zinc) nor the amount of zinc associated with high molecular weight fractions (i.e. soluble zinc) were beneficial in predicting zinc availability in vivo. Chiplonkar et al.,1999 observed that in vivo zinc dialyzability was substantially associated with in vivo human data, contrary to Hunt's claim.

4. Factors Affecting Zinc Absorption

Calcium and Phosphorus

Zinc absorption in adults was lowered when calcium phosphate (1360 mg/day) was consumed (Wood and Zheng, 1997). Zinc absorption may be influenced by calcium sources, processes, and the degree of luminal zinc solubility. Calcium decreases zinc absorption in animals, most likely as a result of co-precipitation with phytate and zinc (Wood and Zheng 1997). Over a wide intake range, phosphorus-containing salts in the diet have no effect on zinc balance (Spencer et al.1984; Abrams & Atkinson 2003; Piskin et al., 2022). Phytate and phosphorus-rich proteins, such as milk casein and nucleic acids, are two more dietary sources of phosphorus that bind zinc tightly and reduce zinc absorption (Dryden 2021).

Iron

The absorption of zinc can be hindered by daily intake of iron at levels comparable to those found in certain supplements (Lonnerdal, 2000). A study on iron-fortified newborn milk formula (40 mg Fe/100 g) revealed lower plasma zinc concentrations compared to a control group that received a commercially available formula with 8.3 mg iron/100 g (SandstroÈm, 2001). Additionally, the presence of other divalent metals such as Ca2+, copper, and phosphorus may also affect zinc absorption (Udechukwu et al., 2016).

Folate

Research suggests link between folate and zinc where they found that monoglutamate is converted to folate by the zinc-dependent enzyme polyglutamate hydrolase, folate bioavailability is improved (Fekete et al., 2010). Folate absorption was reduced as a result of low zinc consumption.

Proteins

Subtle variations in protein digestion can significantly influence zinc absorption (Lonnerdal and Sandstorm, 1989). The lower digestibility of proteins in cow milk, which is rich in casein, compared to human milk leads to higher zinc bioavailability in human milk (Ackland and Michalczyk, 2016). Drago et al. (2004) found that the concentration of casein in infant formulae reduces zinc dialyzability, with the formulas having a higher casein-to-whey protein ratio showing the lowest values. When consuming a diet rich in animal proteins like cheese, beef, and eggs, which have a high phytate content, zinc absorption can be affected (Roohani et al., 2013). However, zinc absorption significantly improved after removing phytate concentration in soy protein isolate through a precipitation procedure (Lonnerdal, 2000). Shilpashree (2015) investigated the bioaccessibility of various mineral complexes in milk proteins and concluded that minerals in complex form exhibited greater bioaccessibility compared to minerals in their free form.

Other food components

Polyphenols, fibre, and saponins are among other food components that impact zinc bioavailability because they generate an insoluble zinc complex (Gibson et al., 2006). According to Bosscher et al.,2001 soluble fibre has a greater inhibitory effect on zinc dialyzability in casein-based formulas than in whey-based formulas. Casein and whey-based formulae supplemented with 0.42g of locust bean gum/100mL had zinc dialyzability of 3.2 and 5.6 percent respectively. The RDA values of zinc for different age groups of people are presented are given below (ICMR NIN 2020):

1) Men: 17 mg/d; Women: 13 mg/d
2) Pregnant women: 14.5 mg/d
3) Lactating women: 14 mg/d
4) Infants (6-12m): 2.5 mg/d
5) Children (1-3y): 3.3 mg/d, Children (4-6y): 4.5 mg/d and Children (7-9 y): 5.9 mg/d
6) Boys/ Girls (10-12y): 8.5 mg/d
7) Boys (13-15y): 14.3 mg/d and Girls (13-15y): 12.8 mg/d
8) Boys (16-18y): 17.6 mg/d and Girls (16-18y): 14.2 mg/d

5. Different Strategies to Combat Zinc Deficiency

Zinc deficiency can be avoided by ensuring that the required amount of zinc is consumed in the diet at the RDA level. To address zinc deficiency in diverse demographic groups, various techniques such as fortification, dietary diversity, and supplementation are applied. Before using these strategies in various food products, it is vital to get understanding of food ingredients, supplement or fortification (Akhtar et al. 2011).

Supplementation

Zinc supplementation is a frequently utilised method around the world because it has several advantages over other approaches, such as rapid improvement, ease of administration, and suitability for people of all ages (Brown et al. 2004). Supplementation programmes are helpful in preventing zinc deficiency in underdeveloped nations.

Dietary Diversification

Dietary variety is a technique for preventing zinc insufficiency. It includes eating zinc-rich foods such as meat, poultry, and fish that are easily available (Temple et al. 2002). For a proper intake of zinc, food selection, preparation methods such as cooking, germination, and fermentation are all important. The phytate content of grains and legumes is reduced during germination and fermentation, which influences zinc bioavailability. Saponins and polyphenols, for example, are dietary components that affect zinc bioavailability and can be lowered through soaking to improve zinc absorption (Davidsson et al., 2004; Hambidge et al., 2004)

Food Fortification

In both developing and developed countries, food fortification stands out as the most efficient approach to address micronutrient deficiencies, including vitamin A, zinc, iron, and iodine. Zinc fortification plays a crucial role in enhancing zinc intake and absorption (Hess and King, 2009). The term “fortification” is defined by the Codex General Principles for the Addition of Essential Nutrients to Foods (Codex, 1987) as the addition of one or more essential nutrients to a food, whether or not it is naturally present in the food, with the aim of preventing or correcting demonstrated deficiencies in the population or specific population groups. Food fortification offers several advantages over other methods, including cost-effectiveness, ease of implementation, and the potential to improve nutritional and health status.

Biofortification

Biofortification is an additional strategy employed to address micronutrient deficiencies in developing nations. This approach involves modifying agricultural, agronomic, or genetic aspects of crops to increase the levels of specific micronutrients. Currently, biofortification is being utilized to enhance the levels of zinc, iron, and pro-vitamin A carotenoids in rice, maize, and wheat, which are crucial staple food crops worldwide (Akhtar et al., 2011).

Table 1 below presents an overview of various zinc compounds used for supplementation and fortification. Among them, only five zinc compounds (zinc sulphate, zinc gluconate, zinc chloride, zinc stearate, and zinc oxide) are approved as Generally Recognized as Safe (GRAS) for fortification purposes (Whittaker, 1998). The solubility of zinc compounds and the intra-gastric pH are factors that influence zinc absorption (Gibson et al., 1998). Despite its limited bioavailability, zinc oxide is utilized in various products in industrialized countries.

Table 1: Zinc compounds used for supplementation and fortification

Zinc oxide	Zinc stearate
Zinc sulfate heptahydrate	Zinc ascorbate
Zinc chloride	Zinc picolinate
Zinc carbonate	Zinc amniote
Zinc citrate	Zinc histidine
Zinc acetate	Zinc methionine
Zinc gluconate	Stabilized zinc gluconate

(Salgueiro *et al.,* 2002)

6. Level of Zinc in Bovine and Human Milk

Milk is a poor source of zinc, and the amount of zinc in human and bovine milk varies greatly (Flynn and Cashman, 1997). Zinc concentrations in human milk range from 0.65 to 5.3 mg per litre, but zinc concentrations in cow milk range from 2 to 6 mg per litre. Blakeborough et al.,1983 found that 95 percent of the zinc in the skim milk fraction was bound and the remaining 5% was present in supernatant. In human milk, casein micelles, whey proteins, and low molecular weight ligand account for 70-75 percent of the zinc, whereas the fat fraction accounts for 29 percent (Singh et al. 1989).

7. Zinc Bioavailability in Milk

The difference in zinc bioavailability between human and bovine milk can be attributed to variations in the ratios of casein and whey proteins, with casein acting as a limiting factor for zinc absorption (Fransson and Lonnerdal, 1983). Studies have shown that zinc bioavailability is higher in human milk compared to bovine milk (Ackland and Michalczyk, 2016; Blakeborough et al., 1983). Krebs et al. (2000) discovered that newborns absorb 54 percent more zinc from human milk. This difference is attributed to the presence of micelles with high phosphorylated casein in bovine milk, which effectively bind zinc and make it less accessible for absorption. In comparison, casein micelles have lower bioavailability than whey protein (Ackland and Michalczyk, 2016).

8. Zinc Fortified Foods

Zinc fortification initiatives have been relatively limited, primarily focusing on specific areas such as infant formula milks (containing zinc sulfate), supplemental foods, and ready-to-eat breakfast cereals in the United States. However, a significant advancement has been made by the Food Safety and Standards Authority of India (FSSAI) through the issuance of the "Food Safety and Standards (Fortification of Foods) Regulations (FSSR) 2016." This comprehensive regulation on food fortification covers five main types of food: wheat flour, rice, milk, oil, and salt. The FSSAI allows the addition of various nutrients including iron, zinc, folic acid, vitamin B12, vitamin A, thiamine, riboflavin, niacin, and pyridoxine to wheat flour and rice. In Indonesia, the addition of zinc to wheat noodles is mandatory, and several Latin American countries have shown interest in fortifying grain flours with zinc (Allen et al., 2006).Studies have demonstrated the potential of zinc supplementation in improving children's growth rates. However, there is a limited number of studies that have evaluated the efficiency and effectiveness of zinc supplementation. For instance, while adding zinc oxide to breakfast cereals increased plasma zinc concentrations in preschool-aged children in the United States, it did not show evidence of improved growth rates or food intake (Hambidge et al., 1979). Conversely, in Turkey, zinc-fortified bread was found to enhance the growth rates of schoolchildren with low plasma zinc levels (Çakmak et al., 1999). Moreover, Hess and Brown (2009) observed that very low birth weight premature infants who were fed zinc-supplemented infant formula containing either 6.7 or 5 mg/L of zinc exhibited significantly faster growth compared to the control group receiving 10 or 11 mg/L of zinc, respectively.

Zinc Fortification in Milk

Goillem et al. (2000) conducted a study on the bioavailability of zinc salts in fortified milk and soy-based formulae. They discovered that the choice of zinc component had an impact on the bioavailability of zinc. Zinc oxide, zinc sulfate, and zinc gluconate were found to be more dialyzable compared to zinc acetate and zinc chloride. Interestingly, the water solubility of the salts used did not affect the dialyzability ratings. In fact, zinc oxide, despite having the lowest water solubility in milk-based formulae, exhibited the highest dialyzability. Sharma et al. (2015) focused on studying the sensory acceptability of lactose-zinc complex fortified milk. They found that fortifying milk with lactose-zinc complex above 30 mg/L resulted in better sensory acceptability compared to milk fortified with zinc salt. Additionally, they observed comparable changes in all physicochemical parameters when compared to the control. The study also revealed that milk supplemented with lactose-zinc complex exhibited better in vitro bioavailability of additional zinc compared to both the control milk and milk fortified with zinc salt.Rana et al. (2017) utilized electrochemical impedance spectroscopy to analyze the physicochemical and electrochemical properties of zinc-supplemented milk. They observed that zinc bound to milk proteins led to a reduction in the current flow of fortified milk. These metal ions have the ability to bind to the active sites of proteins and lipids, thereby reducing the flow of electrons onto the working surface of the electrode and consequently lowering the peak current.

Zinc Fortification in Milk Products

In a study conducted by Gulbas and Saldamli (2005), the impact of selenium and zinc fortification on the quality of Turkish white cheese was investigated. They found that after 60 days of ripening, fortified cheese with a higher concentration of zinc (160 mg/kg) had a significant effect on the organoleptic properties of the cheese compared to the control cheese.

Abd-Rabou et al. (2010) examined the effect of fortifying cheese curd with zinc on the qualities of Edam cheese. They fortified Edam cheese with different sources of dietary zinc, such as zinc acetate, zinc chloride, and zinc sulfate, at a level of 150 mg zinc/kg cheese curd, both in fresh and stored (2 to 12 weeks at 51°C) conditions. During storage, it was observed that Edam cheese fortified with zinc had higher acidity, acid value, and protein fractions (soluble nitrogen) compared to the control cheese. Among the different treatments, cheese fortified with zinc acetate received the highest organoleptic scores, indicating better taste and preference.El-Din et al. (2012) utilized zinc sulfate and zinc acetate at concentrations of 40, 60, and 120 mg Zn/kg of milk to produce zinc-enhanced Domiati cheese. Their findings showed that zinc fortification in cheese (up to 40 mg of zinc/kg of milk) had no effect on the chemical composition or organoleptic qualities when compared to the control cheese sample.Kahraman and Ustunol (2012) focused on making zinc-fortified cheddar cheese and discovered that adding 16 mg of zinc per kilogram of cheese had no impact on the cheese's quality.Shilpashree et al. (2015) developed a lyophilized powder of WPC-zinc/iron complex and compared it to iron-bound WPC. They found that the WPC-zinc complex exhibited higher heat stability, lower pH, and better ionic stability than the iron-bound WPC. They proposed that the produced complex could be used as an organic fortificant with improved solubility in various food applications.

Conclusion of Zinc

Zinc is a micronutrient that is required for optimal human health and development. It has various functions in the human body and is an essential component of several enzymes. To counteract zinc deficiency, several strategies such as supplementation, food fortification, and dietary diversity are used. Phytate, polyphenols, fibres, and saponins inhibit zinc absorption by forming insoluble zinc complexes that reduce bioavailability. Milk is considered a complete food, but its zinc content is very low, at 2-6 mg/liter, and its bioavailability is only 28 percent of total zinc levels. Zinc addition to milk and milk products could be a viable strategy for overcoming zinc insufficiency.

References

Abd-Rabou, N. S., Zaghloul, A. H., Seleet, F. L., & El-Hofi, M. A. (2010).Properties of Edam cheese fortified by dietary zinc salts. Journal ofAmerican Science, 6(10), 441-446.

Abrams, S. A., & Atkinson, S. A. (2003). Calcium, magnesium, phosphorus and vitamin D fortification of complementary foods. The Journal of Nutrition, 133(9), 2994S-2999S.

Ackland, M. L. & Michalczyk, A. A. (2016). Zinc and infant nutrition. Archives of biochemistry and biophysics, 611, 51-57.

Akhtar, S., Anjum, F. M., Rehman, Z. U., Riaz, M., Arshad, M., Basit, A. & Ismail, T. (2011). Effect of zinc and iron fortification of the feed on liver and thyroid function in rats. Biological trace element research, 144(1-3), 894- 903.

Allen, L.H., De Benoist, B.Dary, O. Hurrell, R. & WHO (2006). Guidelines on food fortification with micronutrients.

Blakeborough, P., Salter, D. N. & Gurr, M. I. (1983). Zinc binding in cow's milk and human milk. Biochemical Journal, 209(2), 505-512.

Bosscher, P., Vancallie-Bertrand, M., Robberecht, H., Vandyck, K., Van, callwenbergh, R. & Deelstra, H. (2001). In Vitro bioavailability of calcium, iron and zinc from first age infant formulae and human milk. Journal of Pediatric Gastroenterology and Nutrition, 32: 54-58.

Brown, K. H., Rivera, J. A., Bhutta, Z., Gibson, R. S., King, J. C., Lonnerdal, B. & de Romana, D. L. (2004). International Zinc Nutrition Consultative Group (IZiNCG). Assessment of the risk of zinc deficiency in populations and options for its control. Food and nutrition bulletin, 25(1 Suppl 2).

Chiplonkar, A. S., Agte, V. V., Tarwadi, V. K. &Kavadia, R. (1999). In vitro dialyzability using meal approach as an index for zinc and iron absorption in humans. Biological Trace Element Research, 67 (3): 249-256.

Davidsson, L., Ziegler, E.E., Kastenmayer, P., van Dael P. & Barclay, D. (2004). Dephytinisation of soybean protein isolate with low native phytic acid content has limited impact on mineral and trace element absorption in healthy infants.Br J Nutr, 91:287-294.

Drago, R. S., Binaghland, M. & Valencia, E. M. (2004). Effect of gastric digestion pH on iron, zinc, and calcium dialyzability from preterm and term starting infant formulas. Journal of Food Science, 70 (2): 107-112

Dryden, G. (2021). Fundamentals of Applied Animal Nutrition. CABI.

El-Din, A. G., Hassan, A. S. H., El-Behairy, S. A., & Mohamed, E. A. (2012). Impact of zinc and iron salts fortification of buffalo's milk on the dairy product. World Journal of Dairy & Food Sciences, 7(1), 21-27.

Fekete, K., Berti, C., Cetin, I., Hermoso, M., Koletzko, B. V., &Decsi, T. (2010). Perinatal folate supply: relevance in health outcome parameters. Maternal & child nutrition, 6, 23-38.

Flynn, A. & Cashman, K. (1997). Nutritional aspects of minerals in bovine and human milks. In Advanced Dairy Chemistry Volume 3, 257-302.

Fransson, G. B. &Lonnerdal, B. O. (1983). Distribution of trace elements and minerals in human and cow's milk. Pediatric research, 17(11), 912-915.

FSSAI (2011). Food Safety and Standards (Fortification of Foods) Regulations. Food Safety and Standards Authority of India. Ministry of Health and Family Welfare, Government of India, New Delhi,India

Gibson, R. S., Perlas, L., & Hotz, C. (2006). Improving the bioavailability of nutrients in plant foods at the household level. Proceedings of the Nutrition Society, 65(2), 160-168.

Gibson, R. S., Yeudall, F., Drost, N., Mtitimuni, B. & Cullinan, T. (1998). Dietary interventions to prevent zinc deficiency. The American Journal of Clinical Nutrition, 68(2), 484-487.

Goillem, V. N., Law, A. R. & Leaver, J. (2000). Effect of ph on the thermal denaturation of whey proteins in milk. Journal of Agricultural Food Chemistry, 48 (3): 672-679.

Gulbas, S. Y., &Saldamli, I. (2005). The effect of selenium and zinc fortification on the quality of Turkish white cheese. Journal of Food Sciences and Nutrition, 56(2), 141-146.

Hambidge, K. M., Chavez, M. N., Brown, R. M., &Walravens, P. A. (1979). Zinc nutritional status of young middle-income children and effects of consuming zinc- fortified breakfast cereals. The American Journal of Clinical Nutrition, 32(12), 2532-2539.

Hambidge, K. M., Huffer, J. W., Roboy, V., Grunwald, G. K., Westcott, J. L. & Sian, L. (2004),Zinc absorption from low phytate hybrids of maize and their wild type isohybrides .Am J Clin Nutr 2004; 79:1053-1059.

Hess, S. Y., & King, J. C. (2009). Effects of maternal zinc supplementation on pregnancy and lactation outcomes. Food and Nutrition Bulletin, 30(1_suppl1), 60-78.

Hunt, J. R., Johnson, P. E., & Swan, P. B. (1989). The dynamic nature of zinc availability from foods in vivo. Biological Trace Element Research, 19(1), 119-127.

Institute of Medicine. Food and Nutrition Board. Dietary Reference Intakes for Vitamin A, Vitamin K, Arsenic, Boron, Chromium, Copper, Iodine, Iron, Manganese, Molybdenum, Nickel, Silicon, Vanadium, and Zinc Washington, DC: National Academy Press; 2001

Kahraman, O. &Ustunol, Z. (2012). Effect of zinc fortification on Cheddar cheese quality. Journal of dairy science, 95(6), 2840-2847.

Kapil, U. & Jain, K. (2011). Magnitude of zinc deficiency amongst under five children in India. The Indian Journal of Pediatrics, 78(9), 1069-1072.

Krebs, N. F. (2000). Overview of zinc absorption and excretion in the human gastrointestinal tract. The Journal of Nutrition, 130(5), 1374S-1377S.

Lonnerdal, B. (2000). Dietary factors influencing zinc absorption. The Journal of Nutrition, 130(5), 1378S-1383S.

Lonnerdal, B. O., Sandberg, A. S., Sandström, B., & Kunz, C. (1989). Inhibitory effects of phytic acid and other inositol phosphates on zinc and calcium absorption in suckling rats. The Journal of Nutrition, 119(2), 211-214.

Ollig, J., Kloubert, V., Weßels, I., Haase, H., & Rink, L. (2016). Parameters influencing zinc in experimental systems in vivo and in vitro. Metals, 6(3), 71.

Piskin, E., Cianciosi, D., Gulec, S., Tomas, M., & Capanoglu, E. (2022). Iron absorption: factors, limitations, and improvement methods. ACS omega, 7(24), 20441-20456.

Rana, B., Kaushik, R., Kaushal, K., Arora, S., Kaushal, A., Gupta, S. & Kaushik, P. (2017). Physicochemical and electrochemical properties of zinc fortified milk. Food Bioscience 63.

Roohani, N., Hurrell, R., Kelishadi, R., & Schulin, R. (2013). Zinc and its importance for human health: An integrative review. Journal of research in medical sciences: the official journal of Isfahan University of Medical Sciences, 18(2), 144.

Salgueiro, M. J., Zubillaga, M. B., Lysionek, A. E., Caro, R. A., Weill, R., & Boccio, J. R. (2002). The role of zinc in the growth and development of children. Nutrition, 18(6), 510-519.

Sandberg, A. S. (2002). Bioavailability of minerals in legumes. British Journal of Nutrition, 88(S3), 281-285.

SandstroÈm, B. (2001). Micronutrient interactions: effects on absorption and bioavailability. British Journal of Nutrition, 85(S2), S181-S185.

Sharma, A., ShilpaShree, B. S., Arora, S. & Tomar, S. K. (2015). Lactose–Zinc complex preparation and evaluation of acceptability of complex in milk. LWT-Food Science and Technology, 64(1), 275-281.

Shilpashree, B. G., Arora, S., Chawla, P. & Tomar, S. K. (2015). Effect of succinylation on physicochemical and functional properties of milk protein concentrate. Food Research International, 72, 223-230.

Singh, H, Flynn,A&Fox, P.F(1989). Binding of zinc to bovine and human milk proteins. Journal of Dairy Research,56,235-248.

Spencer, H., Kramer, L., Norris, C., & Osis, D. (1984). Effect of calcium and phosphorus on zinc metabolism in man. The American journal of clinicalnutrition, 40(6), 1213-1218.

Temple, L., Gibson, R. S. & Hotz, C. (2002). Use of soaking and enrichment for improving the content and bioavailability of calcium, iron, and zinc in complementary foods and diets of rural Malawian weanlings. Journal offood science, 67(5), 1926-1932.

Udechukwu, M. C., Collins, S. A., &Udenigwe, C. C. (2016). Prospects of enhancing dietary zinc bioavailability with food-derived zinc-chelating peptides. Food & function, 7(10), 4137-4144.

White, P. J., & Broadley, M. R. (2005). Biofortifying crops with essential mineral elements. Trends in plant science, 10(12), 586-593.

Whittaker, P. (1998). Iron and zinc interactions in humans. The American journal of clinical nutrition, 68(2), 442-446.

WHO (2006). A safer future: global public health security in the 21st century: report of a Joint World Health Organization/Centers, Geneva, Switzerland.

Wood, R. J., & Zheng, J. J. (1997). High dietary calcium intakes reduce zinc absorption and balance in humans. The American journal of clinicalnutrition, 65(6), 1803-1809.

Çakmak, İ., Kalaycı, M., Ekiz, H., Braun, H. J., Kılınç, Y., & Yılmaz, A. (1999). Zinc deficiency as a practical problem in plant and human nutrition in Turkey: a NATO-science for stability project. Field Crops Research, 60(1-2), 175-188.

27

Food and Nutritional Security and Sustainability of Agriculture

J.P. Srivastava

Department of Agril. Extension & Communication, SHUATS, Prayagraj Uttar Pradesh

Introduction

Hunger and all forms of malnutrition continue to cause widespread suffering throughout the world. Malnutrition in children commonly leads to poor growth and poor physical and mental development, increased morbidity and, all too often, death. Among adults, malnutrition leads to lethargy, frequent illness and poor health, reduced physical capacity to do work, poor pregnancy outcomes among women, decreased cognitive function and a loss of learning potential and productivity. Widespread hunger and malnutrition also serve as severe impediments to social and economic development at both community and national levels. After years of steady progress, recent setbacks in the fight against hunger and malnutrition arising from the combined effects of high food prices and the global economic downturn coupled with the continuing problems of underdevelopment, civil strife, inadequate food supplies, social discrimination and environmental stress, jeopardize the earlier hopes for achieving the poverty, hunger and nutrition-related Millennium Development Goals (MDGs).

According to FAO, there were 105 million more hungry people in 2009 than in the previous year and the number of malnourished people now stands at 1.02 billion. In the worst-affected countries, individual productivity losses due to malnutrition are equivalent to 10% of lifetime earnings, resulting in gross domestic product (GDP) losses of up to 3%. Around the world close to 10 million children die before their fifth birthday every year, over one-third of which are associated with under nutrition. One in three developing-country children under the age of five - 178 million children – are stunted due to chronic under nutrition and poor health, and some 148 million children are underweight. Micronutrient deficiencies or "hidden hunger" affects around 2 billion people worldwide or about one third of the world's population. These vitamin and mineral deficiencies, especially of iron, iodine, zinc and vitamin A, lead to poor physical growth and development, lowered mental capacities, reduced productivity, impaired immune systems, blindness and death - all of which are preventable.

Stagnation in agricultural production in India is a cause of serious concern, which needs to be addressed through inclusive development. To address this challenge, the strategy should be to improve irrigation efficiency, ensure balanced plant nutrition management, develop denuded lands for increasing the cropping area, breed drought tolerant high yielding varieties of coarse millets and pulses and promote dairy husbandry, preferably through mixed farming. Active involvement of civil society organisations and farmers' organisations can help in capacity building of small farmers and women, while developing a value chain for the produce. It is necessary to promote sustainable agriculture to safeguard the economic viability of the farmers. Sustainable agriculture is a set of farming practices which can continue to maintain the farm productivity, efficiency and profitability in the long run, without depleting the natural resources and the environment. For ensuring the sustainability of small farmers, it may be useful to encourage the adaptation of indigenous skills, use of internal inputs, preferably from organic sources, least dependence on external inputs, greater emphasis on crop diversity, symbiotic crop rotation, and production focused on local needs and easy marketability.

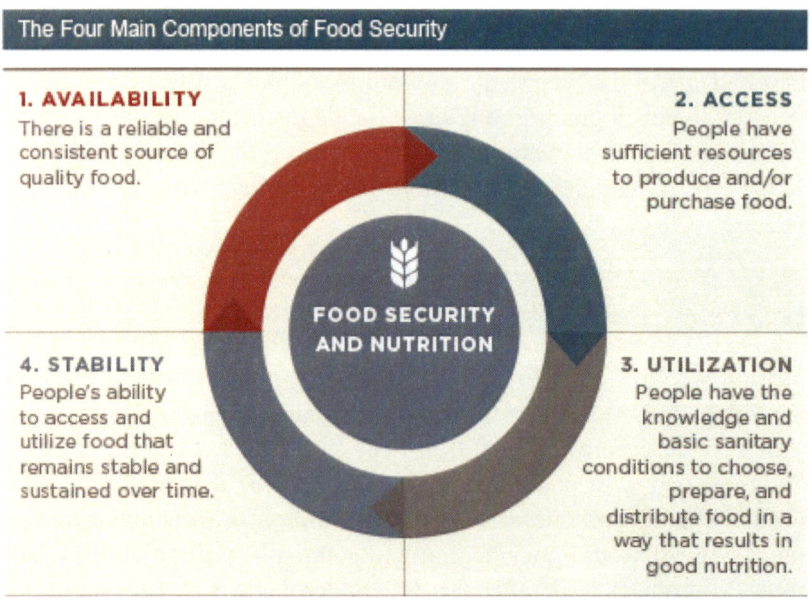

While India has made significant progress in the areas of science, technology and industrial development, food security for the rural poor continues to be a cause of serious concern. This is inspite of the successful Green Revolution, launched in the sixties, which sustained its positive impact for over three decades. Subsequently, since the mid nineties, there has been stagnation in agricultural production, resulting in food and nutritional insecurity.

Problems and Prospects of Food Production

Increasing population and their dependence on agriculture, small and fragmented land holdings, depletion of soil productivity, and inefficient use of water resources, out-dated agricultural production technologies, lack of agricultural credit and infrastructure for post harvest management and marketing of agricultural produce are responsible for low yield and income.

Role of Indian Green Revolution for Food Security

The Green Revolution programme which was launched in a mission mode, with various services for backward and forward integration, enabled the farmers in Punjab, Haryana and parts of Uttar Pradesh and Rajasthan, to increase food production and their income. The food grain production increased from 82 million tons in 1960 to 176.4 million tons in 1990 and to 241.56 million tons in 2010.The yields of rice, wheat, maize and oilseeds had increased by over 200%, but the increase in the yields of pulses and coarse cereals, which are mostly grown in rain-fed areas by small farmers, was insignificant. Increase in land productivity and crop yields generated employment for landholders as well as the landless, while ensuring food security for the country. However, the growth in agricultural production started declining from 10 - 12% in the 80's to 4.7% in the 90's and to 2.3% during 2002-07 (Government of India, 2011). The cost of inputs for cultivation of high yielding varieties increased steeply, reducing the profit margin, but there was no major breakthrough to enhance the crop productivity in rain-fed areas.

"Reduction in the rate of growth in food production reduced the per capita availability of food grains, which accelerated the incidences of malnutrition. Stagnation in agricultural production suppressed employment opportunities in rural areas, thereby reducing the contribution of agriculture to the national GDP from 58% in 1951 to 35% in 1991 and to 17% in 2009."

Considerations for Boosting Agricultural Production

While focusing on the development in agricultural production across the country, it is necessary to categorise the areas which were benefitted through Green Revolution and the other areas which had low impact, and develop a suitable strategy looking at ground realities. For areas like Punjab and Haryana, where the post green revolution growth in agricultural production is stagnant, the strategy should be to change the cropping pattern to rebuild soil productivity, make efficient use of water by adopting micro-irrigation systems, balanced crop nutrition with emphasis on increased doses of farm yard manure, biofertilisers and micro-nutrients and promote intensive large-size dairy farming. This system of farming is already becoming popular in Punjab and suitable policy support can popularise this programme further. With the increase in the yield of food grains in other parts of the country, changed cropping plans in Punjab and Haryana will not affect the food security in the country.

Strategy for Development of Rain-Fed Areas

Over 75% of the rural households in the country are marginal and small farmers, holding less than 2 ha and a majority of them are dependent on rain-fed agriculture for their livelihood. However, as the crop yields are low, 40 – 45 % of the rural families are living in poverty. To ensure food security to them, it is necessary to improve agricultural production, while enhancing their employment generation opportunities. This can be managed by promotion of improved production practices and development of degraded and wastelands through mixed farming. As most of the small holders are unable to earn their livelihood from crop production alone, livestock husbandry should be an integral part of the sustainable agricultural development.

The following priority areas for ensuring food security need special attention:

Water Resources Management: Water is the mostcritical input to boost agricultural production but excessive use of water has not only reduced the irrigation efficiency but has also turned fertile lands into sodic wastelands. Efficient storage of surface water, metered supply of irrigation water with realistic pricing and promotion of micro-irrigation systems, can increase water use efficiency to 70 - 75% while increasing the area under irrigation and reducing pollution of ground water. Therefore, it is necessary to introduce efficient irrigation systems, while covering rain-fed areas under an efficient watershed development programme.

Out of 80 million ha rain fed cropping area, over 60 million ha are located in semi-arid and arid regions where crop yields are low. Although about 50 - 55 million ha have been covered under watershed development so far, in the absence of a holistic development programme, the impact has been low. However, in some watersheds, where attempts were made to launch a holistic programme in convergence with other development activities, the agricultural production has increased by 60 - 80%. Experiences of such successful watersheds can be widely replicated by involving committed NGOs, Panchayati Raj institutions and farmers' organisations. Promotion of Water Users' Groups for efficient distribution and utilisation of water, good incentives for introducing micro-irrigation systems and curb on free power supply, can promote efficient water use.

Improving Agricultural Production: Development of degraded and wastelands for crop production and introduction of eco-friendly practices to improve the production in rain-fed areas should be the priority. Reclamation of 9 million ha sodic wastelands, 2 million ha ravine lands and 30 - 40 million ha of denuded wastelands for agricultural production, involving committed voluntary organisations for backward and forward integration, can easily increase the food production by 90 - 100 million tons per annum while generating year-round employment for millions.

Pulses and millets can be grown in rain-fed areas, even on low fertile soils under moisture stress. There are many pulses which are of 75 – 90 days duration, good for cultivation on residual soil moisture, after harvesting *kharif* crop. Farmers in areas receiving low rainfall, require precise information on weather forecasts, drought resistant varieties of various crops and mechanised implements for carrying out timely operations. This can promote various micro-enterprises for providing farming services.

Promotion of Agri-Horti-Forestry: There are various systems of agroforestry to suit different agro-climatic conditions. While the shallow soils under moisture stress are suited for silvipasture, denuded lands located in areas receiving 800 mm rainfall, can be brought under agri-horti-forestry. With the objective of rehabilitating the tribals on such lands in their own environs, BAIF Development Research Foundation, a civil society organisation, has promoted an unique agri-horti-forestry programme, where 0.4 ha land owned by each family was developed as orchard, by establishing drought tolerant fruit crops, while the inter-space was used for growing food crops. This programme enabled the participant families to earn Rs. 35,000 – 50,000 per annum after 4 - 6 years. Additional income generation from cultivation of food and cash crops in the inter-space, fruit nurseries, collection of minor forest products and medicinal herbs could be generated right from the first year to sustain their livelihood. Presently, over 1.5 lakh families spread over 9 states, have benefitted from this programme, which can be expanded across the country. This is a unique programme where employment generation, food security and environmental conservation can be achieved simultaneously.

Livestock Development: Livestock is a source of supplementary income for over 70 - 80% rural households, who own different species of livestock. However, over 80% of the livestock particularly cattle and buffaloes are uneconomical due to low milk yield. To address this problem, BAIF has demonstrated a unique approach of genetic improvement for producing high yielding cattle and buffaloes at the doorsteps of small farmers and 3 such milch animals can provide sustainable livelihood for a family. Initially, the programme has a gestation period of 3.5 to 4 years. Subsequently, the participant families are able to earn Rs. 35,000 - 40,000 per annum. They can earn additional income by selling surplus animals, which fetch Rs. 30,000 to 50,000 per cow or buffalo. BAIF is presently operating 3500 livestock development centres spread over 16 states covering 4 million families in 60,000 villages, contributing Rs. 3500 crores to the national GDP every year through increase in milk production.

Goat husbandry is another opportunity to provide supplementary income without increasing the herd size without causing any burden on natural resources. This programme, focussing on breed improvement, health care, efficient feeding and direct marketing, could enhance the income by 4 - 5 times within a short span of 18 to 24 months. Piggery, poultry and fishery also have good opportunity in selected areas. Livestock development blended with agriculture, can facilitate an efficient nutrient cycle to boost agricultural production.

Agri-Business Activities: Rural employment can also be generated through production of critical agricultural inputs such as vermicompost, bio-fertilisers, bio-pesticides, mushroom spawn, silkworm eggs, cattle feed, seeds of food and fodder crops, etc. Series of agro service centres as well as livestock breeding centres can be established, by training unemployed youth and supporting their business through banking institutions. Such service centres can also take up the distribution of inputs and organise collection and gradation of farm produce, for marketing or processing. Decentralisedagro-processing units can also generate additional employment.

Off-Farm Employment Opportunities: Promotion of various off-farm activities like production of housing material, generation of alternate energy devices, textile, handicrafts, farm equipment and tools, can open up new employment opportunities for the youth. There is also scope for agro-engineering services, electrical repairs and fittings, automobile servicing, trade and tourism.

Capacity Building for Governance: While focussing on capacity building of the farmers in rain-fed areas, women empowerment should be given priority as women manage 60 to 75% farming operations. This can be done through formation of Self Help Groups of producers, creating awareness through exposure visits, hands-on training on appropriate technologies and establishing linkages with development and financial institutions for easy access to technologies and resources. Development of physical infrastructure such as storage, grading and processing facilities as needed, should be undertaken.

Resources for these activities can be mobilised by converging on-going schemes with various Ministries of the Government of India. To ensure the success and sustainability of the programme, good governance is a critical input. There is also a need for sensitising the members of the local Gram Panchayats, Farmers' Cooperatives and well reputed Civil Society Organisations to take active part in supporting the programme. Such strong grassroot level organisations can make best use of various development programmes while establishing efficient value chains for different commodities.

Sustainable Agriculture

Farmers, particularly the small holders need proper orientation to take suitable decisions on crop selection, investment in various inputs, storage and marketing, based on the information on technical developments, government policies and prices of inputs and outputs. Large and elite land holders should also be encouraged to adapt these eco-friendly measures for profitability and environmental safety. Farming systems which can fulfill these criteria can lead to eco-friendly green revolution. In such situations, we can reduce the consumption and production of agrochemicals. The polluting automobiles need not run up and down between the cities and villages to transport agricultural inputs and outputs. The cost of food production can be lower and the retail price of locally produced agricultural commodities will also come down, when more poor people can afford to buy adequate food to maintain a balanced diet. This is a `win-win' situation for all, except the transport and traders.

In the process of avoiding or reducing the use of external inputs, if the small farmers can adapt organic farming, the consumers can get healthy food and productivity of the soil can also be conserved. With better awareness about benefits of organic food, the consumers will be prepared to pay a higher price for eco-friendly produce. Hence low-external input agriculture can be attractive, particularly for small holders.

Promoting Traditional Wisdom

For ensuring sustainable agriculture, it is advisable to explore the traditional wisdom and promote local practices. The study of local traditions will help in understanding the soil fertility and water management systems developed over years. As these practices have helped in the past to produce enough food grains before the introduction of chemical fertilizers, sustainable agriculture can promoted on these traditions. Thus sustainable farming is the extension of traditional practices with necessary modifications.

With a significant change in the ecosystem, forest coverage, livestock holding pattern and lifestyle in rural areas, it is becoming difficult to generate adequate quantity of farmyard manure which was the major source of nutrients in olden days. With increasing deforestation, diversion of biomass for fuel and commercial use, farmers are not able collect biomass from the forests and community lands. But there has been an increase in agricultural byproducts and domestic garbage, which can be used for composting. With the promotion of agroforestry, live hedges and wind breaks, it is possible to produce biomass useful for green manure. Cultivation of green manure crop of short duration between two crops is another option to enhance soil fertility. Adaptation of dairy farming with high yielding cattle and buffaloes, stall feeding and installation of biogas plants can generate significant quantity of farmyard manure, while generating additional income and energy. Use of earthworms can further improve the quality of the manure.

The soil productivity can be further enhanced through soil and water conservation, improved drainage, contour bunding and establishment of live hedges. In arid and semiarid regions where the use of chemical fertilizers is risky, application of organic manure and bio-fertilizers, adaption of tree-based farming, coupled with improved drought tolerant crop varieties and integrated pest management can be promoted to boost the yields. Replacement of local seeds with certified seeds, change of crops, use of improved farm implements, mineral nutrients, plant protection measures and installation of irrigation system are some of the new initiatives to further enhance the crop yields. Post-harvest processing, storage and marketing are other areas where both small and large holders have been incurring heavy losses. This area need introduction of modern technologies and infrastructure development which can add value to the produce, while reducing the cost of handling.

Need for Support and Coordination

1. **Transfer of Technologies:** Indian agriculture in general is due for modernisation, which has been presently adapted for a few selected crops. These include the use of efficient irrigation systems, biofertilisers, biopesticides, safe agrochemicals and improved seeds, which have direct bearing on crop yields and cost of production. Simple techniques for soil, water and plant analysis to maintain nutritional balance, weed and pest management to reduce nutrient losses are some areas where technology transfer has been lagging behind. Farmers can be oriented to make judicious use of nutrients based on the soil fertility, organic matter content, soil moisture availability and the crop requirement, as the balance between organic manure, macro and micro-nutrients is very critical to induce flowering and crop yields. It is therefore necessary to strengthen the linkage between the Agricultural Research Institutions and State Agricultural Extension Department with the farmers at the village level, preferably through local voluntary agencies. The State Agricultural Extension Department and the Krishi Vigyan Kendras should focus their role on providing such field technical service through training, demonstration and field visits. The Agriculture Department should also play a role in coordinating between various other departments

like Irrigation Department, Food Processing Department, Agricultural Credit and Banking to ensure better services.

2. **Research and Technology Development:** There is good scope to further improve the crop production, through application of growth regulators and hormones. These are available both in natural and synthetic forms useful for control of vegetative growth, floral induction, control the maturity period and to enhance the crop yield. These techniques are also useful to improve the quality to synchronize the harvesting time with peak demand for the produce. Development of safe and inexpensive biological plant protection devices should also be given priority. Production of efficient farm equipment and machinery can improve the productivity and reduce the cost. Presently the tractors are hardly used for 250-300 hours in a year, mainly for pre-sowing operations and for occasional transportation. If the idle time can be used for non-farm operations and to run small scale industries, the farmers can reduce their overheads on crop production and generate additional employment.

3. **Information Support:** Weather forecasts, advice on suitable cropping, information service on market demand and supplies, periodic reports on area under various crops and scope for further expansion and incidence of pests and diseases, through mass media can help the farmers to take suitable decisions. These support services can be established by the government or competent Farmers' Organisations and made available to farmers regularly.

4) **Strengthening of Farmers' Organisations:** Dependency on outsiders has been the cause for rural exploitation. To reduce the exploitation by middlemen, village level farmers' organisation should be developed throughout the country. These organisations can also motivate the farmers to adapt improved farming practice, while procuring critical inputs well

in time. These organisations can arrange loan from financial institutions for their members. With the development of farmers' organisations and improvement in crop production, farmers can appoint their own technical advisors and extension officers, as presently undertaken by the sugar cooperatives. Farmers organisations like Federations of Sugar Cooperatives, Milk Unions, Grape Growers Cooperative, Apple Growers Cooperative, tea producers and many such commodity growers have been very successful in India in organising the growers to improve their production, post harvest handling such as storage, processing and marketing to ensure stable market and remunerative price. In the absence of such organisations for other commodities, there has been tremendous harassment for both growers and consumers. In case of onion, with only 10-15% reduction in the production, the price had increased by 400-600% but the benefits of such price rise has hardly reached the growers. This calls for various kinds of farmers' organisations to cover certain geographic areas and who are specialised in handling specific commodities.

5. **Land Reforms and Consolidation:** There are two kinds of land holding patterns, which are affecting the crop production. The first category of farmers have large holdings but are unable to make efficient use of such lands for agricultural production. Among them are many absentee landholders who are afraid of leasing out for cultivation, keep their land. The other category of farmers is those having small holdings, where improved agricultural practices are not feasible. Farmers belonging to both these categories are not willing to involve others for cultivation, as the present Land Reforms Acts can entitle the tiller to own the land. Hence, there is need for relaxation in the Land Reforms and Land Ceiling Acts. Farmers interested in leasing land for short or long term may be allowed to do so on attractive terms. Corporate sectors interested in agricultural development may be permitted to take land on lease from farmers. With such large scale contract farming, it will be economically viable to introduce new technologies and provide a stable market for farm produce. Hence there is an urgent need to review the present Land Ceiling Acts to encourage corporate farming, while safeguarding the interest of the farmers.

Conclusion

Looking to the prevalent farming practices in India and the scope for coping with advances in agricultural development in other parts of the world, there is good scope for improvement. There is also a wide gap between the elite and backward farmers within the country. The task of achieving sufficiency in food production, involving small farmers, is well within the reach. Small Farmers constitute a great majority of farming community today and they need to be organized into farmers interest group and supported through community managed sustainable agriculture. The Agricultural Extension Department shouldreset its role as a facilitator to coordinate with various departments to ensure better service to small farmers and to reduce the communication gap presently existing between the agencies concerned with sustainable development and farmers.

Selected farmers' organizations and civil society organisations engaged in sustainable agricultural development with good track record can be entrusted to develop suitable agricultural development programmes with a holistic approach. Village level producers' groups can take active part in identifying their needs and undertake suitable income generation activities to enhance agricultural production. Public - private partnership may be promoted to develop value chains to improve the profitability. With such an integrated programme, most of the small farmers can come out of poverty and empower themselves to ensure food security.

References

Government of India. 2011. Faster, Sustainable and More Inclusive Growth – An Approach to the XII Five Year Plan, Planning Commission, New Delhi.

Hegde, N.G. 2012. Water Scarcity and Security in India. 99th Indian Science Congress. KIIT University, Bhubaneswar 3-7 January. 19 pp.

Sustainable Agriculture for Food Security. Indian Farming. 49(12). Mar.: 4-11, 2000.

www.fao.org

28

Home Nutrition Gardens and Enhanced Dietary Diversity of Paniyan of Wayanad, Kerala

Paniyan

M.S. Swaminathan Research Foundation, Wayanad, Kerala

Abstract

Adequate nutrition is considered as the base of health and wellbeing. The availability and access to adequate nutrition is determined by multiple factors, that is power, culture, gender, class, education, caste availability of land, control over natural resources and so on. The deprivation of the factors mentioned above, tend the Adivasis more vulnerable to malnutrition. In India 32% of tribal women are chronically undernourished against 23% of non tribals (Sethi V et al 2019). Kerala a well acclaimed state in social and human development also not an exception in Adivasi malnutrition (Census, 2011). Out of the total tribal population of Kerala 31.23 percent residing in Wayanad. Paniiya the freed bonded labour community are the most vulnerable to malnutrition.

Due to the depletion of biodiversity, land alienation and land use changes led to the decline of food basket items they used to collect from wetlands,forests and other common property resources (CPRs). With a view to bridge the gap, an action research based on Home Nutrition Garden (HNG) establishment have been undertaken among 231 Paniyan households at various regions of Wayanad. As a result of participatory exercise carried out in intervention sites, 30 crop varieties have been promoted in HNG on the basis of community's food preference and crops' nutrition value. The impact has been assessed through baseline and endline survey on homestead food crop diversity and consumption data from the HNG yield. The result shows that the number of household's food crop diversity has enhanced from 5 to 20 and an average 25% increase in the consumption of leafygreens, vegetables, tubers, legumes and fruits. This study result recommends HNGs as an effective tool to improve nutrition status of vulnerable communities.

Keywords: *Malnutrition, Adivasis, Paniyan, Home Nutrition Garden, Food Crop Diversity*

Introduction

Food and nutrition security is the basic requirement of human development. Among the ten global burden of diseases, six are connected with diet related factors. Food and nutrition security is also closely related to other factors like socio economic status, geography and demography. Asia, Latin America and Africa hosts world's most of the undernourished population. Among these regions South Asia is considered as one of the hotspots of malnutrition, in which India is the major contributor. Fifty seven percent of population cannot afford healthy diet in South Asia.

Over the centuries of remarkable economic growth in India concomitant reduction in under nutrition has not been witnessed and this dilemma is called Indian enigma (Gillespie et al 2012; Sengupta 2019). In India among the different socio economic categories, undernutrition has highly been manifested among SC and STs (Gillespie et al 2012; NNMB 2009).The reasons of food and nutrition insecurity are directly linked to remoteness, dire poverty and dysfunctional food security programmes (Khera 2011; UNICEF 2014). Poor and landless among them are mostly falling prey to chronic food insecurity. Coupled with reduction in locally available dietary inputs, rising food expenditure also compel poor to cut back the quantity and resort to low priced calorie dense food (Smyth & Sweetman; 2010). Therefore, poor people's diet assume to fraught with starch, which leads to nutritional imbalances and resulting in to various lifestyle diseases like diabetics among them. This situation necessitates the need of local, sustainable household food production through home nutrition gardens. Many studies testify the impact of household level food crop cultivation in addressing malnutrition. Kuhnlein & Raceveur 1996; Pimbert 2009; Vira et al 2015 argues inadequate diets are associated with the "delocalization of food supply" and disruption in existed food systems. Therefore, in general literatures on agriculture and nutrition elicits the positive correlation between cultivating food crops and nutrition security (Gillespie et al 2012; Hossain et al 2016; Viswanathan et al 2015). In this context, home nutrition garden programmeassumes greater significance in countering malnutrition among Paniyan community in Western Ghats region of Wayanad district of Kerala.

Malnutrition Status of Adivasis in the Kerala

Tribes people are the most marginalized and thus powerless all over the world. Their deprivation of resources is more or less same in all the developing countries. With a population of more than 10.2 crores, India has the single largest tribal population in the world. This constitutes 8.6 per cent of the total population of the country (Census of India, 2011). Majority, ie., 96 percent of Adivasis are residing in rural areas, in forest land, forest fringes and wildlife sanctuaries. This implies that access to land and forest resources is very important to Adivasis, economically, socially and culturally. Among Indian States, Kerala is well recognised for its high level of Social Development Indices. Kerala achieved high literacy, high longevity, low infant and maternal mortality, low population growth and favourable sex ratio (Parayil, 1996). As far as tribes people of Kerala is concerned, the development indicators are paradoxical. Malnutrition deaths of Tribal children have been reported from Kerala, especially from Attappadi and Wayanad districts since 2002. All the remedial measures taken so far from the side of the Government system have not yielded the desired result. Adivasi women are suffering from chronic malnutrition and anaemia which directly lead to maternal mortality, low birth weight and infant mortality. According to the report of a team of medical doctors after the continuous instances of infant mortality in Attappadi of Palakkad district of Kerala, "Malnutrition and anaemia are acute among pregnant and lactating women. In the year 2013 itself about 30 children have died within hours/days of their birth" (Ekbal, 2013). 120 children have succumbed to malnutrition death during the period of three years (Reji, 2015). Similarly, in Wayanad, high rate of anaemia and related morbidity of tribal women and decreasing life expectancy of Paniyan tribal men are also pose critical challenges to the ongoing development initiative of Kerala.

As far as food is concerned, the issues of the vulnerable and marginalized tribes people have been exacerbated due to the reduction in the availability of dietary inputs which they have been used to collect from various locations. Insufficient, imbalanced intake of food and nutrients leading to common nutritional problems like low birth weight, protein malnutrition, micronutrient malnutrition and diet related non - communicable diseases. The absorption of required quantity and quality of food affecting health deterioration. Lack of uptake is an acute issue that tribes are living with improper Water Sanitation and Hygiene facilities.

After these high incidences of malnutrition deaths of tribal children in 2013, Government has been implementing several projects to counter this malady. According to a baseline research study conducted by M.S. Swaminathan Research Foundation on the status of malnutrition of Paniyan in Wayanad, reveals that 90% children of the age group 6- 10 are underweight., 90% children are facing deficient in protein intake, 72% children are facing deficient fat intake (Chandrika et al., 2015). In Kerala 15.98% Adivasi families are malnutrition affected. The largest number of malnutrition affected families are seen in Wayanad, Palakkad and Idukki districts (STDD, 2013). These situations are rooted in the combined impact of the loss of indigenous food, the unavailability of alternate nutritious food and the loss of employment opportunities (Ekbal, 2013). As per the data of health department of Wayanad during the period 2013-15, total 324 infant mortality cases were reported, within this 264 infants belong to Tribal communities. During the period of 2010-16 out of 67 maternal mortality 46 are from Adivasi communities. In 2013, 74 child mortality cases were reported among Adivasis due to nutrition deficiency (Mathrubhumi, 2016). Paniyan community's mortality rate is disproportionately higher compared to other Adivasi communities.

Wayanad District

Wayanad, the biodiversity, climate and tribal hotspot of Kerala, Wayanad district is in the southern tip of the Deccan Plateau which encompasses part of the Western Ghats that situates in the north-east of Kerala state, India. It is a very important part of Nilgiris biosphere. It is at a height of 700 m to 2100 m above mean sea level. It sprawls over an area of 2132 sq km. The district has a varying topography which includes hilly areas, valleys, forests, meadows and paddy fields. The climate also changes drastically depending on the geography and altitude. it has been categorized as bio diversity hot spot since it already lost seventy percent of its original habitat. This loss was highly affected its agrobiodiversity and crop genetic diversity that was the major source of food security. Further, this area is declared as climate hot spot by the Kerala State Action Plan on Climate Change (DoECC, 2022) because of the high degree of vulnerability to natural calamities like drought and flood which affect human life and the entire biodiversity. The KSAPCC predicts an increase in the frequency and intensity of forest fires, loss of biodiversity habitat loss, species migration and increasing human - animal conflict due to climate change could also lead to the outbreak of a host of new diseases. In fact, Wayanad has witnessed extreme drought in 2016 -17 because of the seasonal rainfall deficit of 59% during the southwest monsoon in 2016. As per census 2011, total population of Wayanad district is 8,16,558. Scheduled Tribe population is 1,51,443 which is 18.5% of the total district population Eleven different communities of Adivasis are living in Wayand. These Adivasi populations of Wayanad represents its 'differences' in identities with regard to social hierarchy, cultural practices, economic conditions, linguistic affinities, lineage systems, gender relations etc., within the geographical area. The basic human development indices like education and health are adverse and alarming among both the community. Wayanad has a poverty rank of 75, health rank is 114 and education rank is 112 (TOI 2017). In paradox to the resource richness in the region, District is now turned in to one of the 115 backward districts in India. The District is included in the aspirational district programme of Government of India. Due to the afore mentioned reasons, MSSRF has selected Wayanad district as an intervention area for home nutrition garden. Paniyan community comes 45.12% of the total tribal population of Wayanad. Due to special vulnerability of Paniyan within Adivasis, Paniyan is selected as one of the key beneficiaries of home nutrition garden intervention. The nutrition garden intervention is spread across seven settlements in 3 panchayaths and 1 municipality of Wayanad districts. Details of the intervention site are the following.

Study Villages

SL No	Hamlet Name	Mullukuruma	Paniyan	Katunayaka	Total Families
1	Cheeyampam 73	5	142	164	311
2	Madamkunnu	1	31		32
3	Kalappurakkal		11		11
4	Chenkutty		10		10
5	Thanikkuni	4	17		21
6	Mangavayal		4		4
7	Kuzhimukku		16		16
	Total	10	231	164	405

Methodology

This paper is the outcome of 10 yearlong research cum development project titled "Livelihood Enhancement Agriculture and Food Security" (LEAFS) funded by SEED division of Department of Science and Technology, implemented by Community Agrobiodiversity Center of M S Swaminathan Research Foundation. Food and Nutrition Security Initiative is one of the major core area in the said project, which is specifically focused to reduce malnutrition of the Adivasi communities in Western Ghat region of Kerala, Wayanad, India. Increased incidence of malnutrition among tribes, specifically among certain groups such as Paniyan from Wayanad region is selected as one of the target group for intervention. In prior to the implementation, a random baseline assessment by using anthropometric measures, medical camps and existing literature related Paniyan nutritional status were reviewed to understand commonly existing nutrition deficiency induced illnesses. As a result of this baseline, it was decided to enhance their access to diverse dietary inputs by establishing home nutrition gardens and diversifying their existing homesteads. In prior to the intervention our priority was mainly to understand the existing crop diversity in the homesteads. Though, the crops found before intervention slightly varies across intervention sites. Banana, Papaya, curry leaf, elephant foot yam and colocasia were the major crops found before intervention.

The study adopted both quantitative and qualitative methods of data collection. Qualitative methods like transact walks, observation of homesteads and dietary patterns were noticed. Conducted detailed consultation with the community about what food crop they really wanted to have and what are the bottlenecks to consume/ purchase diverse dietary inputs. In prior to the intervention, random survey among beneficiaries to understand their present dietary diversity by using food frequency table. On the basis of participatory exercise 30 crop varieties were shortlisted and promoted in Home Nutrition Garden in consideration of community's food preference and nutrition requirement. These crops encompass vegetables, leafy greens, tubers, fruits, pulses and banana (See Table 1). Year around nutrition demand of beneficiary households was partially met by this crop combination. The impact has been assessed through interim surveys on dietary diversity and homestead food crop diversity and interim crop harvest data from nutrition garden with an interval of three months. On the basis of interim survey results un survived plants were re-augmented.

Table 1: Diversity of Crops Available Across Seasons

Groups	**Details**		
	Rainy	**Winter**	**Summer**
Leafy Vegetables	Amaranthus, leaves of pumpkin, moringa, Curry leaf, Bashala, Thazhuthama	Amaranthus, leaves of pumpkin, moringa, Curry leaf, Bashala, Thazhuthama	Amaranthus, leaves of pumpkin, moringa, Curry leaf, Bashala, Thazhuthama
Roots and Tubers	sweet potato, white yam, colocasia, elephant foot yam	sweet potato, white yam, colocasia, elephant foot yam	sweet potato, white yam, colocasia, elephant foot yam
Other vegetables	Brinjal, lady's finger, tomato, Bitter gourd, ivy gourd.	Brinjal, lady's finger, tomato, Bitter gourd, ivy gourd.	Brinjal, lady's finger, tomato, Bitter gourd, ivy gourd.
Legumes	Field bean, Winged bean, Lima bean, Sword bean, cow pea, pigeon pee	Field bean, Winged bean, Lima bean, Sword bean, cow pea, pigeon pee	Field bean, Winged bean, Lima bean, Sword bean, cow pea, pigeon pee
Fruits/ Tree species	Passion fruit, Papaya, Guava, Amla, Anona, Banana, Lemon, Avecado	Passion fruit, Papaya, Guava, Amla, Anona, Banana, Lemon	Passion fruit, Papaya, Guava, Amla, Anona, Banana, Lemon
Spice	Bird Eye Chilly	Bird Eye Chilly	Bird Eye Chilly

seasons: Rainy (June-Sept) Winter (Oct-December) Summer (Jan-May)

Intervention to Address Malnutrition

The Food and Nutrition Security Initiative (FANSCI) is focused on food based and knowledge based approach to address malnutrition. As part of Nutrition Garden program, planting materials and seeds were provided to the target group after land preparation. A detailed crop calendar was prepared in consultation with leading farmers in the locality. The crop calendar includes information related to the timing of various crop management operation from seed storage to harvest. The capacity enhancement programs on crop management, maintenance of Home Nutrition Garden were provided to the target groups. Parallel community mobilisers were thoroughly trained on nutrition knowledge which includes nutritional value of crops promoted in the garden, nutrition requirement of different gender and generations, awareness on Water Sanitation Hygiene (WASH) and role and function of different nutrients. Further, upscaling the awareness of training participants, training manual and booklets with infographics were provided. During the project period,55 trainings were conducted with the participation of 2257 individuals (514 men 1743 women).

The well trained community mobilisers lead the community level nutrition knowledge generation activities. Supervision and advisory support related to crop management, data collection related to consumption and time to time documentation of activities are the other assigned tasks of community mobilisers.

Nutrition Garden Utilization; Consumption and Sales

Prior to the project intervention, all the intervention sites, people used to maintain minimum homestead farming. But the diversity was limited to four crops. It was noticed, the produce from nutrition garden utilised for different purposes such as own consumption, gift and sales. Among the crops, leafy greens and pumpkin was most exchanged item among community members. Tubers especially white yam, colocasia and elephant foot yam are the major crops they marketed as it is produced in large volume. Apart from gifting within the hamlet, produces like Papaya and tubers gifted to outside community as well. The household level consumption of average of beneficiary households from home nutrition garden are elaborated in the following table.

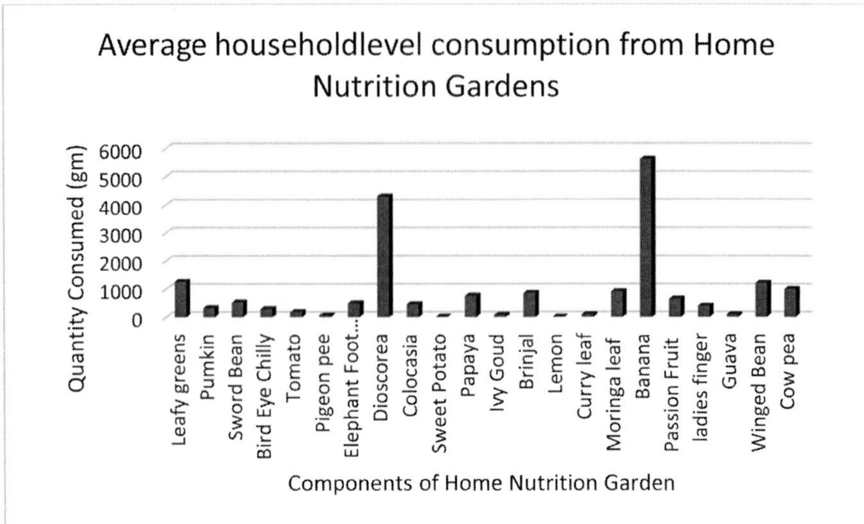

*measures in gram, data collected from the beneficiary household at various points of time during 2019. The local measurements were converted to grams to arrive the figures, still there may be possibility of minor variation in quantity calculation. The category leafy greens consist mainly two varieties of Amaranthus and pumpkin leafs.

As per the study results banana, dioscorea, winged bean, Moringa, Papaya, Brinjal, Cow pea and leafy greens are the major dietary inputs beneficiary households accessed more in terms of quantity. The planted perennial and tree species like Anona, Avecado, Amla, Rose apple and Guava not yet started yielding. Passion fruit and Lemon also not started yielding in majority of the households during data collection. Other parts of banana such as pseudostem and inflorescence were also consumed by beneficiary families are not included in the data. After the data collection period, it was noticed Home Nutrition gardens highly benefited the community during the outbreak of Covid-19.

Challenges

Increased incidence of crop raiding, frequent occurrence of extreme weather events are the major challenges causing significant yield loss and increased incidence of pest and diseases. Another challenge faced was the difficulty to irrigate HNGs as manual watering is required and the lack of modern irrigation facilities is another concern. A critical challenge aligns to the Time poverty of the target group as they are depending 'hand to mouth economy' for survival which doesn't allow the community to engage enough time for proper maintenance of HNGs. Besides, insufficient coverage of sanitation and hygiene facility is also affecting the uptake of nutrients.

Results

1. The activities were implemented in convergence with respective Panchayathi raj institutions. Nutrition knowledge generation initiative were also conducted in collaboration with NRHM, ITDP, ICDS and District Nutrition Mission and other likeminded institutions.

2. The project is being implemented with the active participation of stakeholders. The community participation assured in multiple levels i.e planning, implementation and monitoring level.

3. HNGs also served as a medium for agrobiodiversity conservation by introduction of native food crops and majority of the crops selected for the garden are traditional varieties which can withstand extreme weather events.

4. Community has affinity to local crop varieties which largely requires comparatively less care.

5. The intervention lead to increased homestead food crop diversity of 231 Paniyan families from 4 to twenty crops.

Conclusion

The activity helped to enhance household food crop diversity from 4 to 20 along with reduced expenditure for food consumption among beneficiary families of around 50%. The Home Nutrition Garden concept of FANSCI was shared with officials of State's 'Subiksha Keralam' programme for up scaling as part of food self reliant action plan of the State in the backdrop of covid-19 and agriculture rejuvenation in the State. The main findings of the intervention are focused on food culture of target group, crops resilience to local climatic conditions, nutrition knowledge generation and diversity of crop package must be seriously concerned as part of nutrition security initiatives among Paniyan. The previous studies also pointed couple of issues like displacement from traditional food sources such as forest, wetland, rivers and other common lands, landscape fragmentation, change in cropping pattern and food stigmatization. This all resulted in the considerable reduction in their dietary diversity. Therefore, home nutrition gardens helped significantly to address the reduction in the availability of dietary inputs due to the aforementioned reasons.Therefore, the present article also argues that participatory initiative in cultivation, consumption of locally producing food items is important in nutrition intervention.

Moreover, home nutrition garden initiative helped to address global challenges such as climate change, loss of agrobiodiversity and malnutrition. "Shifting to healthy diets can contribute to reducing the cost of health and climate change and is projected to lead to reduction of up to 97% in direct and indirect health cost of 41-74% of in the social cost of GHG emission in 2030" (FAO 2020). Ultimately the home nutrition garden initiative also played a key role to achieve Sustainable Development Goals at local levels.

Revitalizing cultural practices, traditional knowledge and skill transfer are the indirect benefit of the home nutrition garden initiative. Further, Home Nutrition Garden initiative helped to ensure sustainability of source, health and nutrition literacy and dietary driven solution to malnutrition.

References

Aliza Pradhan, Raju Sathanandhan, Akshaya Kumar Panda, Rupal Wagh. Improving Household Diet Diversity Through Promotion of Nutrition Gardens in India. American Journal of Food Science and Nutrition. Vol. 5, No. 2, 2018, pp. 43-51. B,

C.S, Chandrika; C.P, Haseena; Vipindas; P.T, Shajahan; Bindhu; Joseph ; V.I, Tina;. 2015. "Nutritional status of Paniyan tribes people of Wayanad ." M.S.Swaminathan Research Foundation, kalpetta.

Census of India. 2011. Total Population of Scheduled Tribe and Scheduled Caste. census report, new delhi: Government of India.

DoECC, 2022. Kerala State Action Plan on Climate Change 2.0 (2023-2030), Directorate of Environment and Climate Change, Department of Environment, Government of Kerala. 333pp

Ekbal. 2013. Genocide in Attapadi. medical visit report, attappadi: medical team.

FAO, IFAD, UNICEF, WFP and WHO. 2020. In Brief to The State of Food Security and Nutrition in the World 2020. Transforming food systems for affordable healthy diets. Rome, FAOhttps://doi.org/10.4060/ca9699en

Gillespie, Stuart, Jody Harris and Suneetha Kadiyala (2012): "The Agriculture-Nutrition Disconnect in India: What Do We Know?," IFPRI Discussion Paper 1187, International Food Policy Research Institute, Washington, DC, http://ebrary.ifpri.org/cdm/ref/collection/p15738coll2/id/126958.

Hossain, Mahabub, Nusrat Jimi and Md Aminul Islam (2016): "Does Agriculture Promote Diet Diversity? A Bangladesh Study," LANSA Working Paper Series 11, Brighton: Leveraging Agriculture for Nutrition in South Asia, https://opendocs.ids. ac.uk/opendocs/bitstream/handle/123456789/12884/BRAC%20study%20Hossain%20et%20al%20GG.pdf?sequence=1&isAllowed=y.

Khera, Reetika (2011): "Revival of the Public Distribution System: Evidence and Explanations," Economic & Political Weekly, Nos 44 & 45, pp 36–50.

Kuhnlein, Harriet V and Olivier Receveur (1996): "Dietary Changes and Traditional Food Systems of Indigenous People," Annual Review of Nutrition, Vol 16, No 1, pp 417–22.

Mathrubhumi Daily 2016 "Eemaranathinuaaranuutharavadhi" (Mal) Oct. 18

National Nutrition Monitoring Bureau (NNMB) (2009): "Diet and Nutritional Status of Tribal Population and Prevalence of Hypertension among Adults. Report on Second Repeat Survey,"National Institute of Nutrition, Indian Council of Medical Research, Hyderabad.

Parayil, Govindan. 1996. "The 'Kerala model' of development: development and sustainability in the Third World" ,.Third World Quarterly 17 (5): 941-957.

Pimbert, Michel (2009): "Towards Food Sovereignty: Reclaiming Autonomous Food Systems," International Institute for Environment and Development, London, https://www.iied.org/towards-food-sovereignty-reclaiming-autonomous-food-systems

Reji, Joseph (2015): "Attappadiyilmaranathintethaarattu", (Mal) March 9, Rashtradeepika Daily.

Scheduled Tribes Development Department. 2013. Scheduled Tribes of Kerala Report on the Socio Economic Status. Socio Economic , Scheduled Tribes Development Department, Government of Kerala., trivandrum: Government of Kerala.

Sengupta, Sohini. 2019. "Hidden hunger, burdened women Challenges for food and nutrition security Interventions." Economic and Political Weekly, may 11.

Smyth, Ines and Caroline Sweetman (2010): "Introduction," Gender and Development, Vol 18, No 3, pp 363–72

Times of India. 2017. " Retrieved from Wayanad in list of '115 backward district." times of india, december 5. http://timesofindia.indiatimes.com/articleshow/61921057.cms?utm_source=contentofinterest&utm_medium=text&utm_campaign=cppst.

UNICEF (2014): Nourishing India's Tribal Children: TheNutrition Situation of Children of India's ScheduledTribes, New York: United Nations International Children's Emergency Fund, http://unicef.in/Uploads/Resources/Tribal-low-res-for-view.pdf.

Vani Sethi, Sourav Bhattacharjee, Sonali Sinha, Abner Daniel, Avinash Lumba, Deepika Sharma, Arti Bhanot. 2019. "Sethi, Vani, Bhatacharjee S, Sinha S, Daniel A, Lumba A, Sharma D, Bhanot A (2019): "Delivering Essential Nutrition Interventions for Women in Tribal Pockets of Eastern India",Vol No 54." Economic and Political Weekly, January 26.

Vira, Bhaskar, Christoph Wildburger and Stephanie Mansourian (eds) (2015): "Forests, Trees and Landscapes for Food Security and Nutrition. A Global Assessment Report," IUFRO World Series, Policy Brief, International Union of Forest Research Organizations, Cambridge: Open Book Publishers.

Viswanathan, Brinda, Getsie David, Swarna Vepa and Bhavani R V (2015): "Dietary Diversity and Women's BMI among Farm Households in Rural India," LANSA Working Paper, Series 3, Brighton: Leveraging Agriculture for Nutrition in South Asia, https://opendocs.ids.ac.uk/ opendocs/handle/123456789/12161.

Annexure-1: Scientific names of crops grown in nutrition gardens of Wayanad

Tree/ Fruits/ Vegetables/ tubers/ leafy greens Species	Scientific name	Tree/ Fruits/ Vegetables/ tubers/ leafy greens Species	Scientific name
Amaranthus	*Amaranthus oleraceus, Amaranthus gangeticus*	Lady's finger	*Abelmoschus esculentu*
Amla	*Phyllanthus emblica*	Lime	*Citrus aurantifolia*
Anona	*Annona squamosa*	Lima Bean	*Phaseolus lunatus*
Banana	*Musa sp*	Moringa	*Moringa oleifera*
Beans	*Vigna unguiculata ssp.*	Orange flesh sweet potato (OFSP)	*Ipomoea batatas*
Bashala	*Basella alba*	Papaya	*Carica papaya*
Brinjal	*Solanum melongena*	Passion fruit	*Passiflora edulis*
Bird Eye Chilli	*Capsicum annum*	Pea	*Pisum sativum*
Colocasia	*Colocasia esculenta*	Pigeon pea	*Cajanus cajan*
Cowpea	*Vigna unguiculata*	Pumpkin	*Cucurbita maxima*
Curry leaves	*Murrayakoenigii*	Sword Bean	Canavalia gladiata
Elephant Foot Yam	Amorphophallus companulatus	Tomato	*Solanum lycopersicum*
Guava	*Psidium guajava*	Thazhuthama	*Boerhaviadiffusa*
Indian spinach	*Basella alba*	White Yam	*Dioscorea alata*
Ivy gourd	*Coccinia grandis*	Winged Bean	Psophocarpustert agonolobus

29

Emerging Pedagogy in Higher Education: Lessons Learned Post Covid-19

Aditya Sinha[1], Arindam Nag[1], Kalyan Ghadei[2] Basavaprabhu Jirli[2] and Debabrata Basu[3]

[1]Department of Extension Education, Bihar Agricultural University Sabour, Bihar
[2]Department of Extension Education, Institute of Agricultural Sciences Banaras Hindu University, Varanasi, Uttar Pradesh
[3]Department of Agricultural Extension, Bidhan Chandra Krishi Viswavidyalaya, Mohanpur, Nadia, West Bengal

Abstract

The covid-19 pandemic provided a new challenge to teachers worldwide in retaining their interest in learning online and imparting knowledge and skills in an online environment. The teachers who were novices in using technology in teaching and learning gradually graduated from survival to mastery stages. The expertise in the specific field led the teachers to create new pedagogies in conveying their knowledge in both theoretical and practical fields. The pedagogies enable the teachers to implement more customized, learner-centric experiences in the online environment. However, teachers must not confine their pedagogy in higher education to themselves. On the contrary, a greater sharing of information from teachers of different disciplines will lead to the growth of the concerned subjects. Reflecting on the decisions instructors make in pedagogy related to their subject, diverse educators worldwide could understand their' teaching persona' in a better way leading to the development of multiple philosophies in the same field or interrelated fields. This will again help understand the teaching personality to be effectively implemented in real classroom situations. The current paper provides an overview of different innovative pedagogies in higher education, which the teachers successfully implemented to overcome the digital age challenges. Since learning is a scientific intervention, different evaluation approaches are also necessary to improve learning outcomes and meet the expectations of the present interconnected generation.

Keywords: *Innovative pedagogy, Philosophy, Higher education, Covid-19, Knowledge*

Introduction

Pedagogy in higher education is often referred to as activities of educating, instructing or teaching to impart knowledge and skills (Diaz, 2013). With the onset of the Covid-19 pandemic in the year 2020, instructors worldwide faced a challenge in maintaining the teaching pedagogy suited to the online teaching environments. Many instructors with no prior experience teaching online started developing new approaches to solve the challenges in a changing landscape. The main challenges to solve was concerning creating favourable learning experiences, creating engaging classroom experiences, and meeting the expectations of all the students in the online environment (Teachonline, 2020). The internet has provided a wide range of opportunities, leading to uninterrupted education through various conferencing platforms. Since the year 2000, the proliferation of Web 2.0 was tested in the field of education with several tools such as blogs, wiki articles and use of social media channels. Web 2.0 provided increasing scope to the teachers and students to create, edit and share content which is appealing to the learners. Currently, we have entered the Web 3.0 era in which the technologies can be customized to enhance the learning environment for better facilitation of learning needs.

Pedagogy is referred to as the activity of educating, instructing or teaching the activities that impart knowledge or skill (Korthagen &Kessels, 1999; Crowther, 2013). With the rapid development and proliferation of information and communication technologies (ICTs), education is bound to use the same for enhancement of teaching and learning. Pedagogical innovation is defined as an intentional action to introduce something original in the given context to enhance student learning in situation concerning interaction and interactivity (Bechard, 2000). Although, the use of technology in education is at times challenging for the teachers and learners alike. Sometimes, organizational changes are necessary to bring about a shift in education combining ICTs (example: hybrid learning, blended learning situations, remote learning, etc.). These changes are necessary for the effective integration of innovative teaching and learning practices (Beatty & Gerace, 2009; Heitink *et al.*, 2016). There is a common belief that educational systems should equip students with the knowledge and abilities needed to adapt to a world that is everchanging. The 21st century skills such as critical thinking, problem solving, teamwork, innovation, digital literacy, and flexibility are frequently mentioned. The optimal way to develop these skills under various teaching and learning approaches and under different subjects are often debated. In the current paper, different innovative pedagogies which could be beneficial in the present connected and interactive world are documented from various sources.

Pedagogical Innovations

The role of innovative pedagogies in higher education is of paramount importance to effective cater the learners along with providing an opportunity for enhanced researches from various educators across the globe. There is an urgent need to integrate different approaches in digital learning with traditional teching-learning processes to improve the productivity of both the learners and the educators. The following pedagogies provide an overview of different approaches (both traditional and conventional) used by the educators to effectively provide new experiences to the learners globally.

Experiential Learning

Experiential learning is an old concept which is still relevant in today's interactive world. In traditional classrooms, students learn in a passive manner while remaining uninvolved or unmotivated. Experiential learning situations leads to cooperation and learning with sharing of direct experiences and solving of real-world problems. The role of instructor in experiential learning is more of a facilitator to guide the progress of the students (Northern Illinois University Center for Innovative Teaching and Learning, 2012). The different phases involved in the same are phases of experiencing (doing themselves), reflection (thinking and pondering upon), and application (use in real-life situations). The stages such as reflection and application make this pedagogy different from the traditional "learning-by doing" and "seeing is believing".

Digital Learning

In the current scenario, the wide penetration of internet and low data costs have led to the proliferation of digital gadgets promoting educational opportunities. In fact, due to the widespread effects of digital learning at the local, national, and international levels, it has evolved into a method of education. The learner's response regarding the use of various digital learning platforms demonstrates that the use of digital learning technology increases student engagement in courses, allows for unrestricted access to learning resources, and significantly improves student performance when adaptive technology is used in digital content. However, the effectiveness of a digital learning platform depends on the nature, evaluation style, and accessibility of internet (Singh, 2017). Global trends after the Covid-19 pandemic suggest that the appeal of digital learning to students is driving ongoing expansion in the provision of online education. In fact, the current contact style classroom teaching programmes are finding it difficult to keep up with this digital access to a variety of learning resources.

Many educational startups are focused on producing Massive Open Online Courses (MOOCs) and e-learning for facilitating learning experiences.

Inquiry Based Learning

Inquiry-based learning is a learning process involving high-level questioning and exploration of real-world connections. This approach is helpful in encouraging experiential learning and problem-solving skills among the students. Under the guidance of the facilitator, the students are involved in identifying various issues faced by them, acquiring knowledge on the same and solving the problems. In this way, the students are able to retain more of the knowledge acquired by them to solve real problems on the ground. The benefits associated with this approach is that it offers more flexibility to the students along with their own constraints and solving it. The approach is helpful for higher studies such as postgraduate study and for in-service training among the teaching staff.

Real World Projects

According toBéchard (2000), pedagogical innovation corresponds to a change that he defines as, "an intentional action that aims to introduce something original into a given context, and it is pedagogical as it seeks to substantially improve student learning in a situation of interaction and interactivity." (p. 3), which he later expands upon with: "In a university context, pedagogical innovations are often described as everything which is not lecturing, the method still used by the overwhelming majority of professors." (Béchard and Pelletier, 2001, p. 133).

Real world projects are an innovative pedagogy in higher education which fosters critical thinking, teamwork, real-world problem solving, and communication skills among students is to use real-world group projects as learning activities. The designing of such projects is challenging since it must be relevant for the course's learning objectives, according to instructors or students. Thus, a teacher/ instructor needs to devote lot of time for effective thinking and designing appropriate activities for the same (Queen's University, 2022). Another responsibility of the instructor is to provide necessary feedback to reinforce the learning objectives and deliverables framed for the project. It is observed that real-world projects have high risk of failures associated with several circumstances which are outside the purview of both the facilitator and the students. As such, suitable strategies are necessary to promote critical thinking and new experiences (University of Manchester, 2010). A few strategies associated with the project are finding projects using peer networks, deciding and evaluation of project scope before project start, framing of short time lines for project phases, formation of small teams for group work, weekly activity status and reports, and framing project management methodologies and deliverables (Leih, 2017).

Story Telling-based Pedagogy

Story-telling is a pedagogical tool in which the instructor/ teacher performs the role of a storyteller to inform, engage and entertain the students. The communication of message is compelling and provocative to retain the interest and motivation of the students (Caldwell, 2012). A teacher can be judged by the degree of experience and motivation he/ she puts in narrating the story to the audience. The stories act as a means to widen the horizon of the students along with enhancing the opportunities to learn and practice. This approach could involve a purposive introduction of complexity to meet the desired goals or oversimplification to make the learners understand according to their situations (Landrum *et al.*, 2019). This method is so effective that it involves the emotions which has the potential to stimulate attention and retained in the memory of the students.

Competency Bases Pedagogy

The competency-based pedagogy provides the students an environment to master a skill at their own level. The method could be modified to meet different learning needs of the students leading to different outcomes (Educase, 2022). This approach is similar to learning based on mastery since the students continue to progress as their competencies are proven. The assessment of competencies is important at different stages to foster effective learning. Technology application provides better access to the content along with providing flexibility to the learners to learn at their own pace.

One of the effective strategies for teachers in the digital age is offering courses through digital means (virtual courses). However, though this reduces the digital divide that the students from rural areas used to face earlier due to a lack of qualified teachers, on the flip side, this is causing a loss of interest among the students in attending the structured classes offered in real-time classrooms. Also, virtual learning is leading to the unstructured gaining of information among the students leading to gaps in understanding the entire content better.

Conclusion

Innovative pedagogy leads to the proactive introduction of novel teaching and learning strategies to enhance learning outcomes and meet the challenges of the digital age. Teachers around the world interacted with digital technology post-Covid-19 pandemic. While technology is crucial in shaping future societies, it is also essential to understand the role of society in framing and utilizing future technologies. Technology alone could not lead to better learning experiences among the clientele. The teachers must be accustomed with new pedagogies so that the complexities in usage can be utilized in favour of learners. The ground is still unexplored for designing effective learner-centred pedagogies for a practical approach to engaging learners in the current world driven by social media.

References

Beatty, I. D., & Gerace, W. J. (2009). Technology-enhanced formative assessment: A research-based pedagogy for teaching science with classroom response technology. Journal of Science Education and Technology, 18(2), 146-162.

Béchard, J.-P. (2000). Apprendre à enseigner au supérieur: L'exemple des innovateurs pédagogiques [Learning to teach in higher education: The example of pedagogical innovators]. Cahier de recherche. OIPG.

Caldwell, K. (2012). Storytelling as a pedagogical tool. Retrieved from http://teachingthroughthearts.blogspot.com/2012/07/storytelling-as-pedagogical-tool.html

Crowther, P. (2013). Understanding the signature pedagogy of the design studio and the opportunities for its technological enhancement. Journal of Learning Design, 6(3), 18-28.

Diaz, A. R. (2013). Developing critical languaculture pedagogies in higher education. In Developing Critical Languaculture Pedagogies in Higher Education. Multilingual Matters.

Educase (2022). Competency-based education (CBE). Retrieved from https://library.educause.edu/topics/teaching-and-learning/competency-based-education-cbe

Heitink, M., Voogt, J., Verplanken, L., van Braak, J., & Fisser, P. (2016). Teachers' professional reasoning about their pedagogical use of technology. Computers & education, 101, 70-83.

Korthagen, F. A., &Kessels, J. P. (1999). Linking theory and practice: Changing the pedagogy of teacher education. Educational researcher, 28(4), 4-17.

Landrum, R. E., Brakke, K., & McCarthy, M. A. (2019). The pedagogical power of storytelling. Scholarship of Teaching and Learning in Psychology, 5(3), 247.

Leih, M. (2017). Using Real-World Team Projects: A Pedagogical Framework. Retrieved from https://pillars.taylor.edu/cgi/viewcontent.cgi?article=1014&context=acms-2017

Northern Illinois University Center for Innovative Teaching and Learning. (2012). Experiential learning. In Instructional guide for university faculty and teaching assistants. Retrieved from https://www.niu.edu/citl/resources/guides/instructional-guide

Queen's University (2022). Inquiry-based learning. Retrieved from https://www.queensu.ca/ctl/resources/instructional-strategies/inquiry-based-learning

Singh, O. (2017). Digital Learning in Higher Education: The Indian Perspective. Retrieved from https://www.thehighereducationreview.com/magazine/digital-learning-in-higher-education-the-indian-perspective-NQHZ619177297.html

Teachonline (2020). A New Pedagogy Is Emerging... and Online Learning Is a Key Contributing Factor. Retrieved from https://teachonline.ca/tools-trends/how-teach-online-student-success/new-pedagogy-emerging-and-online-learning-key-contributing-factor

University of Manchester (2010). What is Enquiry-based Learning (EBL)?. Retrieved from http://www.ceebl.manchester.ac.uk/ebl/

30

Blended Learning: A Pedagogical Reform in Educational Process

Janvi Kochar

Agricultural Extension & Communication, Department of Agricultural Communication, G.B. Pant University of Agriculture & Technology Pantnagar, Uttarakhand

Introduction

Education as a process of development deals with three majors teaching, training and instruction. Its objective being to help students to achieve their highest potential and individual development which they are capable. Education needs to be tailored meaning to ensure that the information is relevant. Attitude, Knowledge and Skill "each one of them, with its inner components, prepares students to acquire and obtain different characters as means to produce a professional" Yang *et al.* (2015). Education is a field which is continuously growing from since long and is still growing through a phase of continuous change. To this fast changing world of higher education we need to adapt and adopt the changing environment to meet the circumstances posed by ICT. The educationists and students in particular have witnessed a tremendous change due to the changes brought in by ICT. ICT is what regarded as the efficient management of information, fostering the development of 21st century. The learner in today's world is so much driven by technology that only traditional method of teaching does not bring satisfaction. Today, ICT is nothing but a blend of both which is yielding and having a good impact on the education scenario. ICT applications are undoubtedly improving the education system in all aspects. Escueta *et al.* (2017) the educational landscape has been indelibly altered over the past two decades by technological innovation. These technologies open the potential doors and build bridges by expanding the access to quality education, facilitate the communication between educators and students. Akkoyunlu and Soylu (2006) states that technologies undoubtedly are developing rapidly and have become one of the indispensable elements of the 21st century. They have influenced educational field too like all other fields.

Educational field is one of the major sub-institution of our social structure. They have developed and offered a lot of methods and tools favouring a favourable environment. In today's world mass media, ICT and online platforms are taking the educational stage. They are used to connect to masses all over the globe. Mass media and ICT is what we are exposed to from long whereas online teaching-learning came through MOOCs and also the covid pandemic. Coman *et al.* (2020) during this period it became an essential part hence, now getting it into our habits. Thus, the scenarios changes and also the perspectives. These internet technologies offer many advantages to learners and teachers.

Role of Educational Reform in Development

According to Johnson (2003), if the technology and computer are used correctly, they have the ability to "invoke dream in the minds of visionary educators who saw endless potential for altering traditional notions of teaching and learning." Technology enhanced learning, elearning, blended learning, digital academe and digital literacy have become a very common place in educational world. The National Education policy (2020) is one of the educational reform which constantly talks about implementing the reforms to bring about development. It talks about technology use and integration, online and digital education, digital infrastructure, digital content, etc. This is the world of ICT whereas, on the other hand there is population who thinks that technology's introduction in education will kill its actual traditional charm and essence of chalk and blackboard. As a solution to this, keeping a balance between both traditional and technological method introducing blended learning can serve the purpose.

The idea of blended learning enables the preservation of traditional form of learning along with handing over the learning to new technologies. This XXI century of "informatization" allows us to integrate different wonderful technologies in teaching/learning. Nazarenko (2015) states that Blended learning is a format which is rapidly spreading worldwide.

Recommendations of the NEP

When we talk about upgrading with time, today's technological upgradation brings blended learning into the picture. Talking about the interventions of the New Education Policy which clearly states that the time has come to recognise that a student is the main stakeholder thus, making efforts in the direction to make system responding to their dreams and aspirations. To take up the policy that is undoubtedly student centric. So, when we say blended learning it gives a major advantage to student i.e. the learning becomes self-paced therefore fulfilling the criteria of student-centric. A clear mention of blended learning in NEP makes it a researchable topic and arouses interest to consider it in ongoing educational process hence, a research study is being conducted to check the awareness towards blended learning in agricultural education.

Getting Beyond E-Learning and Classroom Learning

The emergence of blended learning is seen as a pedagogical concept at the beginning of 2000. Guzer and Caner (2013) talking about past, present and future of blended learning gives the statement that with the recent 10 years development it flourished and filled the gap between traditional and distributed learning environments. The use of the term blended learning is relatively new. Earlier the term hybrid-learning was used quite often. Blended learning is now seen as a high road to enhancing student learning experience and has also become important in higher education.

Graham (2006) defines blended learning as follows: "Blended learning systems combine faceto-face instruction with computer-mediated instruction".

Driscoll (2002) clarifies that blended learning refers to four different concepts:

To combine or mix modes of Web-based technology (e.g., live virtual classroom, self-paced instruction, collaborative learning, streaming video, audio, and text) to accomplish an educational goal.

To combine various pedagogical approaches (e.g., constructivism, behaviourism, cognitivism) to produce an optimal learning outcome with or without instructional technology.

To combine any form of instructional technology (e.g., videotape, CD-ROM, Web-based training, film) with face-to-face instructor-led training.

The Concept of Blended Learning

Blended learning emerges from an understanding combining relative strengths of different teaching approaches. Hence, the concept goes beyond the traditional way and evokes a broad range of flexible design possibilities also the challenge of doing things differently. Blended learning: context in which learning is delivered in both ways online and face-to-face. Blended learning is not just a mix of online and face-to-face mode, it also refers to well-planned and meaningful combination of various activities in both modes. Main focusing factor is learning outcomes and learner-centred instructional environment. The resources such as recordings, videos, audios, articles and podcasts are provided prior to transfer the bulk knowledge from teacher to student. This frees up in-class time for teachers to support the activities like discussions, presentations and facilitate engagement. The University Grants Commission describes features of blended learning as:

- Increased student engagement
- Responsibility towards learning
- Time management
- Flexible learning environment
- Enhances student-teacher interaction
- Improved learning outcomes
- Better opportunities for experiential learning
- Self and continuous learning
- Opportunity for collaboration at a distance
- Technology integration

The digital tools which can support blended learning are described by Saliba, Rankine and Cortez (2013) blogs, discussion boards, live internet, etc. Hence, with these technologies we are dawning the age of anytime, anywhere education.

Conclusion

Moving with technological advancements has become the key element in educational process, digital technologies have taken an efficient place in the era. We just cannot imagine a life driven without technology and devices. When we say blended learning, we combine the learning process with technology which pose a lot of advantages and makes learning possible anywhere, anytime. Thus, Indian education system requires blended learning to be adopted with relatively faster pace.

References

Akkoyunlu, B. and Soylu, M. Y. 2006. A study on students views on blended learning environment. Turkish online journal of distance education, 7(3): 43-56.

Coman, C., Tiru, L. G., Schmitz, L. M., Stanciu, C. and Bularca, M. C. 2020. Online Teaching and Learning in Higher Education during the Coronavirus Pandemic: Students'Perspective. Sustainability, 12(24): 1-24.

Driscoll, M. 2002. Blended learning: Let's get beyond the hype. E-Learning, 1(4), 1–4.

Escueta, M., Quan, V., Nickow, A. J. and Oreopoulos, P. 2017. Education Technology: An Evidence-Based Review. National Bureau of Economic Research.

Graham, C. R. 2006. Blended learning systems: Definition, current trends and future directions. The handbook of blended learning: Global perspectives, local designs, 3–21 San Francisco: Pfeiffer.

Guzer, B. and Caner, H. 2014. The past, present and future of blended learning: an in depth analysis of literature. Procedia - Social and Behavioral Sciences 116, 4596 – 4603.

Johnson, D. and Maddux, C. 2003. Technology in education: A twenty-year retrospective. Computers in the Schools, 20(1/2), 1-186.

Nazarenko, A. 2015. Blended Learning vs Traditional Learning: What Works? (A Case Study Research). Procedia - Social and Behavioral Sciences 200, 77 – 82.

Saliba, G., Rankine, L. and Cortez, H. 2013 . Fundamentals of Blended Learning. University of western Sydney.

Yang, M., You, M. and Chen, F. 2015. Competencies and qualifications for industrial design jobs: implications for design practice, education, and student career guidance. Design Studies, 26, 155- 189.

31

Lac Based Intercropping: Viable Option for Livelihood Security in Rainfed Areas

Kiran Kumari and Nirmal Kumar

KVK Dumka, Birsa Agricultural University, Ranchi, Chhatisgarh
Indian Institute of Natural Resin and Gums (IINRG), Ranchi, Chhatisgarh

Abstract

Integration of different component at farm level is the need of hour to enhance the profitability of agriculture production system at farm level. To achieve these, farmers must develop intensive deliberate management systems designed to yield quality integration of farming components to meet emerging need of livelihood options. Integration of agriculture with forest based intervention appears to be viable option to enhance the profits and productivity of upland. In Jharkhand there is abundance of lac host plant in forest upland which potential still underexploited. Ber an important lac host tree in forest upland which interspaces were not utilized. Field experiment was conducted to find out the suitable intercropping system based on (Ber) lac host tree under rainfed conditions to enhance the productivity of forest upland. The shade loving plants Elephant foot yam and Turmeric along with kusumi lac production on Ber tree were studied for their economic viability and compatibility to the existing farming system. Among the two intercrop of Foot yam and Turmeric in interspaced of lac cultivation on ber tree, Foot yam fetched maximum incremental net economic return of Rs.394200 with B: C Rs.7 .16 followed by Kusumi lac cultivation on ber tree with B:C 3.81. Turmeric in interspaced fetched net incremental economic return of Rs. 51804 ih B: C Rs 1.15. It was also observed that in second year lac yield on ber tree improved as the upland become friable and tilth improved due to tillage operations involved in intercropping. It was also observed that eighty percent lac growers of the opinion to adopt intercropping of Foot yam along with kusumi lac cultivation on Ber. In tribal villages where forest upland is abundant and Lac is a traditional enterprise cultivation of Elephant foot yam in interspaced of ber orchard would bring forest upland under sustainable production

Keywords: lac cultivation, intercropping, sustainable livelihood Security

Introduction

Enhancing productivity and profitability are two major factors of agricultural growth for the future. Over dependence on monsoon and limited scope for area expansion made agriculture development is critical to generate employment and additional income for the rural poor. Major part of agricultural population is small and marginal farmers who are economically vulnerable and poor. The consequence of inclement weather on crop production affects ultimately livelihoods of farmers. In the changing scenario improvements in farmers' productivity and profitability, diversification of agriculture have been acknowledged to enhance profit, generate additional employment for rural masses and to conserve the natural resources.

Integration of different component at farm level is the need of hour to enhance the profitability of agriculture production system. Jharkhand state a resource rich state wheremonocroping in rainfed agriculture with small undulated upland is in practice. Rural economy is mixed type where forest produce particularly lac play important role. To enhance rural economy, integration of agriculture with forest based intervention appears to be viable option to enhance profits and productivity of upland which is almost sixty percent of total cultivable land. Even though moisture is available during Rabi free grazing made imperative cultivation of second crop. Continuous aberrant weather demand the inclusion of forest resources to be exploited in diversified and integrated manner as forest interventions require little water.

Diversification and intensification of rice-based system to increase productivity per unit resource is very pertinent. Crop diversification showa lot of promises in alleviating these problems besides, fulfilling basic needs for cereals, pulses, oilseeds and vegetables and, regulating farm income, withstanding weather aberrations, controlling price fluctuation, ensuring balanced food supply, conserving natural resources, reducing the chemical fertilizer and pesticide loads, ensuring environmental safety and creating employment opportunity (Gill and Ahlawat, 2006). Crop diversification has been recognized as an effective strategy for achieving the objectives of food security, nutrition security, income growth, poverty alleviation, employment generation, judicious use of land and water resources, sustainable agricultural development and environmental improvement (Hegde *et al.*, 2003). The crop diversification may enhance profitability, reduce pests, spread out labour more uniformly, reduce risks from aberrant weather by different planting and harvesting times and source of high value products from new crops (Reddy and Suresh, 2009). In the era of shrinking resource base of land, water and energy, resource use efficiency is an important aspect for considering the suitability of a cropping system (Yadav, 2002). Hence, selection of component crops needs to be suitably planned to harvest the synergism among them towards efficient utilization of resource base and to increase overall productivity (Anderson, 2005).

Lac has been a traditional source of livelihood for thousands of tribal families living in the forest fringes. These families primarily depend on agriculture for their livelihood, which is insufficient to provide them with food security and round-the-year income. Poor families living in forests and forest fringe villages have historically suffered from challenges like unfertile or barren land, low asset base, lack of irrigation facilities, small landholding, poor linkages with the market and their own low risk-bearing capacity, forcing them to migrate in search of work. In villages there abundant of ber orchard and interspaces between trees not properly utilized. Although Ber orchard is poorly maintained at farmers field Lac cultivation on Ber tree which is abundantly available within the periphery of their habitat provides appreciable income. The advantage of ber tree is, it host both species of lac i.e. Rangini and Kusumi. So lac cutivationalongwith vegetable in interspaces would be available option to ensure year round income. In the light of above said background Foot Yam and Turmeric being shade loving crop selected to experiment as intercrop in interspaces of Ber with the objective of Profit Maximization of upland which otherwise unutilized and poorly managed.

Methodologies

The experiments were carried out in purposively selected villages (Bandup and Kadla) in Nimdihblock of Saraikela-Kharsawan district in Jharkhand. Fifty respondents were selected purposively on whose field On Farm Trial on Ber Tree based intercropping of Foot Yam and Turmeric was conducted. Ber orchard were traditional and being managed by farmers themselves in upland area. Only intercropping of Foot Yam and Turmeric were intervened in interspaced of ber tree on which rangini lac cultivation were demonstrated. Approximately 800 ber pants were planted in one hectare of upland at 6 m apart.

Foot yam was planted at 2 m apart and turmeric at 40 cm apart was intercropped in interspaces. 1.5 kg kusumi brood lac per tree were inoculated during July and harvested in January as brood lac. Foot Yam and Turmeric were planted in June and harvested during November.

The data were collected by personally interviewing the respondents through a structured scheduled. Apart from the use of schedule, detailed information were collected through informal discussion with the respondents and by critically scrutinizing the practices followed for vegetable production. Use of PRA tool, field observation and non-participant observation techniques were thoroughly used. Field observation was done periodically throughout the growth stage of lac. "Focused Group discussion" of PRA was also followed to gather data on the opinion of respondent's towards the demonstrated technology.

After collection, the data were systematically arranged and tabulated for analysis and interpretation. The statistical techniques used for analysis of data under the study included mean yield, net return, B: C, increase in family income in Rs./year

For economic analysis, economic evaluation data of crops were used. The gross cost of cultivation was calculated on the basis of different operations performed and materials used for raising the crops.

Conceptual Background

Lac is the hardened resin, secreted by the tiny lac insect belonging to a coccid group. The widely known Indian Lac insect is Lacciferlacca. Lac insect settles on the twigs of certain host trees, suck the plant sap and grow, all the while secreting lac resin from their bodies. Since the insects are closely spaced on the twigs, the resin forms continuous encrustations over the twigs of the host trees.

Basically two strain of lac insects are cultivated grown in the district i.e. rangeeni on Butea monosperma(palash),andZizyphusmauritiana (*ber*)kusmi on *ber* and Schlericheraoleosa(*kusum*) tree. So ber tree play important role in production of brood lac as it can play as alternate host for rangini as well as kusmi. Both palash and ber tree are available in plenty and around bund of upland and their economic use had been stopped since many years. Also gravel upland rice undergo climatic stress every years so economic utilization of these appears to be a viable option to integrate the resource to cope up climatic stress also.

Recent studies carried out by Indian Institute of Natural Resin and Gums (IINRG), Ranchi revealed that the income form lac cultivation is about 28% of their total agricultural income. And most of the lac growing families are among the poorest of the poor in the state. The cultivation is done extensively in the remote interior pockets of the state. Further, it was also reported that despite presence of large number of host trees, lac cultivation is not carried only on a few host trees.

Cultivation of lac is simple and investment is very low. It is eminently suited to the farmers living in the vicinity of the forests including women as it demands only their part time attention. In the district, its cultivation provides an important additional income next only to the agriculture. Farmers are also dependent on lac cultivation for their livelihood and Lac is regarded as an important source of cash flow to the marginal, small and large farmers in the district.

Results and Discussion

1. Economics of intercropping system: the economics of interventions carried out were calculated for each economic component of intercrop. It was observed that although the net return (Rs.1980870 per ha) was high in inoculation of ber tree with Kusumilac but highest benefit was obtained over per rupee cost (B:C ; Rs.7.16) in intercropping of Foot yam with ber followed by intercropping of turmeric respectively as Rs. 3.81 and Rs.1.15 Yied (in q/ ha) of economic component of the intervention were recorded as 224, 147, 81, and 48 respectively of elephant foot yam, *kusumi* lac, turmeric and ber fruit respectively.

Although Net Return obtained was maximum in *ber*tree inoculated with *kusumi*lac but due to higher cost incurred in brood lac the component cost incurred was also maximum. As quality brood lac is not available locally in market also its cost is high during inoculation time it plays a critical role in production of *kusumi* lac on *ber.* But as the alternate host of lac insect is available abundant in the area farmers started managing brood lac for inoculation on their own. They started practicing the lac production in coop system of inoculation also they started partial harvesting and using *kusum* tree as alternate host for *kusumi* lac produced on *ber* tree. So the cost of cultivation in consecutive years started decreasing as farmers were able to inoculate the tree with their own brood lac.

Table 1: Economics of intercropping system

Intercropping system	**Economic component**	**Mean Yield (q/ha)**	**Component Cost (Rs/ha)**	**Return from economic component (Rs/ha)**	**Net return of intervention (Rs/ha)**	**B:C**
Ber	Fruit	48.00	56,000	96,000	40,000	0.72
*Kusmi*lac on *ber* tree	Broodlac	147.11	5,20,000	2,50,0870	19,80,870	3.81
ber + Elephant foot Yam	Elephant foot Yam	224.60	55,000	4,49,200	3,94,200	7.16
ber + Turmeric	Turmeric	80.67	45,000	96,804	51,804	1.15

Adoption of Intercropping System

Efforts were made to find out the perception of ac growers on intervened technology for the its application and continuous use in field. It was found that seventy eight percent lac growers opined they will continue the cultivation of foot yam with ber tree. They reported that cultivation of foot yam not need much time and labour. Also it not requires seed tuber every year. They will keep the seed tubr for 2-3 years. The climatic condition also was found to be suitable to cultivation of Foot yam. Cultivating the foot yam improve the tilth of soinl of upland which otherwise was abandoned. It decreses the mortlity of insect gradually. By adopting coop system of inoculation they were able to keep Broood lc on their own tree. It decreased the cost of cultivation of kusumi lac in consecutive year. Out of three intervened technology intercropping of turmeric with ber tree having very less adoption i.e. only 8 percent followed by intercropping of foot yam with ber tree (14%). The profitability of the technology was main concern of lac grower to dopt the technogy and the other aspects were availability of Brood lac at locl level.

Table 2: Inter Cropping system adopted by farmers

Intervention	No. of family adopted the intervention (N=200)	Per cent family adopted the intervention
Kusumi Lac on Ber	28	14
Ber + Elephant foot Yam	156	78
Ber + Turmeric	16	8

Conclusion

Based on findings of this experiment it can be concluded that under conditions of Jharkhand (rainfed monocropping), intercropping of Foot yam with Ber tree was more productive, sustainable, resource-use efficient, and remunerative. Since there was abundant of lac host tree and most of them were on upland which otherwise was non productive, if these resource made productive through this intervention it will change the economic situation of villages. Thousand of tree can be added in rural economy. It will not only add the additional income to farmers pocket but also bring greenery to the village as felling of tree will be checked. It will directly influence the farming ecology and farmers can cope up with aberrant weather. It will ba effective strategy for achieving the objectives of food security, income growth, poverty alleviation, employment generation, judicious use of resources, sustainable agricultural development and environmental improvement.

A climatic resilient combination of crop and forest product will to harvest the synergism among them towards efficient utilization of resource base and to increase overall productivity of upland, throughagro enterpriseconvergence revolves round theinteractiveuse and efficient utilization of land,labour,capital cum available resources. Economic analysisofthe intervention concludesthatlac cultivation in upland on ber host tree which otherwise was abandoned can be advocated for the income security and sustainable livelihood support of small and marginal farmers. Low risk and less labour requirement also make it women friendly. It will raise village economy at house hold level.

References

Aggarwal GC, Sidhu AS, Sekhon NK, Sandhu KS, Sur HS. 1995. Puddling and N management effects on crop response in a rice±wheat cropping systems. Soil Till. Res. 36, 129±139. Ali, M., 1999. Evaluation of green manure technology in tropical lowland rice systems. Field Crops Res. 61, 61–78.

Anderson, RI. 2005. Are some crops synergistic to following crops? Agronomy Journal 97(1): 7-10.

Anonymous. 2009. Agricultural statistics at a glance. Published by Directorate of Economics and Statistics, Department of Agriculture and Cooperation, Ministry of Agriculture, Govt. of India.

Francis CA. 1989. Biological efficiencies in multiple cropping systems. Advances in Agronomy 42:1-36.

Gill MS and Ahlawat IPS. 2006. Crop diversification - its role towardsGill MS and Ahlawat IPS. 2006. Crop diversification - its role towards

Hegde DM, Prakash Tiwari S and Rai M. 2003. Crop diversification in Indian Agriculture. Agricultural Situation in India. August, 2003. pp. 351-354

Hegde DM. 1992. Cropping System Research- Highlights. Project Directorate for CroppingvSystem Research, Modipuram, Meerut, Utter Pradesh. pp. 40.

Kumar Alok, Tripathi HP, Yadav RA and Yadav, SR 2008. Diversification of rice (Oryza sativa) - wheat (Triticum aestivum) cropping system for sustainable production in eastern Uttar Pradesh. Indian Journal of agronomy 53(1):8-21.

Mishra MM, Nanda SS, Mohanty M, Pradhan KC and Mishra SS. 2007.Crop diversification under rice based cropping system in western Orissa. In: Extended summaries 3

Reddy BN and Suresh G. 2009. Crop diversification with oilseed crops for- maximizing productivity, profitability and resource conservation. IndianJournal of Agronomy 54(2): 206-214.

Sharma AK, Thakur NP, Koushal Sanjay, Kachroo Dileep. 2007. Profitable and energy efficient rice-based cropping system under subtropical irrigated conditions of Jammu. In: Extended summaries

Tomar SS and Tiwari AS. 1990. Production potential and economics of different cropping sequences. Indian Journal of Agronomy 35(1, 2): 30-35

Yadav JSP. 2002. Agricultural resource management in India: the challenges. Journal of Agricultural Water Management 1(1):61-69.

32

Restructuring School Curriculum as an Instrument of Entrepreneurship Development: Innovation in the Implementation of NEP 2020

Mishra. Ishita, Vatta. Lalita, Gautam. Anamika and Gupta Roshani

[1]*Department of Home Sciences, MMV, BHU, Varanasi, Uttar Pradesh*

Abstract

The National Education policy 2020 came as pursuit and wisdom .It is all set to change Academic system of India, more focused on steering the education more towards practical and holistic learning. This is concept paper, it will be focusing on connecting national education policy 2020 with the entrepreneurial education model from foundation to middle curricular structure mentioned in NEP 2020.The concept paper is talking about introducing entrepreneurship in school through national education policy that will focus on imparting real-world skill which must be taught to students, it will help the students to live phenomenal lives by thinking innovatively and out-of-the-box not only the students from science, commerce stream but the students from humanities, arts and music background will also learn how to cater real life problems by applying creative thinking skills. Future belongs to incredible innovators and leaders in constantly changing world so it is important for the education system to efficiently adapt to such changes and impart knowledge which would give solution to problem and help in their holistic development on which the paper is primarily focused. Entrepreneurship is an idea and it gives birth to innovation and innovation leads to development. Entrepreneurship will be acting as bridge between holistic development and the national education policy 2020 because education is the driving force for future development. Students and youth should be firmly determined to work for the development of our country in this era of Amrit kaal, Youth today is frontier who can bring revolutionary change. The paper will primarily focus on education of child from the pedagogy level about entrepreneurial education inculcate values and vision about various domains within a child at early stage and make him/ her well prepared for the coming future demand and vast exposure, students should be an enterprise to sustain, national education policy is going to help in changing mindset for innovation in order to sustain long. Create mindset which will generate ideas which are relevant in today's world. Entrepreneurship education has innate capabilities to transform the competitive world into a remunerative place to survive and work.

Keywords: Entrepreneurship, NEP, entrepreneurial skill, Innovation, Self-reliant, holistic development.

Introduction

After so many decades, the new education policy comes in its new form in 21st century. This policy seeks to offer everyone access to an equal, equitable, and fair education. The NEP 2020 represents a significant advancement in the field of education in India, with an emphasis on primary through higher education. It is based on the idea that education must foster the development of not only higher-order cognitive abilities-such as critical thinking and problem-solving skills-but also social, ethical, and emotional capacities and dispositions. These higher-order cognitive abilities include both the foundational abilities of literacy and numeracy and higher-order cognitive abilities. It includes-recognizing, identifying, and fostering each student's individual abilities by educating parents and teachers to support a student's overall growth in both academic and extracurricular areas, by the 3rd grade, all children will possess fundamental skills in literacy and numeracy, it will provide students multidisciplinary and holistic education to ensure the coherence and integrity of all knowledge. The policy manifested 22 guiding concept, some of them are conceptual comprehensive, flexibility and holistic development. These ideas will serve as the foundation for the whole Indian education system (Shamsu, 2021). The New Education Policy 2020 already incorporates vocational training for students' holistic development, A study demonstrates a positive relationship between vocational choice of the subject and variables of personality, traits, parents' involvement, achievements, motivation, and occupation information, (Kaur,1992)but including entrepreneurship in this policy would expose our students to more opportunities because entrepreneurship is the process of taking a risk to launch a new business with the intention of making a profit. Entrepreneurship and entrepreneurial culture are core components to build socially inclusive and participatory economies in highly competitive global world. The paper highlights about development of institutes inclusively promoting entrepreneurship and entrepreneurial culture contributing to pump economic condition of the nation resulting spur growth in wealth of the Nation, (Kant, 2012). It is the capacity and willingness to design, develop, and manage an enterprise while embracing risks and uncertainties in order to turn a profit. Entrepreneurship in economics is the interdependence of land, labor, capital, and natural resources to produce profit. Entrepreneurship is a method of thinking. Study reveals that entrepreneurship positively influences entrepreneurial self-efficacy (ESE) and influences overall task engaged in the venture creation but not significantly (Taneja, 2022). It is about having an obsession with opportunity, a comprehensive approach, and a combination of leadership. It all comes down to viewing challenges as chances to transform the world. Teaching and learning of entrepreneurship education should inculcate knowledge, experience, and action-oriented, learning activities together in one curriculum (Rajput, 2020). Entrepreneurship is the profession that people choose for themselves and steer in the direction they desire. Making the world a better place is the main goal. It has a significant value in the nation development few of them are mentioned below:

The current educational system is grade-oriented, and knowledge is becoming certificate oriented rather than talent-oriented. Instead of making students mark-seekers, the educational ecosystem should develop talent and technology driven experts. Only an exam cannot predict student's future, so they must be given the chance to select a vocation that suits their interests and preferences (Kumari, 2022). The policy seeks to indigenize Indian education. The rich, ancient, and timeless knowledge found in India, the home of spiritual knowledge, serves as the light of policy's compass. Not merely knowledge acquisition, but also preparing students for life in the real world, was the goal of education in ancient India. Education was not limited to the classroom but its major motive was realizing and liberating oneself (Shukla *et al.*, 2022).

Entrepreneurship Education Through NEP-2020 is Bridge to Bring Change

Entrepreneurship aims to make world better place for living. It brings to the world with ideas, products, and innovative businesses. A study address about the major role of business schools in promoting entrepreneurship in India. The paper manifest about loophole in education system of these business schools which may help the students to polish trait and flourish like entrepreneurs (Supriya, 2018). Some products are created which solves burning problem which may bring worldwide change. For example, we can see difference between incandescent light and LED bulbs whereas incandescent bulb had used 50watt energy and LED lights take only 11-12 watts energy. It is revolutionary step in the field of innovation.

Entrepreneurship Promotes Research & Development: Entrepreneur nurture idea and shape them to new form. It strives to discover new idea, new concept and refine the existing ones. Research is a vital tool for one's own growth and improvement, leads to create new enthusiastic mindset for innovation and self-development.

Creates Job: Entrepreneurship is the process of providing job, creating job for man force. New business to grow eventually creates job and promotes sustainable livelihood example- LijjatPapad, the women organization startup started with unit of seven and now to the number of 43,000 sisters resembling "the strength of a woman".

Economic Development: The entire process of entrepreneurship involves procedure like use of innovative idea, maximum use of available resources, identification of strength, development of management skill and improving the standard of living possession of money all these factors contribute in the economic development of country.

Entrepreneurs – Asset for Nation

Entrepreneur is an individual who plays major role in bringing revolution to our society, so it is sensible to cultivate, promote, motivate, and remunerate this greatest asset of country to greatest extent possible. Entrepreneur is an individual who creates new businesses, using skill and ideas, bearing risk and enjoying reward. It is the main source of energy in the entire process of entrepreneurship. They nurture ideas and give them shape, channelize idea and turn into successful business plan. If these budding entrepreneurs can learn entrepreneurship in their academic institutes especially at early years of education. They can do wonders.

School is the most impactful tool for promoting right value, building character belief and attitude, school education can be rightly used to ignite, revamp and radiate attributes of entrepreneurship amongst teenagers. School needs good infrastructure readily available, teacher training and assembling of resources which may be cost effective but will bring tremendous advantages to society. This is one of the best way possible to lead to the epitome utilization of nation human resource for nation building, the study has been done to highlight importance and feasibility of entrepreneurship education at secondary level (Agrawal, 1995).

Development of future entrepreneurs requires not only individuals who are willing to start their entrepreneur but a strong ecosystem that provides them opportunities to groom. Policy makers, academic institutes along with government altogether needs to coordinated their efforts to design intervention around the ecosystem. The entrepreneurial ecosystem face challenges of commoditization. Government of India is providing policy support in form of initiatives like NPSD, NSDC, NEP 2020 etc. to the youth in order to benefitted for establishing their units.

Government Policies, Interventions and Initiatives in the Field of Education and Skill Development

National policy on skill development and entrepreneurship (NPSD) has its commitment for the development of humans as resource. The objective of NPSD 2015 is to speed up and maintain skilling among the human resource of Nation. It is an umbrella framework for all skilling activities being carried out in the Nation. Vision of the policy is "for the citizen of country to create an ecosystem which helps in empowerment by skilling on large scale with quality and high standard, promote entrepreneurship, culture of innovation which ensures wealth and employment leading to sustainable livelihood of citizen. As per NPSD 109.73 millions additional skilled manpower will be required across 24 key sectors by 2022. Skilling of new entrant labors over 2022(both male and female) are 402.87, the other great initiative is the Skill India Mission, under the Ministry of Skill Development. It is on driving seat for implementation of skill development programmes through establishing training centres across India. Till now training centres under this initiative established are near about 717 Pradhan Mantri Kaushal Kendra, adding to the network 14,711 ITI training centres and 309 training centres. The government of India has**"the Vision 2025"**, statement of the Ministry of Skill Development and Entrepreneurship, it adopts an ecosystem for development of nation by enabling lens progressing India to a high skill equilibrium, for positive improvement of individual enterprise and economy. Other mission of government of India is National Skill Development Mission (NSDM), which was launch on July15, 2015 on the World Youth Skill Day by honorable Prime Minister of India. It has provided a strong institutional framework both at the centre and state for strong implementation of skilling activities in the country. There are seven sub mission proposed to achieve objectives of mission, they are Institutional training, Infrastructure, Convergence, Trainers, Overseas Employment, Sustainable livelihood, leveraging public Infrastructure. The next level initiative is National Skill Development Corporation (NSDC) by the government of India, The National Council for Vocational Education and Training (NCVET) dated December 2018, it is overreaching regulatory body for regulating, functioning for vocational education and training entities. Primary function of NCVET includes: Recognition, Ensuring, Discipline, Re-cognizing and regulation of Awarding bodies (ABs), Assessment of agencies (AAs), Skill related Information providers (SIPs). All the above initiatives and mission launched are acting as powerhouse to make India skilled nation, working as powerhouse towards making India self-reliant.

National Education Policy 2020- A Ray of Hope for Entrepreneurship in School Education

The National Education Policy 2020 is a revolutionary step towards transforming India that is BHARAT. This policy is on a mission to provide high quality education to all, moving towards making India global knowledge superpower, "VISHWAGURU". It aims that curriculum structure and pedagogy institutions must develop consciousness toward Nation, respect for Fundamental Duties and Constitutional value, the policy pledges to develop sense of responsibility, commitment towards human right, global well-being.

Why Entrepreneurship is an Important Part of Education System

Introduction of Entrepreneurship Education in Elementary and HighSchool across country with the help of NEP 2020 will help potential value movement in Education system, it will changestudent's motivation and willingness to engage in studies and explore other area of interest. It will for sure have a long-term effect on student potential to perform in various aspects of life. School education will help students to innovate and imagine in different ways most importantly, entrepreneurship as a subject in school education is strongly going to produce army of student ready to face VUCA world – volatile, uncertain complex and ambiguous, it will help students to deal impeccably with VUCA world. Entrepreneurship education will teach children differently, it will provide them tools to be more resilient, it will encourage them to try, play, fail and find opportunity for self-experiment midst of all chaos.

The school education should be focused on teaching pure entrepreneurship as part of new curriculum through NEP2020 because it will produce generation of entrepreneurial thinkers, it will democratize new mindset with dynamic young people who will have different lens on the world. With the rapidly changing time, it will be the new young dynamic mind which will find ways pivoting to find new avenues of growth and on unlocking value in different environment, it should be must for every student. It will help young generation going after their dream probably underrated in school. Mistakes which will be made will help students to be more humble, more disciplined for pursuing dreams to become better and grow putting aside self-doubt. Entrepreneurship as part of curriculum is going to nurture competencies among students like a sound mindset, opportunity recognition, assessment, problem-solving attitude, action orientation, adaptability, resilience.Students learn how to imagine and innovate, mental agility. It is long term process, so entrepreneurship education needs to enter the mainstream schooling through new curriculum being introduced by NEP 2020.

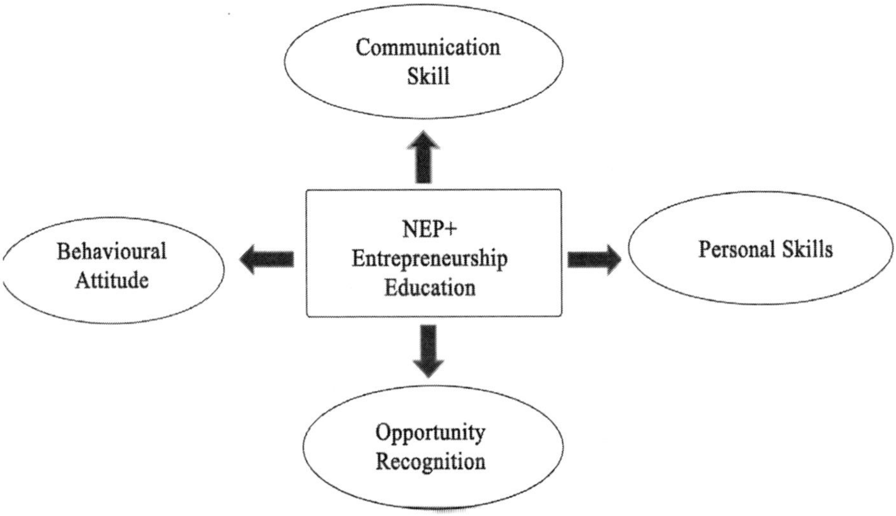

Fig. 1: NEP and Entrepreneurship education resulting in change in behavior

Entrepreneurship is considered as a long-term process not just a mere phenomenon involves individual's behavior and skill as mandatory element. The outcome of entrepreneurship education will be fruitful when the right curriculum is taught to the right student through right method through right teacher in right setting for precisely right duration. Entrepreneurship education in school will be great equalizer, it needs to focus on nurturing student to have hunger for exploration for problem solving and opportunity finding rather than being mental straight jacket. It will help students become technology friendly, help in decision making, to be able to adapt constant changes. Darwin in his theory said that in the long history of humankind those who are able to collaborate and improvise most effectively prevailed. Basically, here Darwin defined about the skills of surviving species right including humans. Today the Entrepreneurship education will do the same to make the upcoming generation well prepared for constant changes the world throws at them. Essence of entrepreneurship education should be teaching students what they can do with what they learn and apply it on projects and assignments they truly care about like Transform problem into opportunity, channel energy and brush up skills students are having. It will help them to perform better, be highly motivated, have high self-esteem.

Entrepreneurship Education in School Interlinked with NEP 2020

Reducing curriculum content to enhance essential learning and critical thinking, there will be more room for critical thinking, more holistic, inquiry based, discovery based, discussion based and analytic based teaching and learning and teaching will be in more interactive manner, question answer session will be encouraged. The entrepreneurial concept is going to work in spectacular way in teaching and learning process. It is surely going to bring revolution in field of education. Opening doors for various career opportunities, creating fresh environment to work in various other field, fresh resolutions, fresh vision and freedom. It will help students to know about their capabilities, strength, weakness, positives and negatives. Because of the revolutionary National Education Policy 2020 the educational institutions are in their transformation phase. Our country is ready to welcome fresh minds, fresh environment ready for revolution. Entrepreneurship education in school will solely support creativity, innovation and collaboration of idea.

Model Educidation about How to Teach Tntrepreneurship to Students at the Foundation and Middle School Levels

Foundational Level

At the foundation level major focus is from age 6 to 8 years, when the child reaches in grade 1 and 2 , introduce Small chapters in books to teach various qualities integrated with entrepreneurial mindset and skills. It will help students to overcome fear. small business model plans through activity should be introduced in which the students would perform assigned character which will initially help them to form set of behavior. This will help them to develop soft skills which include communication skill, flexibility/adaptability, and self-discipline, team work.

Entrepreneurial Skill Development for Kids For early birds, lesson plan through stories and poems through teachers, it will help students identify their strengths through the lessons of variety of entrepreneurial skills. The lesson plan should be structured and taught by 'teacher' and skills learned should be - Risk taking, creativity, courage, believe in self, curiosity, wanting to learn more.

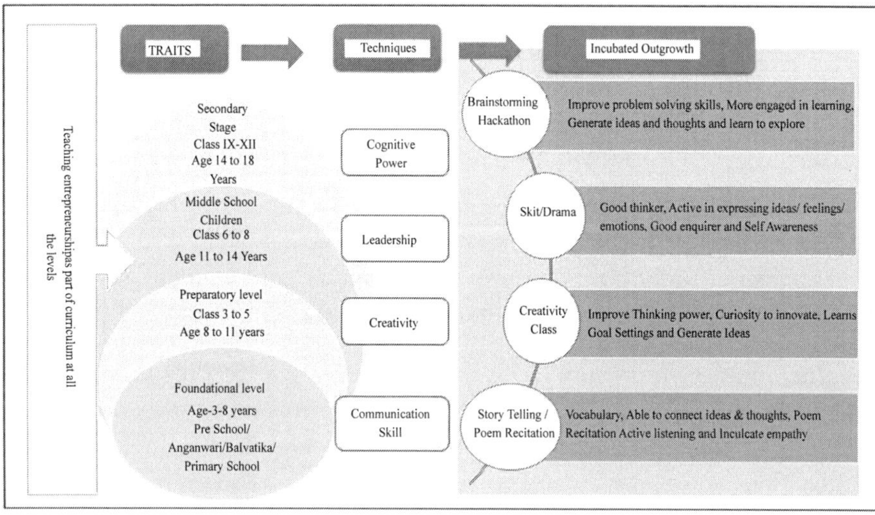

Fig. 2: Model for Introduction of Entrepreneurship in School education: Implementation of NEP 2020

Preparatory and Middle Level

Introduce it as subject solely dedicated to Entrepreneurship which will be more activity based, voluntary activities to perform entrepreneurial activities in school. Project and assignment based learning, navigate future opportunities through interactive talks. Classroom Discussion should be turned in public speaking platform by integrating ignite talk on pop up debates this will help students to build positive pressure for public speaking, regular seminar. Promotion of culture building to bring out positives and negatives, to get a strong hold on roots. To promote student to attend Hackathon and have clarity of thoughts. Promotes student for content writing, publishing about success and failure stories of local entrepreneurs through articles, to change mindset. Encourage student to become 'Innovation ambassador'. Brainstorming session with leaders, entrepreneurs, businessman and learning various skills from them.

Development of Skills Through Various Plans & Activities

For middle school program model should be prepared into various categories which will precisely teach one enriched skill. Model that will be develop should consist of certain cognitive exercises, inspirational video of successful entrepreneur. Articles published of various successful entrepreneur failure stories should be included too. Task for students which will consist various question set which will help student in Brainstorming Hacathon. Students should be encouraged to make project that will safely promote and encourage them for entrepreneurship. Model design should be prepared after pilot study of school students, Edupreneurs, Sociopreneurs and entrepreneur should be associated with the development task of module. It should be structured according to age and grade specific of the student.

Problem/need/desire, solution to problem, model to solve problem, resources used – time, money, knowledge, personnel, skills to perform various task and face challenges, turning difficulty in opportunity, multiple intelligence focus, social values – use of social sciences, meet up with entrepreneur.

Conclusion and Summary

It will encourage children to take risk, encourage them what they want to do and not be afraid of failure. Be future ready and be best prepared for future that is being created in rapidly changing world, students who are tomorrow's adult need to survive and hopefully thrive within.

The development of Entrepreneurial mindset today will be very significant because it opens doors for economic development, personality development, and strong mindset for future problem, innovative thinking. Entrepreneurship education in school when taught integrated effectively to listen, networking, hustle, negotiate, to be a good speaker, to reduce stage fear and be the best version of one. Entrepreneurship education will teach curiosity, it will teach every single part of to be a student to be curious and ask question, encourage imagination that they can dream big and achieve it, it teaches creativity.

Entrepreneurship education should be immediately taught in school as soon as the child reaches in middle to kick start journey to the higher level of studies because there are some core skills which are required to be future ready. School education system should create an ecosystem which should be perfect for teaching entrepreneurship, for teaching valuable skills and creating own mindset. To teach the students way to identify opportunities and way to take risks.

The best way to execute this work is to use NEP 2020 as bridge through which entrepreneurship education will be introduce in school. School will help to learn useful skills like idea generation, pitching, networking, to perform multiple tasks, ready to tackle with issue, engage with the world out of comfort zone. School will encourage them to start working on student's own idea as early as possible which will be extremely valuable, student through this process will came to know about their interest. Students will get to learn a lot and have a swift experience of their own. Students will realize about their potential talent. Introduction of entrepreneurship through NEP will surely ensure more future problem-solving individual dramatically it will help to create more employment, a burning issue which our country is suffering from, it will teach the students to be persistent which will give them higher chance of success.

References

Agrawal, V. K, (1995). Entrepreneurship development through secondary education A case study of CBSE Delhi. S (Ferreira, 2022) Shodhganga, http://hd1.handle.net/10603/99756

Goutam, S. (2018). Role of B schools in Promotion Entrepreneurship in India, A study of select Institutions. Shodhganga, http://hd1.handle.net/10603/270255

Kant, S. (2012). Role of entrepreneurship development institutes in promoting entreneurial culture in India. Shodhganga, http://hd1.handle.net/10603/132534

Kaur, M. (1992). Vocational choices of high school students of Chandigarh as related to personality traits achievement motivation occupational information and parental involvement. Shodhganga, http://hd1.handle.net/10603/81116

Kumari, D. (2022). Roadmap for holistic implementation of the new education policy2020. NEP2020: A way towards build in new India. world lab publication

Ministry of Skill Developmentand Entrepreneurship, National policy on skill development and entrepreneurship (NPSD), Annual Report (2021-2022), page no. 13

Ministry of Skill Developmentand Entrepreneurship, National Skill Development Corporation, Annual Report (2021-2022), Page no. 25

Ministry of Skill Developmentand Entrepreneurship, National Skill Development Mission https://www.msde.gov.in/en/reports-documents/policies/NSDM

Ministry of Skill Developmentand Entrepreneurship, Skill India Mission, Annual Report (2021-2022) page no. 14

Ministry of Skill Developmentand Entrepreneurship, Vision 2025, Annual Report (2021-2022), page no. 15

National Education Policy (2020), Ministry of Human Resource Development Government of India, https://www.education.gov.in/sites/upload_files/mhrd/files/NEP_Final_English_0.pdf, page no. 5

Rajput, V. (2020). Role of Pedagogical Practices in Entrepreneurship Education for New Venture Creation in India. http://hd1.handle.net/10603/311756

Shamsu, S. (2021). Holistic implementation of the National Education Policy 2020-a roadmap. Holistic implementation of the National Education Policy 2020-a roadmap (researchgate.net)

Shukla *et al.*,(2022). Demystifying Approaches of Holistic and Multidisciplinary Education for Diverse Career Opportunities: NEP 2020. Indian Journal of Science and Technology.

Taneja, N. (2022). Influence of Entrepreneurship Education on Entrepreneurial Self Efficacy ESE of the participants. Shodhganga. http://hd1.handle.net/10603/372149

33

A Review
Handicraft as Vocal for Local to Reduce Crop Residue Burning and Its Effects

Neha Agarwal, Madhulika Gautam

Dayalbagh Educational Institute (Deemed University)
Dayalbagh, Agra, Uttar Pradesh

Abstract

India is the 2nd largest populated country and when the population increased its demand for food also increased. At present, the production of the crop is larger, which is why crop residue is also generated in an extended way by cultivation. A huge amount of residue is left without having proper knowledge and management facilities farmers have no option left except to burn it. The existing study centred on the determination of the impact of crop residue burning and finding sustainable ways to use crop residue for income generation. The objective of this study is to present an updated review on the impact of the burning of crop residue on the environment and health and review the possible solution for sustainable use of residue mainly wheat residue and maize husk to reduce the burning residue issues in India. To fulfil the objective of this study, relevant studies were identified from the search engines Science Direct, Pub Med, Scopus, and ResearchGate using relevant keywords crop residue burning, the impact of residue burning on health, the impact of residue burning on the environment, sustainable use of residue, possible income generation activities by use of residue in rural areas, etc. After full-text screening and the use of inclusion criteria, relevant studies were used. With the help of selected studies, it is found that residue burning is doing an additional job of increasing air pollution with vehicles, chimneys, industries, etc. Crop residues release many gases by burning them and these gases are contributing to global warming. Crop residue burning also emits SOx, NOx, NH3, and volatile organic compounds (VOCs). It causes respiratory problems and cardiovascular disease in human health. Cornhusk fibres, wheat straws, rice straws, and other crop residues can be used to make different useful products like carry bags, dresses, etc. by the weaving process and by other techniques. It has a wider scope in international markets. Paper can be produced by residue. These solutions will help local rural people but for this proper knowledge and training is important for them. With this, they can help themselves and this can be successful with the help of extension workers.

Keywords: Environment, Handicraft, Health, Residue burning, Rural employment, Vocal for local, Wheat straw and maize husk

Introduction

India is an agriculture-based country and it contributes most to the economy of the country. Different types of crops are cultivated in a wide range in different ecological regions based on different seasons. So, crop residue is also generated in an extended way by cultivation (Devi et. al 2017). Crop residue is the left-over part of the crop after harvesting. In India, crop residue generates around 600 Mt and from which cereals are 70%, fibres 13%, oilseeds 6%, pulses 3%, sugarcane 2%, and others 6% (Jain *et al.*, 2014). Corn production has increased more than 15 times from 1950-51 to 2016-2017. Rice, wheat, millet, barley, and maize are important cereal crops in Uttar Pradesh (Roy and Ahmad, 2015). Traditionally, Crop residue was used as cattle feed, animal bedding, cooking fuel, organic manure, etc. and some residues were used in making handicraft products by rural women. but still, a huge amount of residue is left without management and farmers have no option left except to burn it. About 90 to 140 Mt of crop residue burning in the field or households as fuels (Jain *et al.*, 2014). In the process of harvesting and threshing, residues are left behind in the field. Farmers need to clear their fields early for the planting of the next crop so they do have not enough time and it is tough to clear the field at that time. Therefore, the farmers opt for the burning of residue in harvested fields due to a lack of access to user-friendly, cost- and time-effective options. The burning of crop residue creates many issues and effects on the environment and health badly. In a report of NAAS in 2017, it was estimated that in November 2016 at the period of rice residue burning, the NASA satellite images showed the intensity of rice residue burning in Punjab, Haryana, Western Uttar Pradesh, and Uttarakhand was very high. It is doing an additional job of increasing air pollution with vehicles, chimneys, industries, etc. Crop residues release many gases by burning them and these gases are contributing to global warming. It was noted that burning one-tonne rice residue released 13 kg particulate matter, 60 kg CO, 1460 kg CO_2, 3.5 kg NOx, and 0.2 kg SO_2 (NAAS, 2017). Crop residue burning also emits SOx, NOx, NH3, and volatile organic compounds (VOCs) (Jain *et al.*, 2014) which are precursors for the formation of particulates in the atmosphere. It causes respiratory problems and cardiovascular disease in human health. It decreases the nutrient value also which is presented in soil (Bimbraw, 2019). Therefore, at present crop residue burning is a major problem and it needed some strong measurements to solve this problem effectively. It needs the effective implementation of sustainable management practices with government interventions and policies. The Indian government took many steps against residue burning. The National Green Tribunal (NGT) established under the National Green Tribunal Act in 2015 banned the burning of residues in Uttar Pradesh, Rajasthan, Haryana, and Punjab and penalties foist on those farmers who break this rule (Datta et. al 2020) and (NAFCC,2018). The use of satellite-based imagery is effective for monitoring crop residue burning, particularly in Punjab, Haryana, and Rajasthan. (Datta et. al 2020). The residue is treated as waste and burns in the field but in actuality, the residues of the crops are also useful for economic generation. It is useful in mushroom cultivation, composting, electricity generation, etc. (Yadav et. al 2018). The present study used a review-based approach to find the local solution by utilization of wheat straw and maize husk so that rural people can use these strategies and make more income and reduce the residue-burning situation.

Methodology

A combinatory review of the literature was carried out before September 2022. The relevant studies were found by using different academic electronic databases such as Science Direct, PubMed, ResearchGate, and websites for relevant data. National and international studies were included in this paper. The comprehensive search was completed based on the keywords of this paper. The papers were reviewed and retrieved by using two steps. The first step was the title and abstract screening and the second was full paper screening and by use of inclusion and exclusion criteria.

Inclusion/ exclusion criteria: Original papers were included for the present paper. Originally observed and intervention-based papers were used to fulfil the objective of this paper. Publications are used only in the English language and qualitative research papers. Referred publications were taken which were fulfilling the need of the present study. No restriction was made for the year/ date of publication.

On the other hand, non-referred publications, duplicate papers, reviews, commentaries, and short communications were excluded and other articles type like conference abstracts, editorials, unofficial reports, and letters to editors were excluded. Furthermore, literature that failed to report the possible income generation solution by the utilization of crop residue. A flowchart of screening and selection of studies is presented in fig 1

Identification	• 414 total citations identified through database searches
	• Science Direct: (n=254)
	• Pub Med: (n= 93)
	• ResearchGate: (n=67)
Screening	• 414 Records were screened and from them 361 records excluded
	• Duplicate: (n= 107)
	• Abstract review: (n=254)
Eligibility	• A total of 53 articles were fully assessed for eligibility
	• 36 articles were excluded after full-text screening
Included	• Articles passed in full-text screening and inclusion criteria (n= 17)
	• Science Direct: (n=8)
	• Pub Med: (n= 6)
	• ResearchGate: (n=3)

Impact of Residue Burning on Human Life

Burning residue created a very crucial situation during the pandemic for patients who were suffering from respiratory diseases because the smoke affects the throat and lungs (Indian Express, 2021). Wheat straws release particulate matter and Green House Gases after burning. More other gases are also released by burning and it contributes to forming black clouds (Fatima *et al.*, 2020; Ravindra *et al.*, 2019)which will exert pressure on crop production and likely increase the agricultural crop residue. The present study estimates atmospheric emissions of various pollutants from crop residue burning using the Intergovernmental Panel on Climate Change guidelines. In India 488 Mt of total crop residue was generated during 2017, and about 24% of it was burnt in agricultural fields. This resulted in emissions of 824 Gg of Particulate Matter (PM2.5. Pollution is caused respiratory problems, heart and lungs problems, eye irritation, skin irritation, blood problems, etc. (Kumar & Singh, 2021) burning of residues also affects soil fertility Loss of nutrients, reduces worms and microbes, etc.

Residue in Making Paper

Corn husk and wheat straws can be used in making paper by using some methods. Previous studies also proved the possibility to complete this task. Fiscal & Dandan (2016) experimented with the combination of corn husk and snake plants for making paper and they found acceptable responses. 75% of corn husk and 25% of snake plant as raw material gave the quality making paper. Fagbemi et. al (2014) also explained that blending of pulps from agricultural wastes such as corn husk and pulps from other non-wood materials such as kenaf contributes to the improvement of the paper strength properties and has the potential of producing paper of differing qualities and can find use in a wide variety of applications.

It is possible to prepare wheat straw pulp after using various processes. The bleached wheat straw pulp can be characterized by good strength properties, reasonable brightness, good printability, etc(Fang & Shen, 2018)i.e., chemical components and anatomy of wheat straw, pulping process and pulping properties, paper and paperboard products from wheat straw pulps, as well as environmental protection issues. Wheat straw is a kind of annual vascular-bundled herbal arthrophyte containing cellulose fibers that are accept- able for pulping and papermaking. The chemical components and anatomy of wheat straw were discussed; practically, soda or soda-AQ chemical pulping processes are com- mon techniques often applied for chemical pulping from wheat straw, conventional and advanced bleaching sequences were introduced, and especially ECF and TCF bleaching techniques have been successfully applied in China's paper mills. Cooking and black liquor extraction equipment and facilities are explained; chemi-mechanical pulping properties of wheat straw, corrugated medium, and linerboard products from wheat straw CMP pulps are evaluated; and chemical recovery from chemical cooking black liquor and effluent treatment processes are discussed. In this chapter, not only labora- tory research results but also some commercial operation experiences are shared. These information and knowledge described in this chapter will help readers to have a good understanding about wheat straw pulping and papermaking; they are useful for pulping and papermaking engineer as reference for design and operation management of wheat straw pulping lines. Proper pulping, cleaning, and screening, pulps from it can be used for manufacturing various writing and printing papers such as light-weighted printing paper, letterpress printing paper, typing paper, writing paper, tissue and sanitary papers, and almost all grades of paper and paperboard grades (Pandita *et al.*, 2015). Wheat straw pulp furnishes must be mixed with some portion of bleached wood pulps or bamboo pulps occasionally (Fang & Shen, 2018)i.e., chemical components and anatomy of wheat straw, pulping process and pulping properties, paper and paperboard products from wheat straw pulps, as well as environmental protection issues. Wheat straw is a kind of annual vascular-bundled herbal arthrophyte containing cellulose fibers that are accept- able for pulping and papermaking. The chemical components and anatomy of wheat straw were discussed; practically, soda or soda-AQ chemical pulping processes are com- mon techniques often applied for chemical pulping from wheat straw, conventional and advanced bleaching sequences were introduced, and especially ECF and TCF bleaching techniques have been successfully applied in China's paper mills. Cooking and black liquor extraction equipment and facilities are explained; chemi-mechanical pulping properties of wheat straw, corrugated medium, and linerboard products from wheat straw CMP pulps are evaluated; and chemical recovery from chemical cooking black liquor and effluent treatment processes are discussed. In this chapter, not only labora- tory research results but also some commercial operation experiences are shared. These information and knowledge described in this chapter will help readers to have a good understanding about wheat straw pulping and papermaking; they are useful for pulping and papermaking engineer as reference for design and operation management of wheat straw pulping lines. (Fang & Shen, 2018. Rural youth can involve in this type of activity so that income can be generated and residue can be used in an eco-friendly way (Devi *et al.*, 2018; Verma, 2014).

Corn Husk and Wheat Straws can be Weaved in Making Useful Products

Hasdiana & Ayuddin, 2017 experimented and explained the process of weaving corn husks and prepared a handbag successfully. More useful and attractive products like mats, coasters, etc. can be developed and rural women can learn them easily. Colourful raw materials can be converted into beautiful products so that women can show their interest to participate in this income-generating activity. Gani, 2020 published an article and mentioned the corn husk weaving project in Sikkim in which more than 20 women were taking the profits earned by making weaved products.

Wheat straws have lustrous, shiny, smooth, flexible, and yellowish in colour(Liu *et al.*, 2013; Rani & Brar, 2020). These can be moved in a cylindrical shape and these straws have woven kinds of abilities so that different products can be developed (Liu et.al, 2013) and it can help in a start-up a micro-enterprise for rural women as the local solution (Kumar *et al.*, 2019). Innovative products can be developed ground on the trends. These environment-friendly products made with wheat residue have much economic value and it is comfortable to use in each type of weather (Yan *et al.*, 2020)

Wheat straws for Making Paintings

The shiny and smooth texture of wheat straw is proved to be advantageous amount because the straws can be used in making paintings by using the cut and pasting method. Attractive paintings can be developed and also the activity of which all ages of people can involve in this craft (Liu, 2016). So, it can be proved as a beautiful and innovative micro-enterprise for all types of aged people (Liu *et al.*, 2013).

Corn Husk can be Used in Textiles

Corn is a renewable resource which used to make corn fibre much more sustainable than regular fibre (www.2wtextile.com). Reddy & Yang (2005) blended the fibres with cotton in various proportions and processed them on the ring and rotor spinning machines and concluded that fibres extracted from cornhusk were suitable for high-value textile applications and had better processability than natural fibres from other agricultural by-products. Sisodia and Parmar (2014) evaluated the response of dyeing corn using disperse dyes and found that the colour strength of the fibres had increased with increased temperature but for the strength, it was important to maintain the temperature. So, it is a chance for rural people who can participate and can be trained in this field.

Sell Wheat Residue Straight for Economic Benefits

Wheat residue gives lots of benefits and opportunities to everyone for income generation. This residue gives a chance to start own micro-enterprise basis on the interest. So, farmers also have an opportunity to deal with these entrepreneurs directly to sell residue in the form of raw material. They need some awareness knowledge and patience to crack a good deal. Threshing the entire crop is not the right way in direction of selling crop residue. Chopped residue cannot be used in all activities so farmers must know all the benefits which can be by wheat residue.

Conclusion

All studies have shown that there is no need to burn residue it can be beneficial as a micro-enterprise. Awareness of farmers regarding modern techniques and technologies and methods through the use of residues can be an outstanding response. Burning straws is the easiest method to save labour costs and prepare the field for the next crop in a short period. This burning solution creates a critical situation that has to deal with every creature and causes many serious health issues and environmental problems. The results suggested some major environmental-friendly solutions which can help in reducing the environmental issues which are created because of residue burning. Corn husk can be used for a different purpose by various processes. Research has suggested that creatively all types of aged rural people can involve in these activities by which rural development can be possible. For this development, government initiatives are very important. It is mandatory to provide proper knowledge and awareness about the benefits of utilization of residue, extension education should be involved for the farmers and other rural people who can participate in these activities. Training should be provided to the participants for that extension workers should be involved because they can easily teach in rural areas by involving Gram Panchayat through various methods like speeches, group discussions, presentations, demonstrations, etc.

References

Bimbraw, A. S. (2019). Generation and Impact of Crop Residue and its Management. Current Agriculture Research Journal.7(3), 5–10. www.agriculturejournal.org

Datta, A., Emmanuel, M. A., Ram N. K. and Dhingra, S. (2020). Crop Residue Management: Solution To Achieve Better Air Quality. New Delhi: TERI

Devi, M.V., Devarani, L., Singh, R. J., Hemochandra,L. (2018) Rice Straw Management in Manipur: Challenges and Finding a Way Forward. Indian Journal of Hill Farming. Identity, 30(2), 181–185. http://www.kiran.nic.in/pdf/IJHF/Vol_30-2/1.pdf

Devi, S., Gupta, C., Jat, S. L., & Parmar, M. S. (2017). Crop residue recycling for economic and environmental sustainability : The case of India. Open Agriculture. 2017; Published by De Gruyter Open. 2: 486–494

Fagbemi, O. D., Fagbemigun, T. K., & Igwe, C. C. (2014). Strength Properties of Paper from Pulp Blend of Kenaf Bark and Corn Husk : A Strength Properties of Paper from Pulp Blend of Kenaf Bark and Corn Husk : A Preliminary Study. British Journal of Applied Science & Technology 4(28): 4124-4129, 2014. 10.9734/BJAST/2014/10745

Fang, G., & Shen, K. (2018). Wheat Straw Pulping for Paper and Paperboard Production. Global Wheat Production. https://doi.org/10.5772/intechopen.77274

Fatima, N., Chandramauli, A. Chaddha. N. (2020). Analyzing Disposal Methods of Crop Residue (Parali). International Journal of Innovative Science and Research Technology. Vol. 5, Iss. 5, May – 2020. 2456-2165

Fiscal, R. & Dandan, K. B. V. (2016). Development and Evaluation of Paper from Corn Husks (Zea mays L .) and Snake Plant Fibers (Sansevieria zeylanica). International Journal of Science and Research (IJSR).10.21275/v5i8.3081601

Hasdiana & Ayuddin. (2018). Quality Improvement of Corn Husk as Raw Material for Textile Products. Advances in Social Science, Education and Humanities Research (ASSEHR), 1st International Conference on Social, Applied Science and Technology in Home Economics (ICONHOMECS 2017). vol. 112

Jain N., Bhatia A., Pathak H. (2014) Emission of air pollutants from crop residue burning in India. Aerosol and Air Quality Research, 2014, 14(1), 422-30

Kumar, P., & Singh, R. K. (2021). Selection of sustainable solutions for crop residue burning: an environmental issue in northwestern states of India. In Environment, Development and Sustainability (Vol. 23, Issue 3). Springer Netherlands. https://doi.org/10.1007/s10668-020-00741-x

Kumar, S., Sharma, D. K., Singh, D. R., Biswas, H., Praveen, K. V., & Sharma, V. (2019). Estimating loss of ecosystem services due to paddy straw burning in North-west India. International Journal of Agricultural Sustainability, 17(2), 146–157. https://doi.org/10.1080/14735903.2019.1581474

Liu, B. (2016). The Research on Culture Inheritance and Development of Straw Art in Northeast China. Advances in Social Science, Education and Humanities Research (ASSEHR), 2016 International Seminar on Education, Innovation and Economic Management (SEIEM 2016) 75(Seiem), 55–58.

Liu, B., Wang, F., Liu, F., & Sun, J. (2013). Study of wheat straw weaving process and decorative effect. Advanced Materials Research, 605–607, 179–182. https://doi.org/10.4028/www.scientific.net/AMR.605-607.179

Pandita, S., Kaula, B., & Passey, S. (2015). Use of Weeds and Agro-Based Raw Materials and their Blends for Handmade Paper Making. DU Journal of Undergraduate Research and Innovation, 1(3), 169–179. http://journals.du.ac.in/ugresearch/pdf-vol3/U17.pdf

Rani, L., & Brar, K. (2020). Physical and tactile properties of woven textile textures from plant and agro-waste materials. The Pharma Innovation Journal 2020; SP-9(4): 104-1099(4), 104–109.2277- 7695; 2349-8242

Ravindra, K., Singh, T., & Mor, S. (2019). Emissions of air pollutants from primary crop residue burning in India and their mitigation strategies for cleaner emissions. Journal of Cleaner Production, 208, 261–273. https://doi.org/10.1016/j.jclepro.2018.10.031

Reddy, N., and Yang, Y. (2005). Properties and potential applications of natural cellulose fibers from cornhusks. Green Chemistry, 7(4), 190-195

Roy, R., Ahmad, H. (2015). State Agricultural Profile of Uttar Pradesh (2014-2015)

Sisodia, N., and Parmar, M. S. (2014). Dyeing behavior and fastness properties of corn (pla) fiber. Journal of Polymer and Textile Engineering. Volume 1(2), 01-07.

Verma, S. S. (2014). Technologies for Stubble Use.Journal of Agriculture and Life Sciences. Vol. 1, No. 2; December 2014 1(2), 106–110. 2375-4214, 2375-4222

Yadav, S., Koli, P., Mina, U. (2017). Crop Residue Burning and Air Pollution. Popular Kheti. Vol. 5, Iss. 2, 2321-0001 https://www.researchgate.net/publication/322750756

Yan, X. C., Wang, X. L., & Zhao, D. L. (2020). Research on 3R design of straw woven living products based on green design concept. E3S Web of Conferences, 179. https://doi.org/10.1051/e3sconf/202017901023

Reports

NAFCC, 2018. Climate Resilience Building among Farmers through Crop Residue Management (For Punjab, Haryana, UP & Rajasthan State). Compiled by NABARD

NPMCR. Available online: http://agricoop.nic.in/sites/default/files/NPMCR_1.pdf. Retrieved on Dec. 20, 2020

Websites

https://indianexpress.com/article/cities/chandigarh/a-silent-contribution-by-farmers-in-covid-fight-no-stubble-burning-6386895/. Written by Anju Agnihotri Chaba Updated: April 30, 2020. Retrieved on March 17, 2022

https://specialtyproduce.com/produce/Corn_Husk_473.php Retrieved on June 2022

https://thenewsmill.com/corn-husk-weaving-supports-destitute-women-in-sikkim-with-sustainability/. Published by Gani, A.in 2020. retrieved on March, 2022

https://www.hindustantimes.com/cities/others/wheat-stubble-burning-starts-in-malwa-experts-warn-covid-scenario-could-worsen-101619552012424.htmlBy HT CorrespondentPublished on Apr 28, 2021. Retrieved on March 17, 2022

Huge spike in Punjab stubble burning cases after wheat harvest https://timesofindia.indiatimes.com/city/chandigarh/huge-spike-in-punjab-stubble-burning-cases-after-wheat-harvest/articleshow/82441945.cmsNeel Kamal / TNN / May 7, 2021, 04. Retrieved on March 17, 2022

www.2wtextile.com Retrieved on March, 2022

34

Women Empowerment Through Self Help Groups in West District of Tripura, North-East India

Daya Ram, M.K. Singh, A. Prasad and M. Deepa Devi

Department of Agricultural Extension, College of Agriculture Central Agricultural University, Imphal, Manipur

Abstract

Self Help Group is a voluntary and self managed group of women who come together to promote savings among themselves and pool savings for activities benefiting either individuals or communities economically. Many Government and non-government organization now have come forward for helping poor communities and to energize their capacity in development activities of which Self Help Group (SHGs) is thought as the best means for improving their position in the society. Women's Self Help Group (SHGs) is emerging as promising and best suited tools in their context. In the Self Help Group (SHGs), usually ten to twenty members having homogeneous socio-economic status, same occupation and same type of needs from a nearly permanent group and use to save some small amount of money (Rs 10-100/-) on regular basis. The present study was conducted in the West Tripura District using the multi stage sampling procedure research design. Total 120 Self Help Group members were selected on the basis of complete enumeration method for the present study. The data were collected during July to August, 2010 with the aid of structured interview schedule. The collected data were then analyze using appropriate statistical tools namely, frequency, percentage, mean, standard deviation simple correlation coefficient and multiple regressions. The study disclosed that the SHGs working in Melaghar block had successfully implemented their planning at the local level, with proper guidance from the Government of Tripura of the West Tripura District. Majority of the SHGs (95%) were having interaction followed by good category (5%) and there were no SHGs members found in the bad category. The contribution of the respondents on the basis of their group motivation show that majority of the SHGs (75%) belonged to very good category were followed by (25%) bad categories and there was no SHGs members found in the good category. Majority (62.50%) of the respondents were in very good category having interpersonal communication as which show their main empowerment. About (25.84%) respondents were categorized under bad category whereas (11.66%) respondents were found under good category. Out of twelve (12) variables, six variables are significant namely age, education, occupation, annual income and socio-economic status and others six variables are non-significant namely size of family, mass media exposure, contact with instructor (SHG), knowledge test and social status. The study also reveals that the average monthly income of the groups due to their empowerment has been enhanced. They have developed abundant self-confidence and self esteem through this SHGs movement. It may conclude that there is a need of sound and sustainable strategy to improve the empowerment of women self help groups.

Keywords: *Women Empowerment, Self help Group and Interaction Group and Variable.*

Introduction

Self Help Group is a voluntary and self managed group of poor people, who come together to promote savings among themselves and pool savings for activities benefiting either individuals or communities economically. Empowerment of women has now become a key issue in the Government's Five year plans by organizing women into Self Help Groups to make the beginning of a major process of empowering women. 320 million people, out of our 1.1 billion, are steeped in abject poverty and due to their asset lessens it is difficult for them to access any kind of aid from banking / financial institutions and this state of the hopelessness and the constraints of an ineffective credit delivery system can be transcended by enhancing the self esteem and raising the feeling of human self worth. The SHGs approach is a reshaping initiative that enhances the livelihoods and ensures sustainability by enabling poor families to access effective economical and social development leading to empowerment.

In Tripura SHG's has been introduced in the year 1999.The main objectives of this programme by mutual beneficial of social and economic activities through women mobilization. SHG's are a way to involve rural women and men actively in the development process by means of increased income Agricultural situation of WTD differs to some extent from the other districts in regard to topography, soil, extent of rain-fall, variation in temperature and humidity which has resulted in difference in cropping pattern also from that of the other three districts. People mainly practice rice and vegetables cultivation. The plain land people also were satisfied to grow single crop preferably paddy without having any knowledge and experience of improved cultivation for growing multiple crops. Animal husbandry is another sector contributing to the economy of WTD. Cattle and poultry are the main livestock wealth of the district. (DRDA office, Agartala, 2010). Many Government and non-government organization now have come forward for helping poor communities and to energize their capacity in development activities of which Self Help Group (SHGs) is thought as the best means for improving their position in the society. Women's Self Help Group (SHGs) is emerging as promising and best suited tools in their context. In the Self Help Group (SHGs), usually ten to twenty members having homogeneous socio-economic status, same occupation and same type of needs from a nearly permanent group and use to save some small amount of money (Rs 10-100/-) on regular basis. They elect one president from their group, who maintains regular meeting of the group is conducted once in a month and they get registration and open an account in any nationalized bank/Regional Rural Bank (RRB). Monthly saving is regularly deposited in the bank. After a year or more when bank satisfies with their regular savings habit, eight to ten times of money of their deposited amount is given as loan. On the basis of decision taken by the member of group, this money is given to the members to establish their own business. Thus either the members of the group can start their needed business individually or in group activities (NABARD, 2000).

The rationale behind the programme was that women's income is known to have positive correlation with the nutritional and educational status of the family and in the building up of a positive attitude towards status of women. Hence, the necessities of ensuring more income are earning avenues for women. (Prof. Mohammed Yunus, 2002).

Self Help Groups are a way to involve rural women and men actively in the development process by means of increased income and resource controls within the household benefits like improved management and technical skills of women, mutual support to enable to manage own affairs and to address constraints. Self Help Groups have the potential of enhancing poor women's access to available opportunities to income enhancement and self reliance. The agenda of SHGs is 'Empowerment' and shift from dormant masses of the rural women to vibrant masses by bringing more income in the hands and control of women so that they can lead a better social life. SHGs aim to improve the quality of life of women by self-reliance, up grading their skills and ensuring sustainability of groups. Self Help Groups are institution for credit delivery to the poor. They are now recognized throughout the country. The movement is supported by Government (an important component of the Swarna Jyanati Gramin Swarozgar Yojana, Rastriya Mahila Kosh) and Non-Governmental organizations. These groups develop and enhance the decision making capacity of the members and instill in them the strength and confidence for solving their problems. They serve as a ideal mechanism for bringing women out of their homes, making them more articulate and developing their leadership qualities and their skills as motivators. SHGs build capacity for prospective entrepreneurs. SHGs move to more transformation agendas, such as finding innovative ways of improving women's situation economically, challenging social inequality, improving women's voice in the public sphere etc. (Singh and Singh, 2005). Keeping these in view, a study was conducted to Women Empowerment of Self Help Groups in West Tripura of Tripura.

Research Methodology

The present study was conducted in the West Tripura District using the multi stage sampling procedure research design. A multi-stage sampling procedure was employed in obtaining the sample for the study. There are 4 districts in Tripura state, one district was randomly selected namely, West Tripura District for the present study. Out of 16 blocks, one block was randomly selected for present study, namely Melagarh (MLG) block (976 number of SHGs are functioning under DRDA). Out of 46 village panchayats, 2 village panchayats were selected randomly, namely Kashchowmani (KC) and Jumerdepha(JD). Out of 255 members of SHGs, 55 members of SHGs were randomly selected under JDP and out of 300 members of SHGs, 65 members of SHGs were randomly selected under KC village panchayats. Total 120 Self Help Group members were selected on the basis of complete enumeration method for the present study. The data were collected during July to August, 2010 with the aid of structured interview schedule. The collected data were then analyze using appropriate statistical tools namely, frequency, percentage, mean, standard deviation simple correlation coefficient and multiple regressions.

Results and Discussion

Table 1: Distribution of selected areas for Empowerment of Women SHGs in Different practices/ working.

N=120

Sl.no.	Selected areas for Empowerment of Women Self Help Groups	Categories	Frequency	Percentages
1.	Group interaction	Very good	114	95.00
		Good	06	5.00
		Bad	Nil	Nil
2.	Group co-operation	Very good	96	80.00
		Good	21	17.50
		Bad	03	2.50
3.	Interpersonal trust	Very good	31	25.83
		Good	61	50.83
		Bad	28	23.34

Sl.no.	Selected areas for Empowerment of Women Self Help Groups	Categories	Frequency	Percentages
4.	Group decision making	Very good Good Bad	77 7 36	64.16 5.84 30.00
5.	Group motivation	Very good Good Bad	90 Nil 30	75.00 Nil 25.00
6.	Interpersonal communication	Very good Good Bad	75 14 31	62.50 11.66 25.84
7.	Group cohesiveness	Very good Good Bad	47 55 18	39.16 45.84 15.00
8.	Manageable group size	Very good Good Bad	29 57 34	24.16 47.50 28.34
9.	Group goal achievement	Very good Good Bad	86 34 Nil	71.66 28.34 Nil
10.	Participation in group activities	Very good Good Bad	23 56 41	19.16 46.64 34.20
11.	Need satisfaction	Very good Good Bad	24 95 01	20.00 79.16 00.84
12	Interpersonal linking	Very good Good Bad	34 70 16	28.33 58.34 13.33
13.	Interdependence of members	Very good Good Bad	01 72 47	00.84 60.00 39.16
14.	Group competition	Very good Good Bad	08 102 10	6.66 8.34 85.00
15.	Group leadership	Very good Good Bad	20 73 27	16.66 60.84 22.50
16.	Conflict management	Very good Good Bad	Nil 118 02	Nil 98.34 01.66

Note: Figures in parentheses show percentage in total.

Group Interaction

The contribution of the respondents on the basis of their group interaction is shown in Table-1 which reveals that majority of the SHGs (95%) were having interaction followed by good category (5%) and there were no SHGs members found in the bad category.

Group Co-operation

The table-1 shows that group co-operation was found very good category (80%) which was followed by good category (17.50%) and bad category (2.50%) respectively.

Interpersonal Trust

It is observed from the Table-1 that majority of the respondent's belonged to (50.83%) very good category followed by good category and very few per cent were in bad categories (22.34%).

Group Decision Making

The same table reveals that about 64.16 per cent of the respondents were under very good category followed by good category (5.84%) and bad category (30.0%) respectively.

Group Motivation

The contribution of the respondents on the basis of their group motivation show that majority of the SHGs (75%) belonged to very good category were followed by (25%) bad categories and there was no SHGs members found in the good category respectively.

Interpersonal Communication

It is also seen that majority (62.50%) of the respondents were in very good category having interpersonal communication as which show their main empowerment .About (25.84%) respondents were categorized under bad category whereas (11.66%) respondent were found under good category respectively.

Group cohesiveness

On further analysis, it is found that 45.84 per cent of the respondents were categorized under bad category followed by very good category (39.16%) and good category (15%) respectively.

Manageable Group Size

As per as "Manageable group size" is concerned, it is seen that majority of the SHGs (47.50 % fell under bad category were followed by very good category (24.16%) and good category (28.34%) respectively.

Group goals Achievement

It is observed that majority (71.66%) of the respondents belonged to very good category 28.34 per cent to good category (28.34%) and there were no SHGs members found in the bad category.

Participation in Group Activities

Table-1 also shows that group activities of SHGs were found under bad category (46.64%) which was followed by very good category (19.16%) and good category (34.20%) respectively.

Need Satisfaction

After assessing "Need Satisfaction" of the Group members as evidenced by Table -1 show that majority of the respondents 79.16 per cent were under good category followed by very good category (20.00 %) and bad category (0.84) respectively.

Interpersonal Linking

After being analyzed the "Interpersonal linking" it was found that about 58.34 per cent of the respondents were under good category followed by very good category (28.33%) and bad category (13.33%) respectively.

Interdependence of Members

Table-1 also shows that good category was adhered by 60% per cent of the respondents which was followed by 39.16 per cent members under bad category and only 0.84 per cent by very good category respectively.

Group Competition

It was observed that majority of the respondents i.e., 85.00 per cent were under good category 6.66 per cent were under very good category and 8.34 per cent under bad category.

Group Leadership

The majority of the respondents (60.84%) belonged to very good category 16.66 per cent belonged to good category and (22.50%) belonged to bad category.

Conflict Management

Data presented in the Table concerned reveals that highest i.e., 98.34 per cent were found having "good conflict management" ability and 1.66 per cent had "bad conflict management" ability whereas, nobody was found under "very good category".

Table 2: Correlation coefficient between independent variables and Empowerment of Women SHGs (dependent variables)

Sl.No.	Independent Variables	Correlation coefficient (r)
X_1	Age	0.327**
X_2	Education	0.532**
X_3	Type of family	0.258**
X_4	Size of family	-0.162 NS
X_5	Primary occupation	0.269**
X_6	Annual income	0.248**.
X_7	Mass media exposure	-0.053 NS
X_8	Contact with concerned instructor/SHGs	-0.110 NS
X_9	Social participation	0.106 NS
X_{10}	Socio – economic status	0.394**
X_{11}	Knowledge test	-0.023 NS
X_{12}	Social status	0.020 NS

** Significant at the 0.01 level * Significant at the 0.05 level and NS = Non Significant

The results on correlation coefficient between socio-economic characteristics and empowerment of women SHGs are presented in the Table 2. It is observed that the age, education, type of family, primary occupation, annual income and socio-economic status had positive and significant correlation with empowerment of women SHGs. These findings are in conformity with results of Kala (2004), Mohamed(2004), Vadivoo and Sekar (2004) who observed that age, education; type of family, primary occupation, annual income and socio-economic status had positive and significant correlation with empowerment of women SHGs. Where, size of family, mass media exposure, contact with concerned instructor/SHGs, social participation, knowledge test and social status were showing positive but no significant correlation with empowerment of women SHGs. Similar findings were also reported by Mandal (2005), Sarma (2000) andRajendrakumar C. and Kaptan, S. S. (2004).

Table 3: Regression analysis of independent variables and dependent variable (Empowerment of members of SHGs)

Sl. No.	Independent variables	Regression co-efficient 'b' value	't' value
X_1	Age	0.209	2.17*
X_2	Education	0.175	1.744NS
X_3	Type of family	-0.305	-2.630*
X_4	Size of family	0.014	0.133NS
X_5	Primary occupation	0.034	0.398NS
X_6	Annual income	0.296	2.830*
X_7	Mass media exposure	-0.055	-0.562NS
X_8	Contact with concerned instructor	-0.230	-1.924NS
X_9	Social participation	-0.005	-0.068NS
X_{10}	Socio – economic status	0.276	2.282*
X_{11}	Knowledge test	0.106	0.806NS
X_{12}	Social status	-0.212	-2.015*

** Significant at the 0.01 level R^2= 0.519 F = 9.637

* Significant at the 0.05 level R=0.721

NS = Non Significant Degree of freedom=119

The above table-3 reveals that the twelve (12) independent variables taken together explain to the extend 51.90 per cent of variation in the empowerment of women of SHGs. The R^2 value (0.519) suggests that 51.9 per cent was contributed towards the change in empowerment of women SHGs due to independent variables. The F-value (9.637) was also found 0.05 per cent level of probability at 12 and 119 degree of freedom. Thus, the results explained that all the independent variables would account for a significant amount of variation in the empowerment of women SHGs. From the above observation, "t" test of significance indicates that the co-efficient of regression (b-value) was significant for the five variables, age (x_1), family type ($x_{3)}$, annual income (x_6), socio- economic status (x_{10})and social status (x_{12}) .This means that variables were found most important in predicting the empowerment of women SHGs. The regression co-efficient were not significant for the variables viz, education (x_2), family size (x_4), occupation (x_5), mass media exposure (x_7), contact with concerned instructor (x_8), social participation (x_9) and knowledge test (x_{11}).

Conclusion

The study disclosed that the SHGs working in Melaghar block had successfully implemented their planning at the local level, with proper guidance from the Government of Tripura of the West Tripura District. Out of twelve (12) variables, six variables are significant namely age, education, occupation, annual income and socio-economic status and others six variables are non-significant namely size of family, mass media exposure, contact with instructor (SHG), knowledge test and social status. The study also reveals that the average monthly income of the groups due to their empowerment has been enhanced. They have developed abundant self-confidence and self esteem through this SHGs movement. It may conclude that there is a need of sound and sustainable strategy to improve the empowerment of women self help groups.

References

DRDA (2010). District rural development agency (DRDA), Annual report, Agartala West Tripura District.

Kala (2004). A hand book on forming Self Help Groups. Regional office, Guwahati, pp, 11-22.

Mandal, A. (2005).Swarna Jayanti Gram Swarozgar Yojna, SHG's.Kurukshetra,53 (3) :4-6.

Mohamed,S. M. (2004).Self Help Groups for the success of women entrepreneurs. kissan world, 31(12),30-31.

NABARD (2005).A handbook on farming self help groups. Regional office, Guwahati, pp.11-12

Rajendrakumar, C. and Kaptan, S. S. (2004). SHGs strategic partner (SAP) to rural market. Kisan World.28 (10):24.

Sarma, S. (2000).Microfinance in rural development. A study of the impact of SHG's –Bank linkage programme in Assam. Project Report of NABARD, pp, 14-15.

Singh P. & Singh B. (2005). Self Help Group for Empowerment: Concept, Best Practices and Impact Assessment.Division of Agricultural Extension, IARI, New Delhi.

Vadivoo, S. K. and Sekar, B. (2004).Self Help Groups:-A moment for women empowerment. Kisan World,31 (7): 13-14

Yunus, Muhammad, (2002). Grameen Bank ii: designed to open new possibilities. At http:// www.Grameen info: organization/ bank/ bank2.html.

35

Impact of NEP 2020 on Higher Education: A Theoretical Analysis

Jyoti*[1] *and Avadhesh Singh2

[1]*Department of Sociology, Faculty of Social Science, Banaras Hindu University, Uttar Pradesh*
[2]*Faculty of Commerce, Banaras Hindu University, Varanasi, Uttar Pradesh*

Abstract

Education is a crucial prerequisite for the movement of a country. Education is important to meet the essential human potential, for each country, to make an equivalent and fair society, and to lead us towards a sensible answer for genuine issues. Since Independence, different commissions and strategies have been set up for the development and improvement of the instructive area. NEP 2020 is a creative and cutting edge proposition with both positive and negative perspectives, outlined with the target to give a quality school instruction and advanced education to every body with an assumption for comprehensive and examination situated progress. Education has a key and definitive job in this situation of possibilities. NEP-2020 gives for quality advanced education through multidisciplinary colleges and independent schools. The research study's primary goal is to determine how new education policy can influencelearner'sineducationalinstitutions.This study in vestigates the relationship between new education policy and higher education. This research study is descriptive in nature and it isaqualitative study and this research study uses secondary data, which it gathers from existing literature, articles that have been published in various journals, periodicals, newspapers, and online domains during the research period and analyse through content analysis.Thistopichas the potential to be expanded upon and has a number of theoretical and practical applications . This article can be considered as a source of perspective to the strategy execution groups of Govt of India.

Keywords: Education system, Higher education, National education policy, Right to Education, Quality assessment.

1. Introduction

The Indian government's New Education policy (NEP 2020) is apositive improvement in this covid-19 scourge. The main NEP for India happened in

1968, trailed by NEPs in 1986 and 2020 under Narendra Modi. India passed the NEP 2020 on July 29, 2020. It tries to reform education and make the countrymoretaught. K Kasturirangan, the previousheadofISRO, led- the NEP 2020 arrangement council. The decision Government of India vowed to modernise Indian schooling in it selection pronouncement (Saha Mushkan, 2020). Educationassumes a fundamental part in building a country; it chooses the nation's future and its kin's fate. The effect will be sturdy concerning the development and advancement of the stateandsubject. In this day and age, schooling is urgent. Pre-freedom and post-autonomy periods show the extension and progress Indian Government. It will alter the way India studies; this is the third correction to the instruction strategy. National education policy 2020 plans changes for all gatherings (Kalyani Pawan, 2020).

World wide occasions and in clinations have previously influenced India' s advanced education. In ongoing years, India has drawn in countless global understudies, including a rising number of non-occupant Indians, who come to participate in different academic and professional programmes. Various abroad colleges license Indians to acquire degrees without leaving the country (Kugiel, P., 2017). Open college frameworks are on the ascent, making it simpler for private colleges to close the hole. Higher Educational Institutions currently shows ought after authority level abilities and capabilities. Future work valuable open doors will no doubt draw inpeople (Hiremath, S.S., 2020). Rebuilding and internationalising Asia'sadvanced education. This renationalised Singapore's advanced education. Worldwide understudies gain personalities, desires, and perspectives through intercultural learning. Global understudies join in Singapore (Hannah Soong, 2020). Institute of Higher Education is filling in an unrestricted economy time. Many countries' Higher Education strategies adhere to this guideline. Contrasting Higher Education methods in Chi globalisation illustrated globalization's effect. Education globalisation (MinHong, 2018). InternationalisationInterAmericanOrganisations for Higher Education will fortify the nation, organisation, and individual connections. It's againstpublic and social interests, saycritics. Singapore and Japan' sschool ssho wthis(Justin S. Sanders, 2018).

The National Education Policy 2020 imaginesan India focused schoolingsystem by integrating its custom, culture, values, and ethos to change the country into an evenhanded, manageable, and energetic information society. The NEP-2020 is created by thinking about the wide and profound authentic legacy of the nation and the commitments of numerous researchers to various subjects as the establishing stone to construct excellent multi-disciplinary liberal education at both school and higher and proficient instruction level. With the target to expand the Gross Enrolment Ratio of school instruction enlistment and higher and proficient training enlistment to increment from 28% and 05% to half what's more, 20% individually by 2030, through revolutionary changes in the current training approaches and administration frameworks by presenting responsibility in each stakeholder (Prahlada, 2020). A new educational method to see tremendous developments in schools also, high level illumination. He has pointed the void between vision and mission would require more than the activityplans also, the system for execution. In this way, it takes the right measures to guarantee that execution lives up to assumptions (Sunil Kumar et al, 2020).

In January 2015, acommittee under former cabinet Secretary T.S.R. Subramanianbeganthe meeting process for the New Education policy. In view of the panel report, in June 2017, the draft NEP was submitted in 2019 by a board drove by previous Indian Space Research Organisation chief Krishnaswamy Kasturirangan. The Draft New Education policy (DNEP) 2019, was subsequently delivered by Ministry of Human Resource development, trailed by various public discussions. The Draft NEP had 484 pages. The Service embraced a thorough meeting process in figuring out the draft strategy: "More than two lakh ideas from 2.5 lakh gram panchayats, 6,600 blocks, 6,000 Metropolitan Neighborhood Bodies, 676 areas were gotten." The vision of the Public Schooling Strategy is: " National Education Policy 2020 envisions an India-centric education system that contributes directly to transforming our nation sustainably into an equitable and vibrant knowledge society by providing high-quality education to all" (Kurien & Chandramana, 2020). Therefore the research problem for the research study is how educational sector was impacted by new education policy 2020.

2. The Oretical Background of the Study

The theoretical system directs and ought to reverbe rate with every-aspect of the research cycle from the meaning of the problem, literature study, philosophy, show and conversation of the findings as well as the ends that are drawn. Eisenhart (1991) contends that the hypothetical system helps the specialist in considering elective hypotheses that could challenge his or her perspective, in this way advancing the qualities of the review. Simon and Goes(2011)aswell as Maxwell (2004) affirm that theoretical frameworks develop the substance of the review. Subsequently, research proposals that require financing must obviously show the hypothetical system that pivots the planned examination. It persuades subsidising offices that the search request merits supporting. For researchers in the field and readers, the legitimate choice and presence of a hypothetical framework convinces them that the review did not depend on the individual impulses of the specialist yet rather is immovably established in a laid out theory selected by means of solid examinations (Akintoye, 2015).

2.1 New Education Policy 2020

NEP 2020 will work on Indian advanced education. It will tackle worldwide communities experiences. Advanced education under NEP 2020 will use Information and communication technology to fulfils the post covid era's instructive requirements. An obsolete education systemisfutile. India should change its higher education educational plan to match worldwide expertise requests. Intricacy encompasses NEP2020 implementation. The strategy will prod advancement. Rather than different controllers, schools will-have one. Scholarly opportunity, research expansiveness, and advancement benefit from liberal autonomy. Foreign school scan assemble grounds in India, and Indian colleges can do a similar abroad. NEP 2020 would further develop expertise improvement in India by giving under studies numerous passageand exit-options and the opportunity to seek after advanced education subsequent to working. NEP 2020 is a somewhat advanced paper that figures out the ongoing financial circumstance and future difficulties. No policy is viable without legitimate execution.

This approach is all around considered and honestly advances Indian training. This technique in corporates proficient training into higher education to foster abilities and make occupations. NEP 2020 has likely delineated India 2.0. If implemented, it will make India worldwide education community by 2030. Only a select gathering of establishments will be allowed to give the B.Ed. DigitiseMary'sschool in struction should satisfythisguideline.Digitalisedeach library. National Research Foundation supporting promotes undergraduate and graduates research. Teachers will have independence in course delivery assessment and examination advancement.Online degrees ought to support Gross Enrolment Ratio. Indian Universitiesand establishments can unreservedly work together on instructing and research with global University (Gaur, 2022). Through this New Education Policy, the kids who are in the age gathering of 3 to 6 years approaches free, protected, top caliber, formatively fitting consideration, and training by 2025. Since right now there is an extreme learning emergency in India since kids are enlisted for pre-essential however they neglect to accomplish fundamental abilities.

India, being a developing liberal country for instructive changes, as of now has around 845 colleges and roughly 40,000 advanced education establishments, mirroring the general high discontinuity and many little estimated Higher Educational Institutions in the country which are subsidiary to these colleges.

It is seen that as more than 40% of these little estimated organisations are running single program me against the normal change to a multidisciplinary style of advanced education which is a fundamental necessity for the instructive changes in the country for the 21st hundred years. It is additionally noticed that more than 20% of the schools have yearly enrolment under 100 understudies making them nonviable to work on the nature of instruction and just 4% of universities enlist in excess of 3,000 understudies every year because of provincial irregularity as well as the nature of training they offer. A portion of the reasons found for the fracture of the higher education framework in India are: It is anticipated that India will be the third biggest economy on the planet by 2030-2032 with assessed Gross domestic product of ten trillion bucks. It is obvious that the ten trillion economies will be driven by information assets and not by there gularassets of the country. To help the development of the Indian schooling area, the current government chose to patch up it by presenting thorough New Education Policy 2020.

This is in accordance with the State leader's new approach utilising the Fourth Modern Upheaval to take India higher than ever. The right now presented New Education Policy 2020 imagines an India focused school system that contributes straight forwardly to changing our country reasonably in to a fairand lively information society, by giving great instruction to all. The NEP looks to present a shift from 10+2 construction to 5+3+3+4 design, where youth instruction will be a section of formal training. What's more, the NEP 2020 spotlights on lessening the educational program content to account for decisivereas on in gandinturn, foster people with 21st-century abilities imparted in them. Subsequently, all parts of the educational plan and teaching method should be rebuilt to achieve these objectives. The difficulties in effectively executing these progressions incorporate changing the educational program as per the Public Educational plan System. Likewise, instructors need to reexamine the learning content rubric and adjust the reading material appropriately (Venkateshwarlu, 2020).

2.2 Higher Education

It's critical to see the approach in setting of what has been occurring in state funded colleges, and ongoing calamity of colleges of prominence. There has been persistent disintegration of college independence by the state. Unreasonable states avagery released endless supply of the best state funded colleges in India didn't occur in that frame of mind past. Political arrangements of college pioneers who are best case scenario, the instruments of state, rather than being centered around instructing, learning, exploration ororganisation.

Howeverthe archive features administrative independence, it would be troubling assuming the archive additionally implied monetary independence. This 'envisioned' independence is imagined through substitution of UGC (University Grant Commission) and AICTE (All India Council for Technical Institutions). New body Advanced education Commission of India depends on the possibility of division of capabilities what's more, division of exercises.

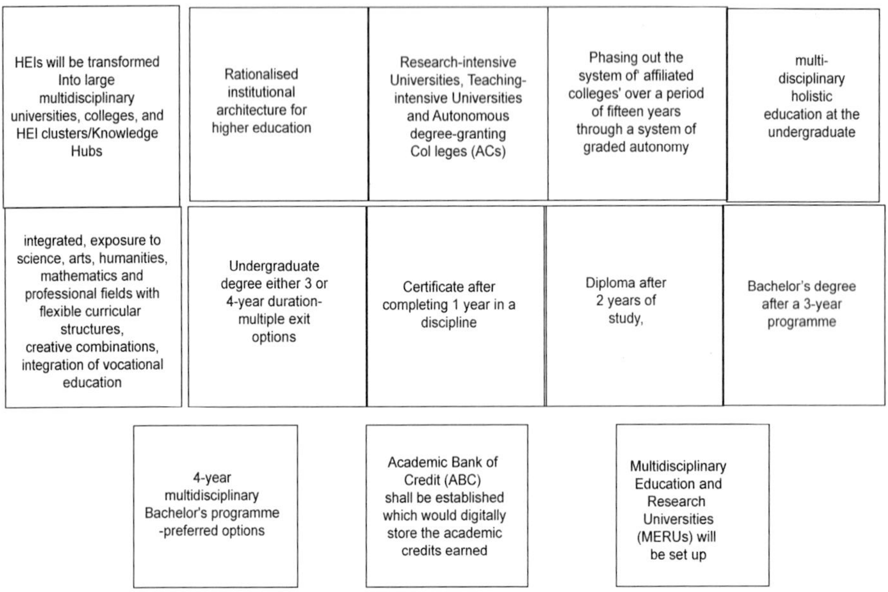

Source: https://news.careers360.com/nep-2020-what-does-it-say-about-schools-and-universities

NEP's principal target concerning advanced education is to raised a ruckus around town of a half Gross Ratio by 2035 with arrangement for various passage/exit. Incases during numerous sections and leaves, declarations would be given after the main year, for recognition, it's after the subsequent year, and for the degree, it's following three-four years. Stopping of MPhil courses. Just a single controller expert for HE in the country with isolated verticals for endorsement and money. The controller will chip away at the 'Online Self Exposure Based Straight forward Framework'. All Higher Educational Institutions, other than Regulation and medical universities will be controlled by similar specialists. Shared rules for both private and public higher instructive establishments.(Anubavam, 2020).

Strategy likewise contends against commercialisation of training. Nonetheless, in a similar expansiveness consider sun familiar colleges to come to India.

There has been critical expansion in number of private colleges by Indian suppliers. Assuming the thought was to increment contest, it appears to be legit. Nonetheless, inclusion of the assertion doesn't. Center around cutting edge educational plan seems OK, and a different body committed to zero in on coordinating innovation in foundations is a fundamental bearing. National Research Foundation is another good thought.Not with standing, in the event that these spaces get filled by people who are driven by philosophical plans, little could be anticipated. Indian Colleges will be permitted to set up grounds somewhere else on the planetthere is major areas of strength for a for this to create in gulf- markets. There is an enormous interest for quality training by Indian diaspora (Venkateshwarlu, 2021).

As of now an undergraduateand post graduate confirmation or in light of the placement test directed at college level or state level with the exception of National Institute of Technology and medical universities. In NEP 2020 all undergraduateand post graduate affirmation of Public Higher educational institutionsdepend on National Testing Agency scores directed at public level (Dhimmar, 2021).

2.3 Impact of New Education Policy 2020 on Higher Education

Prof. Sheth expressed that NEP 2020 is aground breaking and aggressive strategy made by the public authority. That's what he featured: The NEP 2020 makes ready towards not so much guideline but rather more independence for foundations through a viable self-administrative structure. Over-guideline has forever been drilled in Indian instructive foundations not at all like the western organisation's who practice self-guideline. Through NEP 2020 continuous independence and authorisations will be conceded to all advanced education establishments. Foreign Colleges will be permittedtosetupgroundsin India. It will advanceexaminationand bring adaptable methodology. The unfamiliar colleges would get program and organisation versatility. The possibility of international isation of advanced education depends on the versatility of understudies, employees, projects, and organisations across nations. The NEP 2020 targets setting up multidisciplinary organisations offering programs with excellent instructing, exploration, and local area commitment. This will helpund erstudies with expanded adaptability and selection of subjects across different surges of expressions, humanities, sciences, sports, and other professional subjects.

Featuring the National mission on training which is to zero in on fundamental education and essential numeracy. The significant reorganisations in educational design won't have any unbending partition between streams. The divisions between professional, scholarly, curricular, and extra-curricular investigations will be removed. Board tests will test the information procured rather than repetition learning (Anubavam, 2020). The new strategy centres around on the web and e-learning. Instruction has forever been a greater amount of physical methodology instead of innovation based. Online training will assume abroad part to overcome any barrier between the understudies and the organisations. In schooling what is reciprocal today will become center. The virtual world will turn out to be more genuine than this present reality with advanced and video driven approaches. New Education Strategy isn't just about the degree. Everything revolves around center around fundamental abilities and professional courses. It will be understudy driven approach where moderateness and availability will be there and where the understudies can use the most out of instruction. Expanded admittance, value, and incorporation through open tutoring, online training and Open Distance Learning will be advanced (Sheth, 2020).

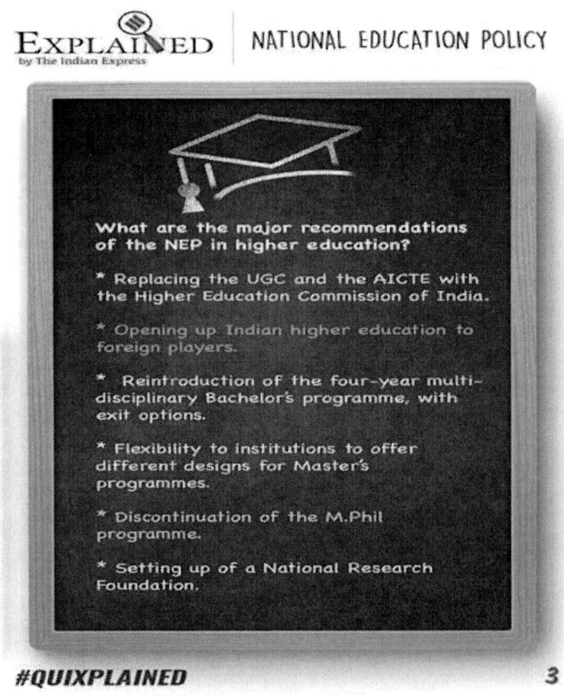

Source: https://indianexpress.com/article/explained/ reading-new-education-policy-india-schools-colleges-6531603/

Higher education assumes a significant part in working on human prosperity and creating India and through this NEP they plan to build their Gross

Enrolment Ratio from 26.3% to half by 2035, and around 3.5 cr. new seats will likewise be added to advanced education establishments. Under this, UG schooling could be of 3 or 4 years with quantities of passage and leave choices. Multidisciplinary instruction and Exploration College will be set up of worldwide norms. HECI (Higher Education Commission of India) will be the main body for whole advanced education (except medical and clinical reason) HECI will have 4 free verticals like NHERC (National Higher Education Regulatory Council) foranorder, GEC (General Education Curriculum) for standard setting, HEGC (Higher Education Grant Council) for supporting, and NAC (National Accreditation Council) for recognization. Globalization of advanced education. As cholastic bank of credit will be laid out in which credit acquired by the understudies during their scholastics from various Higher Educational Institutions could be put away and moved at the time last degree. Public examination establishment will go about as a pinnacle body that encourages serious areas of strength for a culture and constructs research limit covering advanced education (Negi, 2022).

3. Objective of Study

1. To examine the relationship between New Education Policy and Higher Education.
2. The study also highlights of salient features of NEP and analyses how they influence the existing education system.
3. To determine how new education policy can influence learner'sin Educational Institutions.

4. Methodology

Research can be characterised as "a movement that includes finding out, in a pretty much precise way, things you didn't have the foggiest idea" (Walliman and Walliman, 2011, p.7). "Methodology is the philosophical system inside which the examination is led or the establishment where upon the exploration is based" (Brown, 2006). The Research Methodology part of the examination portrays research techniques, approaches, and plans exhaustively featuring those utilised throughout the review, legitimising my decision depicting thebenefits and drawbacks of each methodology and configuration considering their functional materialness to our exploration.

The research design of the research study is descriptive in nature and is a qualitative study. Data is collected from various secondary sources like journals, periodicals, newspapers, various websites, and books and analyse through content analysis. Descriptive research can be made sense of as an explanation of issues as they are at present with the scientist having zero commands over factor. Also, "descriptive investigations might be portrayed as essentially the endeavour to decide, depictor distinguish what is, while logical examination endeavours to lay out why it is like that or how it became Descriptive research is pointed toward illuminating recent concerns or issues through a course of information assortment that empowers them to depict what is going on more totally than was conceivable without utilising this strategy" (Manjunatha, 2019).

5. Analysis and Discussion

Quality advanced education should plant of oster people who are magnificent, insightful, well- adjusted, and imaginative. It should empower an individual to concentrate on at least one particular are as of interestat a top to bottom level and foster person, moraland protected values, scholarly interest, logical attitude, inventiveness, administration soul, and the abilities of the 21st 100 years across a scope of fields, including sciences, sociologies, artistic expression, humanities, dialects, individual, mechanical the professional subjects. The new instruction strategy brings some essential changes to the ongoing framework, and the key features are multidisciplinary colleges and universities, with somewhere around one in or close to each locale, redoing understudy educational programs, teaching method, assessment, and backing for improved understudy insight, laying out a Public Exploration Establishment to help great companion assessed work and actually seed learn at colleges furthermore, schools.

The fundamental issues looked by the Indian advanced education framework incorporates implemented partition of capabilities, early specialisation and understudy spilling into confined research regions, less center around research all things considered colleges and schools, and absence of serious companion assessed scholarly exploration financing and enormous associated colleges prompting low degrees of undergrad training. Point of NEP 2020 inregard of advanced education can be determined as follows: The strategy visualizes for anothervision and design for advanced education with huge, well-resourced, dynamic multi-disciplinary foundations driving towards greatness. The strategy accommodates expansive based liberal instruction through thorough yet adaptable curricular designs, imaginative blends of study, coordination of professional schooling and various section as well as leave focuses. The strategy works with for deliberate and independent administration as institutional administration will be founded on independence in terms of scholastic, authoritative and monetary. The approach tries to guarantee great guideline, as guideline will be light however close, as the guideline will be in the possession of free bodies to dispense with irreconcilable situations.

Another significant change the NEP 2020 proposes that the National Testing Agency will serve as a chief, master, independent testing association to lead placement tests for Undergraduate and graduate affirmations and associations in Advanced education Foundations. The superior grade, reach, and adaptability of the NTA testing administrations will empowermost colleges to utilise these normal selection tests - as opposed to having many colleges each formulating their own selection tests in this manner radically lessening the weight on understudies, colleges furthermore, schools, and the whole school system. It will be surrendered to individual colleges and schools to involve NTA appraisals for their confirmations. It additionally definitely helps the understudies to without any problem move their certifications and credits to colleges abroad.

6. Conclusion and Findings

India is equipped to carry out the rules of National Education Policy 2020 all through the country to change and roll out extremist improvements in school training and higher schooling with a target of making another instruction framework which ought to enable the adolescents and lift their certainty to make new information, new abilities, along with human qualities to tackle current and future issues and difficulties of the humanised society through their improved imaginative capacity and technical education. It is known that in novation which is a use of logical reasoning has the abilities to work on the personal satisfaction of everybody in this universe and quality schooling is an ideal starting point for it. With the target of offering some benefit based, information based, and expertise based advance deducation for every body in the country, the new instruction strategy has numerous natural recommendations to work on thenature of school and higher training making interest in their picked region to find challenges and changing over them interests valuable open doors by finding creative answers for make life agreeable also, fruitful with anticipated satisfaction.

The quality higher instruction has the goal of creating individuals who-are answerable for making better society through further developed human worth based discipline, and regarding each other for development and thriving. Quality advanced education additionally makes everybody to add to finding new innovation, embracing new innovation, or advancing new innovation which can add to the advancement of society. It is normal that the new instruction strategy which is research focussed, will speed up the achievement of the above goals and makes each partner as pioneer. NEP 2020 will work on Indian advance deducation. It will saddle worldwide networks'encounters. Advanced education under NEP 2020 will use ICT to fulfils the post Coronavirus period's instructive necessities. An obsolete education's system is futile. India should adjust its higher instruction educational plan to match worldwide ability requests.

Intricacyen compasses NEP 2020 execution. The arrangement will spike advancement. Rather than numerous controllers, universities will have one. Scholastic opportunity, research broadness, and advancement benefit from liberal independence. Unfamiliar schools can construct grounds in India, and Indian colleges can do the equivalenta broad. NEP 2020 would further develop ability improvement in India by giving under studies numerous passage and exit choices and the opportunity to seek after advanced education subsequent to working. NEP 2020 is a to some degree high level paper that figures out the ongoing financial circumstance and future difficulties. No strategy is powerful without legitimate execution. This approach is very much though to fandsincerely advances Indian training. This procedure coordinate sproficient instruction into higher training to foster abilities and make occupations. NEP 2020 has likely outlined India 2.0. On the off chance that executed, it will make India worldwide schooling place by 2030.

The policy presents an entire range of changes and peruses generally as an exceptionally moderate report, with astrong handle on the current financial scene and the possibility of future vulnerability. Instruction for another age of students needs to basically draw in with the expanding dematerialisation and digitalisation of economies, which requires a totally new put of capacities together to have the option to keep up. This is by all accounts a much more fundamental perquisite presently, with the pattern towards digitalisation and problematic mechanisation being stimulated by the pandemic. Generallyspeaking, the NEP2020 addresses the need to foster experts in an assortment of fields going from Farming to Man-made consciousness. India should be prepared for what's to come. What's more, the NEP 2020 prepares ahead for the overwhelming majority youthful hopeful understudies to be furnished with the right range of abilities.

The new education policy has a praiseworthy vision, however its solidarity will rely upon whether it is ready to really coordinate with the other strategy drives of government like Advanced India, Expertise India and the New Modern Strategy to give some examples, to impact an intelligent underlying change. Consequently, strategy linkages can guarantee that instruction strategy locations to and learns from Ability India's involvement with connecting all the more powerfully with the corporate area to shape professional schooling educational program to make it a triumph. There is likewise a need for more proof based independent direction, to adjust to quickly developing changes and interruptions. NEP has reassuringly provisioned for constant assessment frameworks and a consultative observing and audit system. This will enable the school system to continually change itself, rather than expecting foranother schooling strategy consistently for a change in educational program.This,init-self, will be a momentous accomplishment. The NEP 2020 is a pivotal turning point for advanced education. Compelling and time-bound execution will make it really way breaking.

References

Akintoye, A. (2015). Developing Theoretical and Conceptual Frameworks. Jedm. oauife.edu.ng>uploads>2017/03/07 (accessed 2017 February 22) DOI: 10.1080/08841241.2020.1728465Kugel, P. (2017). India's soft power: A new foreign policy strategy. Taylor &Francis.Kumar, K., Prakash, A., & Singh, K. (2020). How National Education Policy 2020 can be a lodestar to transform.

Eisenhart, M. (1991). Conceptual Frame works for Research Circa 1991: Ideasfroma Cultural Anthropologist ; Implications for Mathematics Education Researchers.Virginia: Blacksburg Press

Hannah Soong (2020) Singapore international education hub and its dilemmas: the challenges and makings for cosmopolitan learning, Asia Pacific Journal of Education, 40:1, 112-125, DOI: 10.1080/02188791.2020.1725433

Hiremath, S.S.(2020). Education In India: Impact of Globalisation And ICT. Blue DiamondPublishing.Pawan(2020).Anempirical study on NEP 2020 with particular reference to the future of Indian education and its effects on stakeholders Managing .

https://iimranchi.ac.in/p/nep-2020-impact-on-higher-education.html

https://nvshq.org/article/new-national-education-policy-nep-2022/

https://www.creatrixcampus.com/blog/The-National-Education-Policy-NEP-2020

https://www.educationworld.in/nep-2020-school-education-service-environment-and culture.

https://www.researchgate.net/publication/346654722_Impact_of_New_Education_Policy_2020_on_Higher_Education

https://www.researchgate.net/publication/366177195_A_Roadmap_for_India%27s_New_Education_Policy_NEP_2020.

Justin S. Sanders (2019) National international is ationcontext and competition, Comparative and International Education, 49:3, 413-429, DOI: 10.1080/03057925.2017.1417025.

Maxwell, J. (2004). Qualitative Research Design: An Interactive Approach (2nded.). Thousand Oaks, CA: Sage.

SahaMushkan(2020).InternationalJournalofResearch(IJR)

Simon, M.K. & Goes, J. (2011). Developinga Theoretical Framework. Seattle, WA: Dissertation Success, LLC.

Sunil Kumar Saroha, & Uttam Anand(2020). New in structionprocedure 2020 Highlights:To see huge movements in schools and advanced edification. IOSR Journal of Humanities and Social Science (IOSR- JHSS), 25 (8), 59-62.

36

A Study on Consumer Behavior and Environmental Consciousness: Implications for Sustainable Consumption in a Contemporary Context

Shabana Gufran Khan

SVKM's Mithibai College of Arts, Chauhan Institute of Science & Amrutben Jivanlal College of Commerce & Economics, Autonomous

Abstract

The study delves into understanding what drives consumers to embrace eco-friendly products and sustainable consumption habits. As global concern for the environment rises, grasping consumer behavior in this realm becomes paramount for businesses and policymakers alike.

Starting with a thorough review of existing literature, the study pinpoints key influencers such as environmental awareness, perceived value, media impact, and demographic factors on consumer choices. Despite existing research, there remain gaps in understanding how these factors interact to shape sustainable consumption habits.

To bridge these gaps, the researcher proposes conducting descriptive research through surveys and data analysis. By examining consumer attitudes, purchase intentions, and demographic profiles, the study seeks to uncover insights that can guide companies and policymakers in promoting sustainable consumption effectively.

The study's findings hold significant implications. Businesses can use them to develop eco-friendly products and tailored marketing strategies. Meanwhile, policymakers can craft targeted interventions to foster sustainable consumption on both individual and societal levels. Overall, the study aims to contribute practical insights to enhance eco-conscious behaviors and product adoption among consumers, thus advancing the discourse on sustainable consumption.

Keywords: *Eco-friendly products, Sustainable consumption, Consumer behaviour*

1. Introduction

Sustainability embodies an ecological vision aimed at fostering enduring human coexistence on Earth. Over time, its definition has evolved in literature and context, generating ongoing debate. Generally, experts frame sustainability around three pillars: social, economic, and environmental. While environmental concerns often dominate discussions, sustainability encompasses broader challenges like pollution, land degradation, biodiversity loss, climate change, and ecosystem service decline. This concept influences decision-making globally, politically, and personally.

Sustainable development, often used interchangeably with sustainability, is distinguished by UNESCO as a long-term objective versus the diverse processes and pathways to achieve it. Within the economic dimension, the tension between welfare and environmental preservation is explored through concepts like "weak and strong sustainability," acknowledging the necessity of trade-offs. Ideal policies would aim to decouple environmental degradation from economic growth.

Environmental sustainability is pursued through various means such as ecosystem preservation, collaborative resource management, reducing food waste, advocating plant-based diets, managing fertility rates to curb population growth, and transitioning away from fossil fuels towards clean energy and innovative green technologies.

Sustainable consumption entails utilizing goods and services with minimal negative environmental impact, considering the needs of present and future generations. It encompasses both individual and collective choices, influenced by political institutions, social norms, and cultural values. Sustainable consumption and production are interconnected, aiming to enhance quality of life while minimizing resource use, waste, and pollution to safeguard future generations' needs.

Eco- friendly Products

Environmentally friendly things have no negative impact on the ecosystem, whether they are manufactured, utilized, or discarded. Despite the fact the raw materials are natural, precautions are taken to avoid damaging the environment. Natural materials are employed in its manufacture.

Importance of Environmentally Friendly Products

Preservation of Nature: Natural resource depletion, pollution, and overexploitation can disrupt the balance of nature. Given humans' integral connection to the environment, adopting eco-friendly products is crucial to prevent further degradation.

Human Well-being: Human health is intricately linked to environmental health. Negligent manufacturing and disposal practices contribute to the introduction of new diseases into society. Utilizing eco-friendly products can mitigate these risks and safeguard human health.

Climate Improvement: Accelerated global warming poses significant threats to various aspects of life on Earth, including agriculture, polar ice caps, and biodiversity. Eco-friendly products help mitigate climate change impacts by reducing emissions and promoting sustainable practices.

Healthier Lifestyle: Switching to all-natural products eliminates exposure to harmful chemicals, promoting a healthier, chemical-free lifestyle for individuals. Additionally, it contributes to broader environmental conservation efforts.

2. Research Framework

2.1 Research Objectives

1. To investigate the current level of consumer adoption of environmentally friendly consumption habits.
2. To analyse how consumers' decisions about eco-friendly items are influenced by marketing tactics (such as product labelling and green advertising).
3. To explore the connection between consumer sentiments regarding eco-friendly items and demographic traits including age and gender.
4. To investigate the main variables affecting customers' acceptance of environmentally sustainable products.
5. To offer analysis and suggestions on how companies and legislators can encourage environmentally friendly consumer behaviour.

2.1 Significance of Study

- This research holds significant value for companies embarking on the integration of eco-friendly and sustainable materials into their production processes and product lines.
- The findings of this study have the potential to raise environmental consciousness among individuals, shedding light on the positive impact eco-friendly products can have on the environment.

- The insights gathered from this investigation offer valuable guidance to businesses seeking to align their operations with eco-conscious practices, fostering sustainability across industries.
- By illuminating the environmental benefits of eco-friendly products, this study serves as a catalyst for informed consumer choices, driving demand for more sustainable alternatives in the marketplace.
- For policymakers and advocacy groups, the implications of this research underscore the importance of promoting eco-friendly initiatives and regulations to mitigate environmental degradation and promote long-term ecological health.

2.3. Limitations of the study

- One potential limitation of the research is individual bias, where participants' responses and actions regarding eco-friendly products and sustainable consumption may be influenced by their personal opinions, preferences, or past experiences.
- Additionally, the study was constrained by a small sample size.

3. Literature Review

1. **Alghamdi, O. A., & Agag, G. (2024):** The researchers state that the Consumer behavior has been significantly impacted by the COVID-19 pandemic. In the context of acquiring and consuming commodities, there has been a shift from conventional to ecologically friendly shopping. Notwithstanding the fact that these aspects had not previously been thoroughly investigated, our research aimed to identify the variables that impact consumers' conscious green purchasing behavior (GPB). The findings of the study imply that, in comparison to media influence, peers have a larger effect on awareness on GPB through a variety of factors. The results of this analysis have a number of important theoretical and managerial ramifications for marketers operating in the context of green consumerism.

2. **Grover, A., & Hareesh, K. T. (2023):** Global overconsumption of goods and services has skyrocketed, resulting in the decline of natural resources and associated climate catastrophes like scarcity of food and water, water pollution, and air pollution, among others. Anxiety over climate crises has clearly increased, and this has caused a corresponding shift in people's consumption habits. Customers are beginning to change their reckless consumption habits into more conscientious and

sustainable ones, and they are displaying a positive attitude towards "green" or ecologically friendly products. This study intends to establish a conceptual framework on the antecedents of environmentally friendly consumption behaviour in order to expand the theoretical knowledge of sustainable consumption behaviour. The study looks at how environmental, economic, and external factors affect consumers' intentions to buy green items and how they behave sustainably when making purchases.

3. **Cao, Y. (2023):** This study attempts to summarise current data, identifies a number of reasons why consumers aren't buying as many green products, and offers a thorough grasp of the actual consumer decision-making process. This study conducted a systematic assessment of 73 trustworthy publications regarding green purchase behaviour posted between 2010 – 2022, utilising the framework of the Consumer Buying Process. The findings of the study indicated that there can be a "loop tendency" in the green procurement process that involves the creation and adoption of assessments and their circulation. Furthermore, the minority of consumers' purchasing behaviours might not be powerful enough to establish a legitimate societal norm, and the products that are now on the market are still insufficient to meet the needs of customers.

4. **Dantas, R., Sabir, I., Martins, J. M., Muhammad, B. M., Rafiq, M., Jéssica, N. M., & Rana, K. (2023):** This study's main goal is to clarify the significance of green packaging as a crucial component of a sustainable ecosystem. It has been shown that using green packaging has a remarkable impact on the environment. Green packaging is one workable response to the severe environmental damage and pollution caused by the misuse of carbon goods. Two independent factors have been used to study the phenomenon: multisensory packaging and green packaging. Ecological sustainability is the dependent variable, whereas consumer perception and environmental knowledge have been employed as moderating and mediating variables, respectively. The results of the study show that multimodal packaging has a negative influence on environmental sustainability, whereas green packaging has a large and favourable impact.

5. **Wong, W. M., Qu, M., Nunualvuttiwong, C., & Ngamcharoe-nmongkhon, K. (2023):** This study, which is conceptual in nature, looks at the complex interactions that exist between consumers' sustainable purchase patterns and their environmental orientations (egoistic, altruistic, and biospheric), with a focus on the pervasive impact

of COVID-19 dread. The research also clarifies how the COVID-19 epidemic has altered the relationship between environmental concerns and concrete sustainable consumption practices. The study adds to the body of knowledge in a post- pandemic context by emphasising the relationship between sustainable consumption and environmental awareness.

6. **Yunita Budi, R. S., & Sukresna, I. M. (2023):** The purpose of this study is to examine how green knowledge and inhibiting factors, such as habits, skepticism, and lack of availability, affect consumers' decisions to buy environmentally friendly products. The moderating influence of green knowledge on the connection between the factors that limit the purchase of green products and the actual purchase of green products is also explained by this research.

7. **Kim, N., & Lee, K. (2023):** This study looked at situational factors that cause reluctance to buy eco-friendly products as well as consumer awareness of the environment as a variable influencing the buying of eco- friendly products. The Researchers looked examined how these variables affected the convenience of purchasing and the trustworthiness of eco-labels. The following are the precise verification results. First, the intention to purchase eco-friendly products was not significantly impacted by environmental interest. Second, the desire of consumers to purchase eco-friendly items was significantly influenced by their assessment of consumer effectiveness and level of environmental understanding. Third, the aim to buy environmentally friendly products had a big influence on the way people bought them.

8. **Anwar, S. S., & Cheah, I. (2023):** This paper seeks to investigate how consumers' reactions are affected by environmentally friendly packaging. In order to ascertain the relative significance of the essential precursors of eco-friendly packaging behavior—environmental responsibilities, understanding, perspectives, and message framing—a study framework is studied. Two studies are conducted to test a set of hypotheses. Research 1 (n = 160) demonstrates that when it comes to consumers' aspirations and awareness about the environment while implementing eco-friendly packaging, attitude has a significant mediating role. Additionally, research 2 (n = 132) discovers that in order to promote pro-environmental behavioural intention in relation to eco-friendly packaging, the advertising message must be presented with a certain emotional appeal congruent to a particular psychographic feature.

9. **Jaiswal, D., & Singh, B. (2018):** The authors in their study states that Green purchasing practices can encourage businesses and society to pursue sustainability while addressing the positive effects on the environment and society, all the while balancing the supply and demand sides of the business cycle. In the conditions of sustainable consumption at India & Asia, this research aims to comprehend and evaluate the holistic approach of green buying behavior. The findings confirm that "green purchase intention," which is strongly and favorably influenced by "attitude towards green products," considerably determines "green purchase behavior."

4. Research Methodology

4.1 Research Type

This research employs a descriptive methodology to describe and characterize the consumer behaviour in relation to eco-friendly products and sustainable consumption practices.

4.2 Sample design

The researcher has used Convenience Sampling Method to collect the data from the sample of Respondents.

4.3 Sample size

Responses for research were collected from total of 109 respondents.

4.4 Sample Location

Respondents for the research were from Tier-1 and Tier-2 cities across India.

4.5 Data Collection Method

Primary data for the research has been collected through Survey using Questionnaire &

Secondary data is collected from different websites and reputed journals.

4.6 Data Analysis Techniques: -

Chi-Square test has been applied to analyze the association between variables. Graphical analysis is done using pie charts and bar graphs.

5. Hypothesis Testing & Findings

Hypothesis – 1

Null Hypothesis (H_{01}): Gender does not have a significant effect on the level of consumer adoption of eco-friendly consumption.

Alternative Hypothesis (H_{11}): The level of consumer adoption of eco-friendly consumption varies significantly depending on gender.

Purchase of Eco-friendly products

Calculation	Observed (O)	Expected (E)	O - E	$(O - E)^2$	$(O - E)^2 / E$
Rarely, Male	32	35.84	-3.84	14.7456	0.41
Always, Male	17	13.16	3.84	14.7456	1.12
Rarely, Female	47	43.16	3.84	14.7456	0.34
Always, Female	12	15.84	-3.84	14.7456	0.93
Total	-	-	-	-	**2.8**

Chi-Square Value:2.8083

Degrees of Freedom: 1

P-value: 0.093776

"The Chi-square statistic is 2.8083. The p-value is .093776. Not significant at $p < .05$."

Result: - With a p-value greater than 0.05, the alternate hypothesis is rejected, and the null hypothesis is accepted. This suggests that the level of consumer adoption of eco-friendly consumption does not significantly vary based on gender.

Hypothesis – 2

Null Hypothesis (H_{02}): There is no significant association between marketing tactics (such as product labeling and green advertising) and consumers' decisions regarding eco-friendly items.

Alternative Hypothesis (H_{12}): There is a significant association between marketing tactics (such as product labeling and green advertising) and consumers' decisions regarding eco-friendly items.

Purchase of Eco-Friendly Products

Calculation	Observed (O)	Expected (E)	O - E	$(O - E)^2$	$(O - E)^2 / E$
Always, Always	19	10.83	8.17	66.75	6.16
Rarely, Always	20	28.17	-8.17	66.75	2.37
Always, Rarely	11	19.17	-8.17	66.75	3.48
Rarely, Rarely	58	49.83	8.17	66.75	1.34
Total	-	-	-	-	**13.34**

Chi-Square Value: 13.3423

Degrees of Freedom: 1

P-value: 0.000259

"The Chi-square statistic is 13.3423. The p-value is .000259. Significant at $p < .05$."

Result – The p-value is less than 0.05, indicating a significant association between marketing tactics (like product labeling and green advertising) and consumers' decisions on eco-friendly items. Therefore, the alternate hypothesis is accepted, and the null hypothesis is rejected.

Hypothesis- 3

Null Hypothesis (H_{03}): There is no significant association between consumer sentiments regarding eco-friendly items and their age.

Alternative Hypothesis (H_{13}): There exists a significant association between consumer sentiments regarding eco-friendly items and their age.

Willingness in changing their lifestyle for reducing their carbon footprint

Calculation	Observed (O)	Expected (E)	O - E	$(O - E)^2$	$(O - E)^2 / E$
Less than 25 years of age, Willing	48	48.69	-0.69	0.4761	0.0097
26 years of age and above, Willing	39	38.69	-0.69	0.4761	0.0124
Less than 25 years of age, Not willing	13	12.31	0.69	0.4761	0.0386
26 years of age and above, Not willing	9	9.69	-0.69	0.4761	0.0491
Total	-	-	-	-	**0.1094**

Chi-Square Value: 0.1094
Degrees of Freedom: 1
P-value: 0.740823

"The Chi-square statistic is 0.1094. The p-value is .740823. Not significant at $p < .05$."

Result – With a p-value greater than 0.05, the alternative hypothesis is rejected, and the null hypothesis is accepted. This indicates that there is no significant association between consumer sentiment regarding eco-friendly items and demographic traits such as age.

6. Data Analysis and Interpretation

1. Age?

From the above figure it can be seen that maximum number of respondents, i.e., 47.7% were from the age group 18-25. And the least number of respondents were from the age group less than 18.

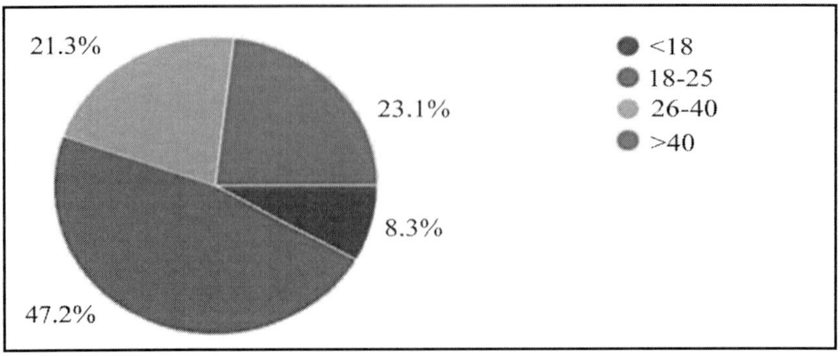

2. Gender?

In the above figure it can see that out of total 109 respondents 60 are female which contribute a total of 55% and on the other hand 49 are male which contribute a total of 36.8%.

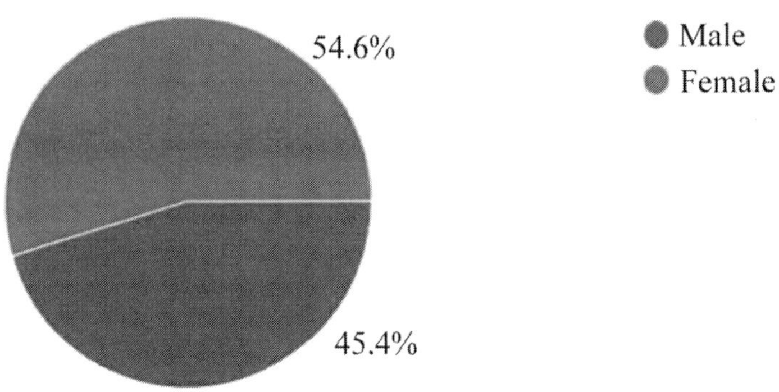

3. What factors influence your decision to buy eco-friendly products?

The bar graph indicates that among 109 respondents, the majority, 66, consider environmental impact as the primary factor influencing their decision to purchase eco-friendly products. Following closely, 53 respondents prioritize quality in their buying decisions. Additionally, 26 respondents prioritize price, while 25 and 14 respondents prioritize brand reputation and convenience, respectively.

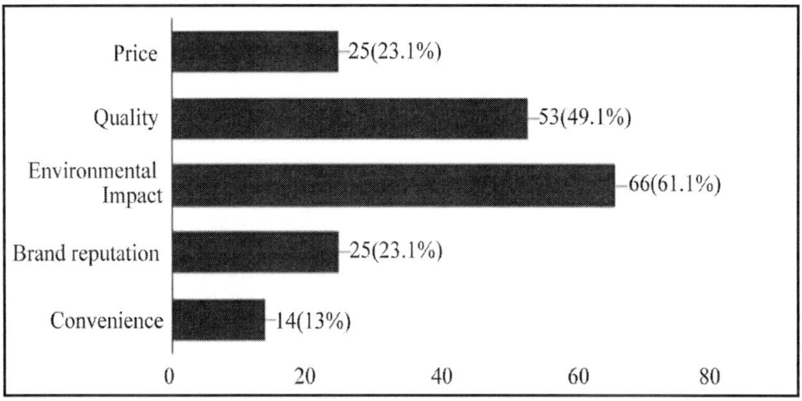

4. How important is environmental sustainability to you when making purchasing decisions?

From the above figure it can be seen that maximum number of respondents i.e., 35 out of 109 respondents says that environmental sustainability is important for them while making a purchase decision. And 30 out of 109 says that it is very important. This indicates that maximum number of people says that environmental sustainability is important for them.

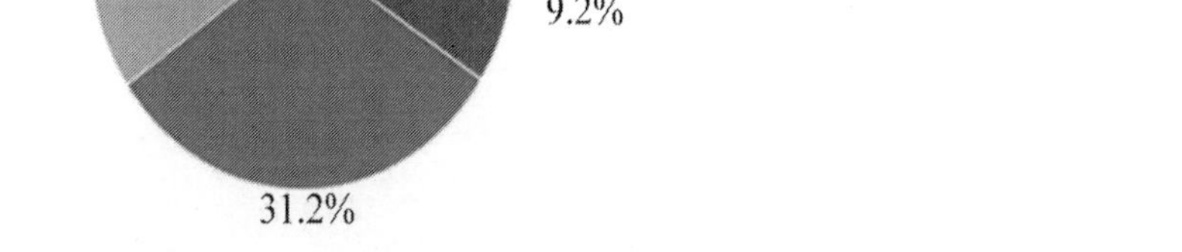

5. Have you ever switched from conventional products to eco-friendly alternatives?

In the provided figure, it's evident that the majority of respondents, comprising 78.9%, have switched from conventional products to eco-friendly alternatives. Conversely, 21.1% of respondents state that they have not made the switch from conventional products to eco-friendly ones.

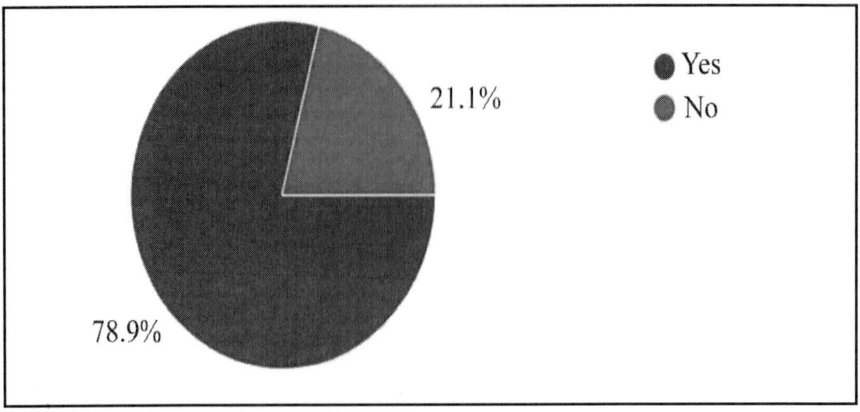

6. How likely are you to pay a premium price for eco-friendly products?

The bar graph reveals that out of 109 respondents, 22 are inclined to pay a premium for eco-friendly products. The largest group, consisting of 46 respondents, holds a neutral stance towards paying a premium. Conversely, only 13 respondents out of 109 are very unlikely to pay a premium for eco-friendly products.

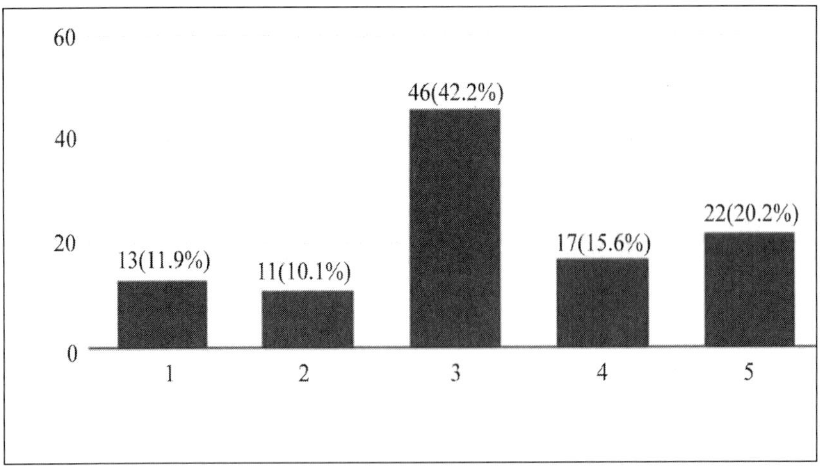

7. How often do you consider product labelling when making purchasing decisions?

The pie chart illustrates that out of 109 respondents, the largest portion, comprising 43 individuals, occasionally take product labeling into account when making purchase decisions. Furthermore, 18 respondents consistently factor in product labeling when making their purchasing choices.

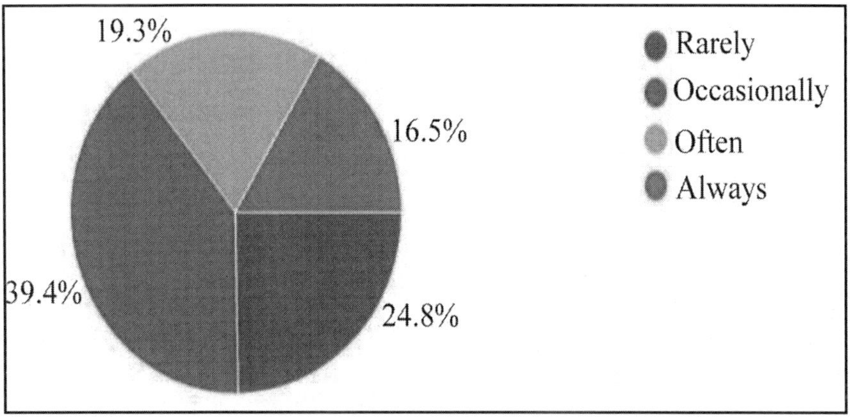

8. Are you willing to make lifestyle changes to reduce your environment footprint?

From the above figure, it can be seen that maximum number of respondents are willing to make lifestyle changes to reduce their environmental footprint i.e., 71 respondents and only 6 respondents are not willing to make a lifestyle change to reduce their environmental footprint.

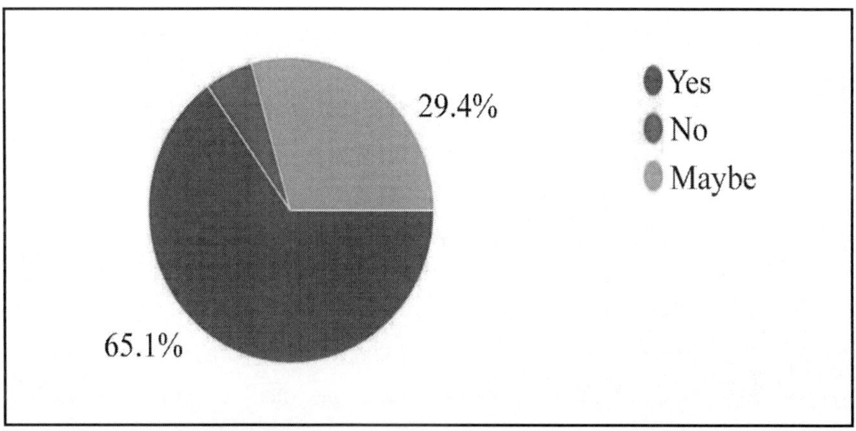

Chapter 8. Suggestions and Conclusion

- **Suggestions**

  1. Governments ought to incentivize the adoption of eco-friendly products through policies and awareness campaigns.
  2. Investing in research and development for innovative eco-friendly materials and manufacturing processes can drive sustainability efforts further.

5. Collaborating with suppliers and partners to establish sustainable supply chains can enhance the overall eco-friendliness of products.

• Conclusion

This study indirectly addresses environmental concerns, a topic of heightened sensitivity in the contemporary landscape. By exploring various facets of sustainable consumption, the research underscores its significance as a substantial issue. Findings indicate a noteworthy level of awareness among individuals regarding sustainable products and their environmental impact. Moreover, demographic variables such as age and gender emerge as influential factors shaping awareness and attitudes toward sustainable products. Consequently, tailoring sustainable products according to demographic segmentation and considering the frequency of product diffusion becomes imperative.

The study reveals a widespread willingness among participants to transition to sustainable products, despite potential incremental costs. Participants recognize the value proposition of such products, citing benefits such as cost savings and reduced resource wastage, which ultimately contribute to environmental preservation.

Furthermore, there is consensus among participants regarding the necessity for heightened consumer awareness concerning sustainable products, indicating an opportunity for advocacy and educational initiatives.

In conclusion, the research aims to offer mutual benefits to both companies and consumers engaged in the use of eco-friendly and sustainable products. Through its findings and recommendations, it aspires to foster a positive impact on environmental conservation efforts while meeting consumer needs and preferences.

References

Alghamdi, O. A., & Agag, G. (2024). Understanding factors affecting consumers' conscious green purchasing behavior. Sustainability, 16(2), 705.

Grover, A., & Hareesh, K. T. (2023). Exploring the Drivers of Sustainable Consumption Behaviour: A Path towards a Greener Future. Indian Journal of Sustainable Development, 9(1), 43-51.

Cao, Y. (2023). "Green Products": A Review with the Consumer Buying Process Framework. Journal of Environmental Management & Tourism, 14(1), 52-66.

Dantas, R., Sabir, I., Martins, J. M., Muhammad, B. M., Rafiq, M., Jéssica, N. M., & Rana, K. (2023). Role of green and multisensory packaging in environmental sustainability: Evidence from FMCG sector of pakistan. Cogent Business & Management, 10(3)

Wong, W. M., Qu, M., Nunualvuttiwong, C., & Ngamcharoenmongkhon, K. (2023). Environmental concerns, sustainable consumption, and COVID-19 fear in online consumers: A research exploration. Theoretical and Practical Research in Economic Fields, 14(2), 246-257.

Raut, G., & Taneja, U. (2023). Adoption of green information systems for sustainable business processes in india: A review. IUP Journal of Business Strategy, 20(4), 20-36.

Gohoungodji, P., & Amara, N. (2024). Greening live shows: What factors matter? A systematic review of factors affecting sustainability practices in music festivals. International Journal of Arts Management, 26(2), 69-90.

Kim, N., & Lee, K. (2023). Environmental consciousness, purchase intention, and actual purchase behavior of eco-friendly products: The moderating impact of situational context. International Journal of Environmental Research and Public Health, 20(7), 5312.

Yin, M., & Lee, E. (2023). Planet earth calling: Unveiling the brain's response to awe and driving eco-friendly consumption. Frontiers in Neuroscience.

Anwar, S. S., & Cheah, I. (2023). Consumers' preference for eco-friendly packaged products: Pride vs guilt appeal. Marketing Intelligence & Planning, 41(2), 186-198.

Jaiswal, D., & Singh, B. (2018). Toward sustainable consumption: Investigating the determinants of green buying behaviour of Indian consumers. Business Strategy and Development, 1(1), 64–73

37

A Study on Persistence of Gender Discrimination and the Glass Ceiling Effect on Women's Advancement In Leadership Positions

Shabana Gufran Khan [2]Mrunal Deo [3]Aditi Anand and [4]Parvin Munshi

SVKM's Mithibai College of Arts, Chauhan Institute of Science & Amrutben Jivanlal College of Commerce & Economics (Autonomous)

Abstract

This study meticulously explores the enduring challenge of the Glass Ceiling in today's corporate environment.

With a focus on three key objectives—measuring prevalence, analysing corporate approaches, and assessing personal factors—the study systematically investigates gender-based barriers obstructing career progression to leadership roles. Employing empirical data and thorough analysis, the research quantifies the contemporary prevalence of the Glass Ceiling, scrutinizes strategies adopted by large corporations, and evaluates the impact of personal factors on its acceleration. By revisiting this phenomenon, the study aims to contribute substantive insights to the ongoing discourse on gender equality in leadership, offering valuable considerations for corporate policies and societal initiatives aimed at dismantling persistent barriers to equitable professional advancement.

Keywords: Glass Ceiling, Gender Discrimination, Leadership Roles, Corporate Bias, Career Progression, Organizational Strategies

Introduction

In the modern corporate landscape, the Glass Ceiling remains an enduring challenge, obstructing the path to leadership positions for qualified individuals based on gender. This study embarks on a comprehensive analysis of the persistent gender discrimination inherent in leadership roles within corporations.

With a primary focus on measurement, analysis, and assessment, the study seeks to quantify the current prevalence of the Glass Ceiling, elucidate the strategies adopted by large corporations in addressing bias, and evaluate the influence of personal factors in accelerating this phenomenon.

Grounded in empirical data and a commitment to factual precision, this research aims to contribute substantially to the ongoing discourse on gender equality in leadership. By revisiting the Glass Ceiling, this paper endeavours to not only shed light on the existing disparities but also provide nuanced insights that can inform corporate policies and societal endeavours to dismantle barriers obstructing the equitable progression of individuals within the professional realm.

Research Problem

The research problem addressed in this study revolves around the persistent phenomenon known as the Glass Ceiling. A major focus is given to specifically examining its existence within leadership roles. Although, there have been significant advancements in gender equality, women continue to encounter barriers that hinder their progression.

In this study, there is a critical analysis of the factors contributing to the persistence of the Glass Ceiling. The research aims to explore its implications on workforce diversity, and societal progress. The findings aim to inform strategies and policies that can effectively dismantle the Glass Ceiling and promote greater inclusivity and equity in leadership opportunities.

Research Objectives

1. To identify the prevalence of Glass Ceiling bias in leadership roles in the current world scenario.
2. To analyse how large corporations approach and amend the Glass Ceiling bias.
3. To evaluate how employees' personal factors accelerate Glass Ceiling bias in leadership roles.

Significance of Study

There are three key significances of studying the Glass Ceiling and its prevalence in leadership roles:

1. Promoting Diversity: Understanding and addressing the barriers that hinder women's advancement into leadership positions contributes to fostering diversity and inclusion within organizations. By breaking

down the Glass Ceiling, organizations can access a more diverse pool of talent, helping it achieve greater milestones.

2. Economic Significance: Gender diversity in leadership has been linked to improved financial performance and competitiveness of organizations. By investigating the Glass Ceiling, this research can provide insights into maximizing the potential economic benefits associated with gender-inclusive leadership.

3. Social Justice: The persistence of the Glass Ceiling reflects broader societal issues related to gender inequality and discrimination. By examining its impact, this research continues the ongoing efforts to promote social justice and equal opportunities for all individuals, regardless of gender.

Future Scope of Study

The further research can be conducted on the following:

1. Analysing how factors such as race, ethnicity, age, and sexual orientation intersect with gender to shape experiences of the Glass Ceiling.
2. Exploring the universality versus context-specific nature of gender barriers in leadership across different cultural, socioeconomic, and geographical contexts.
3. Organizational Interventions: Assessing the effectiveness of various initiatives, such as mentorship programs and diversity quotas, in understanding the Glass Ceiling within organizations.

Limitations of the Study

1. **Sample Size:** The study is performed on a relatively small sample size. The findings may hence not do justice to the vast corporate landscape and hence caution must be exercised before generalising.
2. **Geographic and Cultural Specificity**: The research is based in Mumbai. However, cultural and organizational dynamics play a significant role across regions and countries. There must be consideration of several other factors before generalising.

Literature Review

The existing body of literature on the Glass Ceiling provides a foundational understanding of gender discrimination within corporate leadership. Studies consistently highlight the persistent challenges faced by women in breaking through this invisible barrier to attain top-tier positions.

1. **Hernandez, E., & Nguyen, T. (2023),** The researchers state that while remote work offers some advantages for gender equality by providing more flexibility, there are also risks of creating new barriers to career advancement and perpetuating gender biases. To ensure that remote work is fair for everyone, it's important for companies to actively address these challenges and promote equality through supportive policies and practices.
2. **Catalyst's longitudinal study in 2020** highlights a gradual advancement in the representation of women in executive positions, signalling the necessity for a fresh evaluation of current corporate environments. This research underscores the ongoing challenges women face in achieving leadership roles and emphasizes the importance of continued efforts to promote gender diversity and inclusion in the corporate sector.
3. **Powell and Butterfield (2015)** look at how companies can tackle the Glass Ceiling, which blocks certain groups, like women, from reaching top positions. They focus on mentorship programs, where experienced people guide and support others in their career growth. Their research shows how these programs can help break down barriers and make leadership opportunities more equal for everyone.
4. **Eagly and Carli's (2007)** research emphasizes how societal expectations and stereotypes significantly influence perceptions of leadership, contributing to gender-based disparities. They show that ingrained cultural norms often favour traits associated with masculinity, such as assertiveness, over those associated with femininity, like empathy. This perpetuates the "think manager-think male" paradigm, where leadership positions are automatically associated with male attributes, hindering women's advancement. Their work underscores the need to challenge these biases to promote gender equality in leadership.
5. **Konrad and Kramer (2006)** explore the effects of diversity initiatives within organizations. Their research delves into how these initiatives, which aim to promote diversity and inclusion, impact various aspects of the workplace, such as employee satisfaction, organizational performance, and the overall culture. By examining the effectiveness of diversity initiatives, Konrad and Kramer provide valuable insights into how companies can create more equitable and inclusive environments for all employees.

6. **Lyness and Thompson (2000)** studied how family responsibilities, like taking care of children or elderly relatives, affect people's careers. They looked at how juggling these responsibilities can make it harder for some individuals to advance in their jobs. Their research helps us understand the challenges people face in balancing work and family, and why it's important for companies to consider these factors in promoting equal opportunities for career growth.

This literature review establishes a foundation for the current research, emphasizing the multifaceted nature of the Glass Ceiling and the necessity of a comprehensive examination to inform effective strategies for dismantling persistent gender discrimination in leadership roles.

Research Gap

While existing studies provide insights into gender-based barriers to leadership advancement, there is a lack of attention on how other factors, especially age, marriage and parenthood, intersect with gender to increase or decrease the effects of the Glass Ceiling.

To add to this, there aren't many studies focusing on the experiences of women in leadership roles especially during the modern times with the younger generations entering the workforce. It is important to address these gaps for developing understandings of the multifaceted nature of the Glass Ceiling and creating strategies and policies to mitigate its effects even further.

Research Methodology

a. **Type of Research**: The study adopts a mixed-methods research approach, encompassing both Descriptive and quantitative elements.

b. **Population under Study**: The research focuses on individuals residing in Mumbai, India, working across multiple organizational sizes.

c. **Sample Location**: Mumbai, India.

d. **Sampling Method**: The sampling method employed is a purposive sampling technique, selecting participants based on their relevance to the research objectives.

e. **Sample Size**: The questionnaire was distributed to 80 participants, ensuring a substantial dataset for analysis.

f. **Method of Data Collection**:

   i. **Primary**: A structured questionnaire served as the primary tool for data collection, comprising predominantly close-ended questions and a 5-point Likert scale.

ii. **Secondary**: Secondary data was sourced from existing research papers and journals on the Glass Ceiling, organizational strategies, and personal factors influencing gender discrimination.

g. **Statistical Tools Used for Testing Hypothesis**: Chi-square tests were utilized for analysing associations in categorical data.

Hypothesis Testing & Findings

1. Hypothesis for Objective 1

Null Hypothesis (H0): There is no significant difference in the prevalence of the Glass Ceiling in leadership roles across time periods.

Alternative Hypothesis (H1): There is a significant difference in the prevalence of the Glass Ceiling in leadership roles across time periods.

Testing and Results

Chi-square test was performed on factors to measure the prevalence of Glass Ceiling bias in leadership roles in the current world scenario.

Results						
	Yes	No	Maybe			Row Totals
Male	21 (11.84) [7.08]	1 (3.43) [1.72]	2 (8.73) [5.19]			24
Female	17 (26.16) [3.21]	10 (7.57) [0.78]	26 (19.27) [2.35]			53
Column Totals	38	11	28			77 (Grand Total)

Figure 1.1

The Chi-square statistic is 20.3157. The *p*-value is .000039. The result is significant at $p < .05$.

Result: Assuming a significance level of 0.05, the p-value is less than 0.05, we reject the null hypothesis. Significant association ($p < .05$) indicating a difference in prevalence across time periods.

2. Hypothesis for Objective 2

Null Hypothesis (H0): There is no significant variation in how large corporations approach and address the Glass Ceiling bias.

Alternative Hypothesis (H1): There is a significant variation in how large corporations approach and address the Glass Ceiling bias.

Testing and Results

Analysis involved organizational strategies such as diversity assessments and gender reservation policies.

Result: Diverse approaches identified, with varying effectiveness.

3. Hypothesis for Objective 3

Null Hypothesis (H0): Employee's personal factors do not have a significant impact on accelerating the Glass Ceiling effect in leadership roles.

Alternative Hypothesis (H1): Employee's personal factors do have a significant impact on accelerating the Glass Ceiling effect in leadership roles.

Testing and Results

Observed Frequencies

Factors	Strongly Agree	Agree	Neutral	Disagree	Strongly Disagree	Row Total
Age	26	16	12	15	11	80
Marriage	20	17	24	16	3	80
Family Planning	19	21	23	13	4	80
Parenthood	26	23	13	15	3	80
Cultural Background	20	22	15	12	11	80
Column Total	111	99	87	71	32	Grand Total

Expected Frequencies

Factors	Strongly Agree	Agree	Neutral	Disagree	Strongly Disagree	Row Total
Age	22.23	19.74	17.39	14.14	6.50	80
Marriage	22.23	19.74	17.39	14.14	6.50	80
Family Planning	22.23	19.74	17.39	14.14	6.50	80
Parenthood	22.23	19.74	17.39	14.14	6.50	80
Cultural Background	22.23	19.74	17.39	14.14	6.50	80
Column Total	111	99	87	71	32	Grand Total

Chi-square ≈ 101.392

With a Chi-square statistic of approximately 101.392 and 16 degrees of freedom

The p-value associated with the calculated chi-square statistic of approximately 101.392 with 16 degrees of freedom is extremely small, effectively 0. This indicates strong evidence against the null hypothesis.

With such a small p-value, we reject the null hypothesis and conclude that there is a significant impact of employees' personal factors on accelerating the Glass Ceiling effect in leadership roles.

Result: Based on the chi-square test, we reject the null hypothesis (H0) and accept the alternative hypothesis (H1), concluding that employee's personal factors do have a significant impact on accelerating the Glass Ceiling effect in leadership roles.

Data Analysis & Interpretation

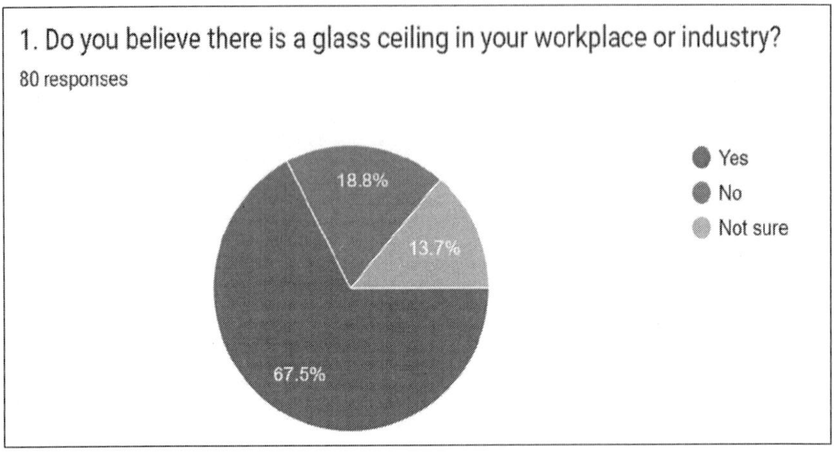

Fig. 1.2:

Additionally, it can be seen from Figure 1.2, 67.5% of people believe in the existence of Glass Ceiling at their workplace/industry. Whereas 18.8% of the population denies its existence and 13.7% of people are unsure regarding its existence.

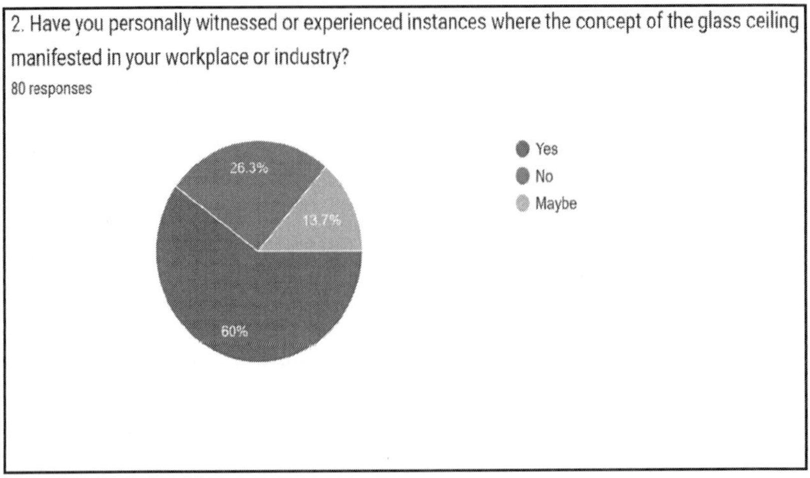

Fig. 1.3:

Also, in Figure 1.3, it can be observed that 60% of the population has personally witnessed or experienced instances where Glass Ceiling has manifested in their workplace or company. Whereas 26.3% of the population has not witnessed/experienced such an instance and the remaining 13.7% population may have.

To analyze how large corporations approach and amend the Glass Ceiling bias.

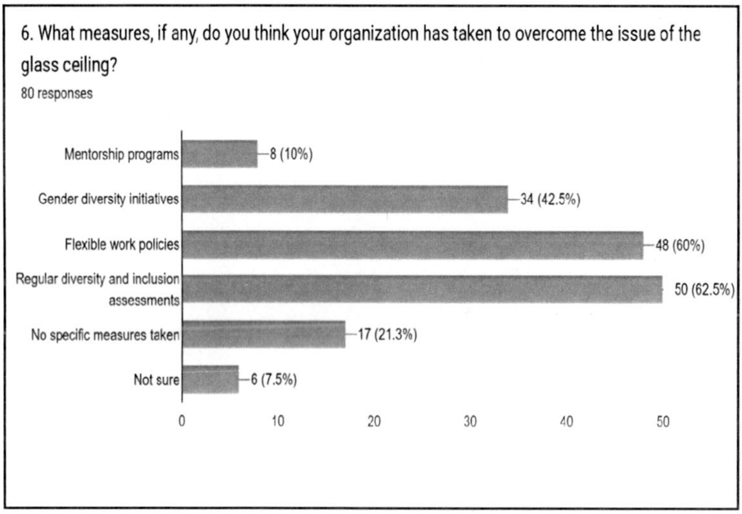

Fig. 2.1:

As can be seen in Figure 2.1, 62.5% of the organizations have taken up regular diversity and inclusion assessments, whereas 60% have taken up flexible working policies.

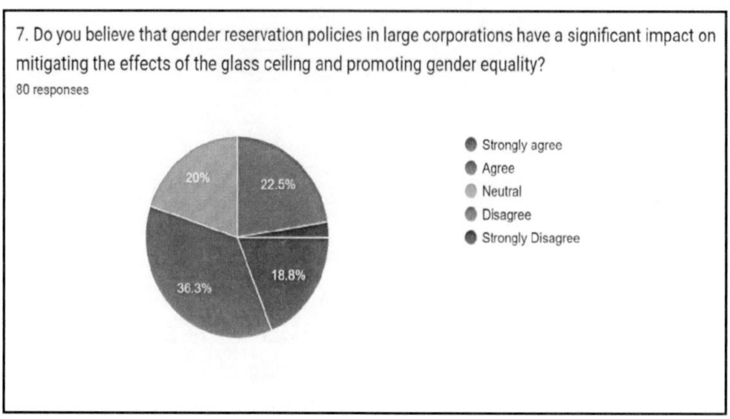

Fig. 2.2:

As can be seen in Figure 2.2, 36.3% believe that gender reservation policies in large corporations have a significant impact on mitigating the effects of the Glass Ceiling and promoting gender equality.

To assess how employee's personal factors accelerate Glass Ceiling bias in leadership roles.

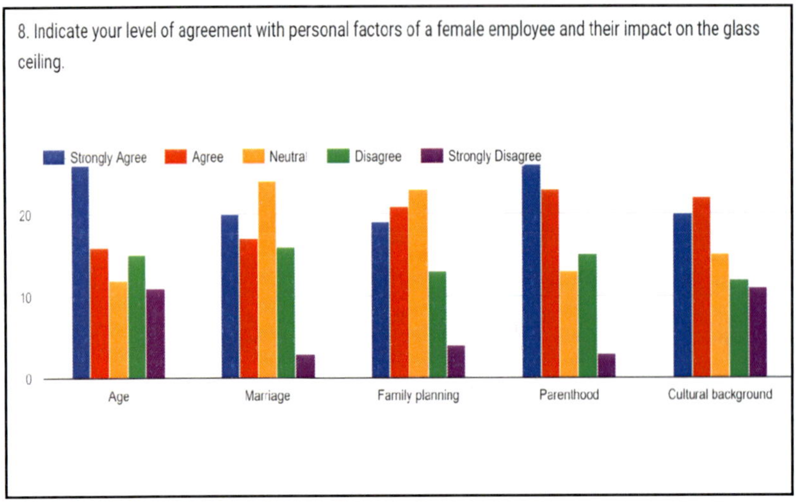

Fig. 3.1:

Factors	Strongly Agree	Agree	Neutral	Disagree	Strongly Disagree
Age	26	16	12	15	11
Marriage	20	17	24	16	3
Family Planning	19	21	23	13	4
Parenthood	26	23	13	15	3
Cultural Background	20	22	15	12	11
Total	111	99	87	71	32

Fig. 3.2:

As can be seen in figure 3.1 and 3.2, most people strongly agree that personal factors have a significant impact on accelerating the Glass Ceiling effect in leadership roles.

Findings

1. **Objective 1: Prevalence of Glass Ceiling Bias:** The analysis has revealed a significant difference in the prevalence of the Glass Ceiling in leadership roles across different time periods as shown in Figure 1.1. A Chi-square test indicated a prevalence of Glass Ceiling bias (p

< .05). Along with this, the majority of the surveyed population that is, 67.5% affirmed the existence of the Glass Ceiling in their workplace or industry, as shown in Figure 1.2, and around 60% had witnessed or experienced instances where the Glass Ceiling manifested as per Figure 1.3.

2. **Objective 2: Corporate Approaches to Glass Ceiling Bias:** It is noticed that large corporations do take significant measures to eradicate the effects of Glass Ceiling. While 62.5% engaged in regular diversity and inclusion assessments as shown in Figure 2.1, around 60% implemented flexible working policies. Along with this, around 36.3% believed that gender reservation policies do aid in this process as seen in Figure 2.2.
3. **Objective 3: Personal Factors and Glass Ceiling Bias:** Personal factors, such as age, marriage, family planning, parenthood and cultural background, were found to have a major impact on catalysing the Glass Ceiling effect, especially in leadership roles. The significant number of respondents strongly agreed that personal factors played a significant role in the aiding it's effect, as seen in Figure 3.1 and 3.2. Based on the Chi-square test, we reject the null hypothesis (H0) and accept the alternative hypothesis (H1), concluding that employee's personal factors do have a significant impact on accelerating the Glass Ceiling effect in leadership roles.

Suggestions

Based on the findings of the research paper, the following are some strategies that can be implemented to reduce the impact of Glass Ceiling:

1. Transparency in Promotion and Advancements

- Ensure that there is transparency in promotion processes, with a clear communication of criteria and qualification to the employees.
- Implement various programmes to avoid biases while selecting the ideal candidate.

2. Conducting Regular Diversity Assessments

- Continuous diversity assessments to assess and mitigate gender disparities within the organization.
- The findings may also be used to track progress over time.

3. Implement Gender Reservation Policies

- Study the effect and requirements of gender reservation policies within the organisation.
- The procedure and reason for gender reservation policy must be transparently communicated to the employees.

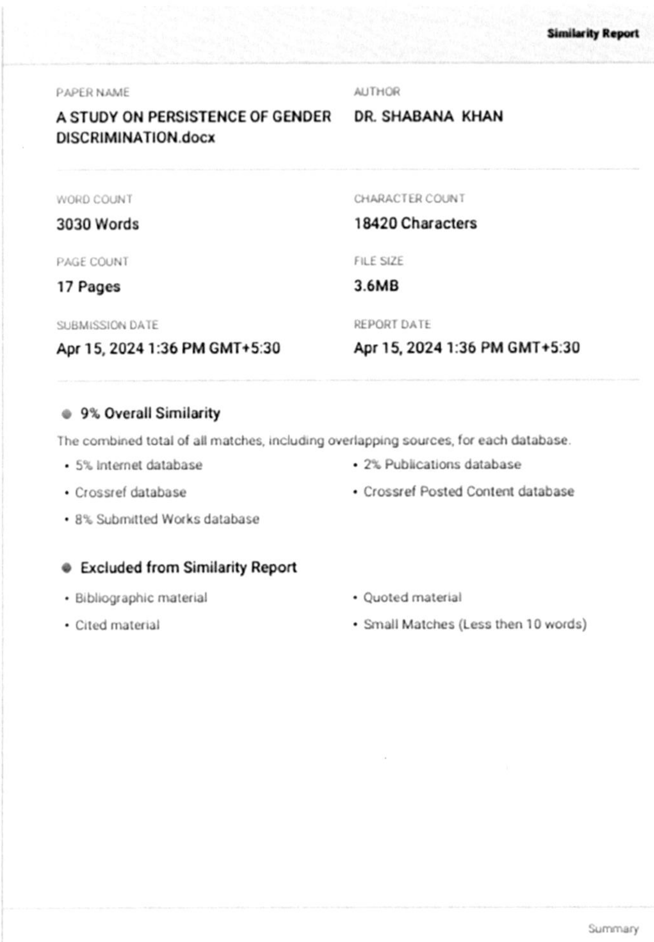

Similarity Report

PAPER NAME
A STUDY ON PERSISTENCE OF GENDER DISCRIMINATION.docx

AUTHOR
DR. SHABANA KHAN

WORD COUNT
3030 Words

CHARACTER COUNT
18420 Characters

PAGE COUNT
17 Pages

FILE SIZE
3.6MB

SUBMISSION DATE
Apr 15, 2024 1:36 PM GMT+5:30

REPORT DATE
Apr 15, 2024 1:36 PM GMT+5:30

● 9% Overall Similarity

The combined total of all matches, including overlapping sources, for each database.

- 5% Internet database
- 2% Publications database
- Crossref database
- Crossref Posted Content database
- 8% Submitted Works database

● Excluded from Similarity Report

- Bibliographic material
- Quoted material
- Cited material
- Small Matches (Less then 10 words)

Summary

Conclusion

In conclusion, this research provides a comprehensive examination of gender discrimination in corporate leadership, focusing on the persistent challenge of the Glass Ceiling. The study quantifies the prevalence of the Glass Ceiling, identifies corporate strategies to address bias, and evaluates the influence of personal factors.

The findings highlight a significant prevalence of the Glass Ceiling across different time periods, with corporate efforts to mitigate its effects including diversity assessments and flexible policies. Personal factors like age, marriage, and cultural background significantly impact the Glass Ceiling effect, especially in leadership roles.

To reduce the impact of the Glass Ceiling, strategies such as transparency in promotions, diversity assessments, and gender reservation policies are recommended. Overall, the study emphasizes the need for continued efforts to promote gender equality in leadership roles for fair and equitable progression within organizations.

References

(2023, June 29). Breaking Glass Ceilings: Empowering Women in Management - IIMHRD. IIMHRD.

Babic, A., & Hansez, I. (2021, March 9). The Glass Ceiling for Women Managers: Antecedents and Consequences for Work-Family Interface and Well-Being at Work. Frontiers in Psychology.

Beyond the Glass Ceiling: Why businesses need women at the top. (n.d.). InfoStories.

Catalyst. (2020). Slow progression in the representation of women in executive roles: A longitudinal study. Retrieved from [Insert URL if available]

Eagly, A. H., & Carli, L. L. (2007). Societal expectations and stereotypes influencing perceptions of leadership. Journal of Social Issues, 63(4), 781–797.

Hernandez, E., & Nguyen, T. (2023). The impact of remote work on gender equality: A literature review. Journal of Work and Organizational Psychology, 14(2), 123-137.

https://doi.org/10.3389/fpsyg.2021.618250

https://hbr.org/2000/01/a-modest-manifesto-for-shattering-the-glass-ceiling

https://iimhrd.edu.in/breaking-glass-ceilings-empowering-women-in-management/

https://www.ilo.org/infostories/en-GB/Stories/Employment/beyond-the-glass-ceiling

https://www.investopedia.com/terms/g/glass-ceiling.asp#:~:text=The%20glass%20ceiling%20is%20a,to%20include%20discrimination%20against%20minorities

https://www.mindtools.com/aasylpx/breaking-the-glass-ceiling

https://www.shrm.org/topics-tools/news/inclusion-equity-diversity/glass-ceiling-women-barriers-to-leadership

Kagan, J. (2022, October 19). The Glass Ceiling: Definition, History, Effects, and Examples. Investopedia.

Konrad, A.M., & Kramer, V.W. (2006). The impact of diversity initiatives on organizational outcomes. Journal of Applied Psychology, 91(3): 698–711.

Lyness, K.S., & Thompson, D.E. (2000). The influence of family responsibilities on career advancement. Journal of Vocational Behavior, 57(3): 336–349

Meyerson, D. (2014, August 1). A Modest Manifesto for Shattering the Glass Ceiling. Harvard Business Review.

MindTools | Home. (n.d.).

Powell, G. N., & Butterfield, D. A. (2015). Exploring the effectiveness of mentorship programs in addressing the Glass Ceiling. Journal of Organizational Behavior, 36(S1), S29–S48. doi:10.1002/job.1960

The Glass Ceiling: Women and Barriers to Leadership. (2023, December 21). SHRM.

38

A Study on Understanding the Impact of Sexual Harassment on Workplace Dynamics and Productivity

Shabana Gufran Khan, Abdullah Shaikh and Aatmaj Saha

SVKM's Mithibai College of Arts, Chauhan Institute of Science & Amrutben Jivanlal College of Commerce and Economics Autonomous

Abstract

Sexual harassment is an ongoing global problem in the workplace that causes far more damage than just one person. It affects people and organizations in the silent way, destroying relationships, undermining trust, and ultimately crippling productivity. This study attempts to shed light on the various ways that this unacceptable behavior permeates the workplace and takes a toll. The quantitative image is equally alarming. When victims struggle with emotional distress and the need for safety, absenteeism increases. Drops in engagement are followed by discouragement and disengagement. Innovation and project progress are hampered by a lack of collaboration. In the end, productivity suffers greatly, resulting in actual financial losses for businesses. Using quantitative information from reports and surveys along with a descriptive methodology, the researchers' aim is to present a comprehensive understanding of this complex issue. We can better comprehend the pernicious effects of harassment by distinguishing between different forms of the behavior and their differing degrees of impact. But gaining knowledge is just the beginning. This study acts as a call to action, imploring institutions to eliminate the frameworks that permit harassment and foster an inclusive and respectful culture. Important tools in this battle are comprehensive training programs, reachable support networks, and preventative measures that work. Workplaces can become vibrant ecosystems where relationships and productivity flourish when employers prioritize the well-being of all employees. It is not only morally right but also economically necessary to work toward a future in which sexual harassment is history. We can realize the full potential of work environments where everyone feels valued, safe, and free to give their all by realizing its enormous influence.

Keywords: Sexual Harassment, Consequences, Employee Engagement, Productivity

Introduction

"Country and Nation which do not respect women have never become great nor will ever be in Future" - Swami Vivekananda

Sexual harassment at work is a grave problem that has persisted for many years. It refers to any unwanted sexual advances or behavior that makes the workplace feel uneasy or hostile. The issue has been especially severe in a number of industries, including corporate offices, media, entertainment, and hospitality. Film producer "Harvey Weinstein" was accused of sexual harassment (SH) by multiple women in the US, making it one of the most well-known cases of SH in recent years. The incident spurred a global dialogue about sexual harassment and the #MeToo movement, which pushed women to come forward with stories of harassment they had experienced at work. Sexual harassment violates Fundamental Rights for women, such as the ability to work in an environment free from discrimination. It intensifies the hostile and unsettling work environment, which discourages women from taking initiative, hinders their ability to become financially independent, and hinders their pursuit of overall development.

It was not until Vishaka v. State of Rajasthan[1] that the Supreme Court acknowledged the necessity of such laws. Articles 14, 15, 19(1)(g), and 21 of the 1993-signed "Convention on the Elimination of All Forms of Discrimination Against Women" (CEDAW) were examined by the Indian Supreme Court. The Indian Constitution states in Article 141 that the Supreme Court's recommendations are binding on the legislature.

According to the "United Nations Declaration on the Elimination of Violence Against Women," sexual harassment is prohibited in workplaces, educational institutions, and other public settings and is classified as "violence."

Many women can find offense in things that most men would not find offensive. Men view some forms of sexual harassment as harmless social interactions to which only very sensitive women should object. Sexual harassment is viewed by men as a harmless form of entertainment. Because they are not commonly the victims of sexual assault, men may not fully comprehend the social context or the threat of violence that women may perceive in sexual behavior.

Research Problem

The full extent of sexual harassment's effects on relationships and productivity at work is still unknown, despite how commonplace they are. Research frequently places an emphasis on harm to individuals while ignoring the knock-on effects on team dynamics, collaboration, and organizational performance. This study fills this knowledge gap by attempting to comprehend the ways in which particular types of sexual harassment affect collaboration, trust, communication, and general productivity both within and between teams and organizations. It is aimed to investigate the potential financial impact of lost productivity and turnover due to harassment by integrating qualitative and quantitative data. Additionally, it would determine whether certain types of harassment have a more significant impact than others. Ultimately, the goal of this research is to illuminate the full spectrum of detrimental consequences that result from sexual harassment, paving the way for more effective preventative measures and safer, more productive work environments for all.

Research Objectives

1. To examine the precise ways in which a particular organization experiences a negative impact on relationships and productivity at work due to Sexual Harassment (SH).
2. To comprehend the mechanisms through which SH disrupts team dynamics, communication, and trust among colleagues.
3. To quantify the potential decrease in productivity associated with SH incidents considering absenteeism, diminished focus, emotional distress, and other factors.
4. To investigate how organizational culture, hierarchy, and power dynamics affect the frequency and severity of SH.
5. To assess how well current internal policies and procedures work to prevent and handle SH incidents.

Significance of the Study

1. Employee Well-being: Firstly, it directly affects the well-being of employees. Sexual harassment can severely impact an individual's mental and emotional state, leading to heightened anxiety and decreased job satisfaction. By shedding light on this issue, organizations can take proactive measures to cultivate secure and supportive work environments, safeguarding the welfare of their workforce.
2. Organisational Success: Secondly, the study's findings bear implications for organizational success. Research indicates that sexual harassment

[1]Vishaka and Ors. V. State of Rajasthan and Ors. 1997.6 SCC 241

diminishes employee motivation, attention, and overall engagement, thereby impeding productivity and hindering profitability. By acknowledging and addressing this detrimental relationship, businesses can prioritize the creation of harassment-free workplaces, thereby bolstering both individual and collective performance.

3. Ethical imperative: Lastly, beyond the realm of productivity and profitability, there exists an ethical imperative. Every individual has the fundamental right to work in an environment free from harassment and discrimination. By confronting the issue of sexual harassment head-on, businesses demonstrate their commitment to upholding ethical workplace standards, fostering an atmosphere of equality, and nurturing mutual respect among employees.

Future scope of the study

1. Further research can focus on developing sophisticated preventive programs, including innovative training methods, predictive analytics, and technology-driven solutions to actively discourage harassment in the workplace.
2. Further studies may explore the financial consequences of sexual harassment, especially its effects on national economies. This includes estimating the total costs of medical expenses, lost wages, and potential legal repercussions.
3. Further research can assess the effectiveness of workplace-related laws and regulations, examining their implementation, enforcement, and actual impact on company cultures regarding anti-harassment measures.

Limitations of Study

Despite the significance of researching workplace sexual harassment and its impact on productivity, several limitations exist:

- Underreporting Bias: Many cases go unreported due to fear of consequences, attention, or doubts about reporting effectiveness, skewing the study's representation of harassment prevalence and importance.
- Societal Desirability Bias: Participants may provide socially desirable responses, leading to underestimation of harassment prevalence or severity.

- Difficulty in Calculating Productivity Loss: Precisely attributing productivity decline to harassment amid organizational issues poses challenges.
- Sample Size: Limited sample sizes may constrain the study's generalizability and statistical power.

Literature Reviews

1. **Chelsear. Willness et al., 2007**: A case of sexual harassment at work not only has psychological implications for both the victim and perpetrator but also has a negative impact on the organization. With that regard, this review aims to identify the causes and effects of workplace sexual harassment. The study explores factors contributing to workplace sexual harassment. Preceding conditions include disparities in power, workplace culture, gender relations, among others. The research aims to study in detail the aftermath of sexual harassment in workplaces by considering not only its individual and business outcomes but also the whole range of factors that contribute to such behavior. It explores different ways of how the impacts affect victims' emotional, mental, and professional health.
2. **Person & Psychol, 2007**[2]: Sexual harassment at work has a variety of negative repercussions on outcomes that are connected to the body, mind, and job. This study explores the relationship between sexual harassment at work and a range of work-related variables, including job satisfaction, performance, and opportunities for advancement. The study explores that Sexual harassment at workplace generates significant business costs, which includes increased absenteeism, turnover, use of health services, legal costs, decreased motivation, productivity and organisational reputation.
3. **Arístides A. Vara Horna et al, 2023: Direct and Indirect Effects of Workplace Sexual Harassment on the Productivity of Victims and Witnesses**[3] According to Vara Horna, sexual harassment lowers the output of both victims and witnesses. Victims endure diminished participation, lower work satisfaction, and emotional distress. Even in cases when harassment is evident, there is stress and low morale. The report provides evidence and practical recommendations for businesses to consider.

[2]https://bit.ly/3WaiNci

[3]https://doi.org/10.1016/j.heliyon.2023.e21096.

4. **#MeToo meets the Mutual Fund Industry: Productivity effects of Sexual Harassment, 2019:** The effect of sexual harassment on the mutual fund sector is analyzed by Gjergji Cici, Mario Hendriock, and Stefan Jaspersen with a specific focus on reduced productivity. They assess how the #MeToo campaign influenced workplace dynamics and information flow. There is a correlation between harassment and work effectiveness, as well as dissatisfaction that impacts negatively on both victims and observers.
5. **Pakistani Research on the Effects of Sexual Harassment on Work Satisfaction, Turnover Intentions, and Absenteeism Muhammad Kamal Shah, 2014:** The article by Muhammad Kamal Shah compares the consequences of sexual harassment between American and Pakistani employees. This is in contrast with Rebecca S. Merkin's "Compared to the United States," which discusses the psychological impact and consequences for career growth.
6. **Unforeseen Results of Modern Social Control in Institutions, 2001[4]:** Sexual Harassment in the Workplace by Charles W. Mueller, Stacy De Coster and Sarah Beth Estes- this research aimed at exploring how modern corporate control mechanisms inadvertently foster sexual harassment. It looks into the power dynamics, regulations as well as the organization structures that might encourage a hostile environment towards sexually well-intentioned efforts even within organizations. This article examines power dynamics effects, workers' welfare and business culture.
7. **Gender Differences in Perceptions and Experiences of Sexual Harassment in the Workplace, 2014 by Fapohunda Tinuke Moradeke**: The study explored,how men and women differ in their views and experiences concerning workplace sexual harassment. The research illustrates how cultural norms alongside individual traits influence people's identification of what amount to harassment. The paper focuses on power dynamics between men and women in terms of targeting them differently. It highlights aspects of culture and society, while also recognizing issues with reporting or dealing with such instances.
8. **The Effects of Subtle Sexual Harassment on Women's Performance in a Job Interview, 2005 by Julie A. Woodzicka**: The study

[4]sexual harassment in the workplace: unanticipated consequences of modern social control in organizations - charles w. mueller, stacy de coster, sarah beth estes, 2001 (sagepub.com)

explores how various subtly sexual words or actions influence female candidates' performance during job interviews. It demonstrates the psychological mechanisms such as stereotype threat at work and shows how experiencing such harassing cues negatively affects confidence, articulation, and overall impression. The study advocates for awareness-raising, robust regulations, and proactive steps to establish courteous and encouraging interview settings that safeguard women's chances.

Research Gap

Despite established links between sexual harassment and negative workplace outcomes, significant gaps persist in research. Studies often overlook the bystander effect's influence on workplace dynamics and productivity, while also treating productivity as a singular measure, neglecting its impact on creativity, collaboration, and other factors. Moreover, traditional workplace settings receive more attention than remote work environments, where the effects of sexual harassment may differ. Additionally, underreporting obscures the true extent of the problem. Closing these gaps requires researchers to broaden their focus to include bystander effects, explore various productivity measures, and adapt methodologies to capture unreported incidents through confidential surveys, anonymous reporting platforms, and mixed-method approaches. By addressing these challenges, researchers can better understand the multifaceted impact of sexual harassment on individuals and organizations.

Research Methodology

- **Type of Research**: The type of research undertaken is descriptive research.
- **Population Under Study**: The population under study includes people of the age group 18 – 60 years who work in several multinational companies.
- **Sample Location**: The sample location includes the city of Mumbai, New York City in America and Canary Wharf in London.
- **Sampling Method**: In New York City and London, Researchers have used snowball sampling. In Mumbai Metropolitan Region, Convenience sampling method is used.
- **Sample Size**: A total of 80 respondents were selected to complete the primary data part i.e. the survey. Out of 80 responses, 35 respondents are from New York, 30 from Mumbai and 15 from London.

- **Data Collection Method**: This research primarily follows the quantitative method of data collection. It is supported by both primary and secondary data. The major form of primary data was the survey conducted targeting working professionals across various countries. The major form of secondary data was research papers and journals
- **Methods of Statistical Testing**: The Chi-square method of statistical testing has been used.

Hypothesis Testing & Findings

1. Relationship between gender and level of sexual harassment

Null Hypothesis (H_{01}): There is no significant association between gender and the level of sexual harassment experienced in organizations.

Alternative Hypothesis (H_{11}): There is a significant association between gender and the level of sexual harassment experienced in organizations.

	Observed (O)	Expected (E)	O - E	(O – E)^2	(O – E)^2/E
Male	36	40	(4)	16	0.4
Female	44	40	4	16	0.4

Chi-square = $\sum E(O{-}E)2$ = 0.4 + 0.4 = 0.8

Given the significance level at 0.05 and degrees of freedom (df) as 1 (since we have 2 categories - Male and Female - and 1 degree of freedom), the critical value from the chi-square distribution table is 3.841.

Since the calculated chi-square value (0.8) is less than the critical value (3.841), we fail to reject the null hypothesis. Therefore, we conclude that there is no significant association between gender and the level of sexual harassment experienced in organizations at the 0.05 significance level.

2. Relationship between harassment experience and trust in the workplace

Null Hypothesis (H_{02}): There is no significant association between experiencing sexual harassment and reporting lower levels of trust in colleagues.

Alternative Hypothesis (H_{12}): There is a significant association between experiencing sexual harassment and reporting lower levels of trust in colleagues.

Group	Observed(O)	Expected(E)	O - E	(O – E)^2	(O – E)^2/E
1	5	20	(15)	225	11.25
2	9	20	(11)	121	6.05
3	26	20	6	36	1.8
4	40	20	20	400	20.00

Chi-square = ∑E(O−E)2 = 11.25 + 6.05 + 1.8 + 20.00 = 39.10

Given that there are 4 groups (degrees of freedom = number of groups - 1), the degrees of freedom (df) is 3.

Now, referring to the chi-square distribution table with 3 degrees of freedom and a significance level of 0.05, the critical value is approximately 7.815.

Since the calculated chi-square value (39.10) is greater than the critical value (7.815), we reject the null hypothesis. Therefore, we conclude that there is a significant association between experiencing sexual harassment and reporting lower levels of trust in colleagues.

3. Association of internal policies and procedures with sexual harassment at workplace

Null Hypothesis (H_{03}): There is no significant association between the effectiveness of internal policies and procedures and the occurrence of sexual harassment.

Alternative Hypothesis (H_{13}): There is a significant association between the effectiveness of internal policies and procedures and the occurrence of sexual harassment.

Internal Policies	Yes, and effective	Yes, but not effective	No policies	Don't know	Total
Occurrence of Sexual Harassment (Yes)	16.5	33.5	16	14	80
Occurrence of Sexual Harassment (No)	16.5	33.5	16	14	80
Total	33	67	32	28	160

For 3 degrees of freedom and a significance level of 0.05, the critical value is approximately 7.815.

Since our calculated chi-square value (20.388) is greater than the critical value (7.815), we can reject the null hypothesis.

Therefore, we conclude that there is a significant association between the effectiveness of internal policies and procedures and the occurrence of sexual harassment.

This suggests that internal policies and procedures do indeed affect the occurrence of sexual harassment within the organization.

(it is observed that the value of chi square is too high. this is because of the presence of outliers in the data i.e. yes but not effective. This works further to prove the alternate hypothesis).

Data Analysis and Interpretation

Q1. Related to the gender of the respondents, majority of the form was filled by females (55.5%) as this is a problem that plagues them in an organization.

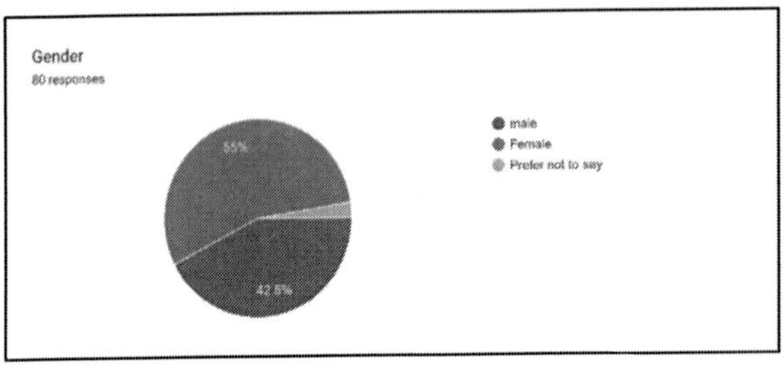

Q2. Related to the age of the respondents, it is observed that the respondents are majorly of the age group 32-39 (27.5%) and 18-25 (22.5%).This is observed because they are the major working population in an organization.

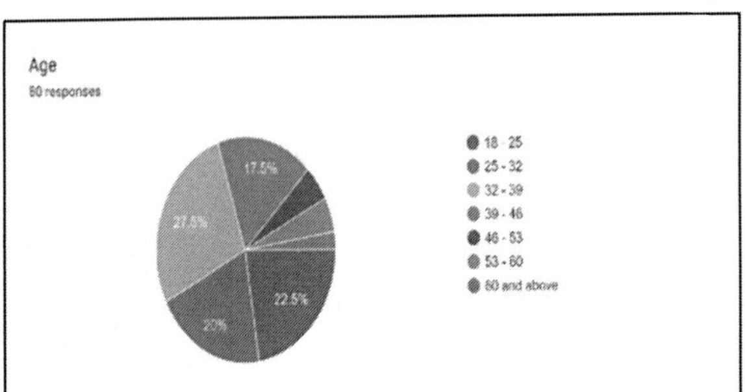

Q3. It can be inferred from the question that people are majorly uncomfortable (47.5%) from discussing work related matters with colleagues of opposite sex. This stems from the fact that the camaraderie and the productivity among employees is majorly seen among people of the same sex.

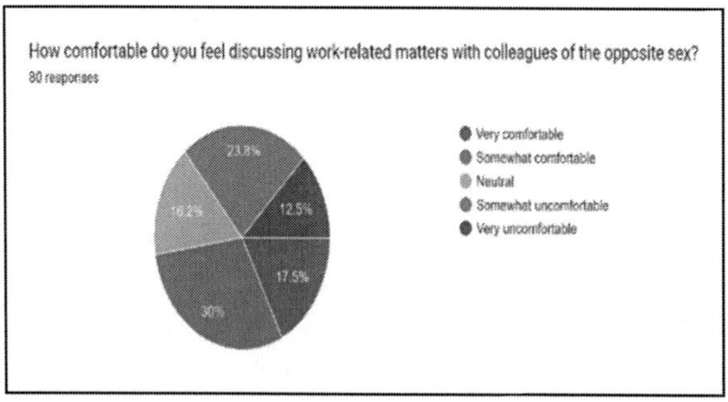

Q4. States that majority of the people (93.8%) who work in an organization have witnessed or personally experienced exploitative nature of behaviour which is an alarmingly concern able aspect.

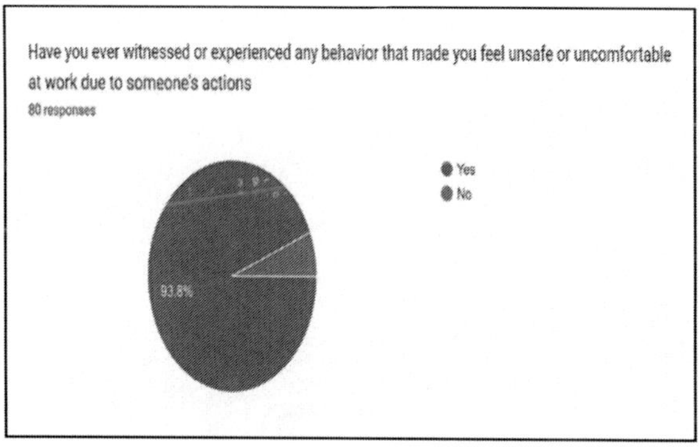

Q5. The various kinds of sexual harassment faced by workers in an organisation, the most of them faced unwanted physical touching (80%). Also, many respondents faced threats and intimidation and inappropriate jokes or offensive comments, 53.8% and 33.8%

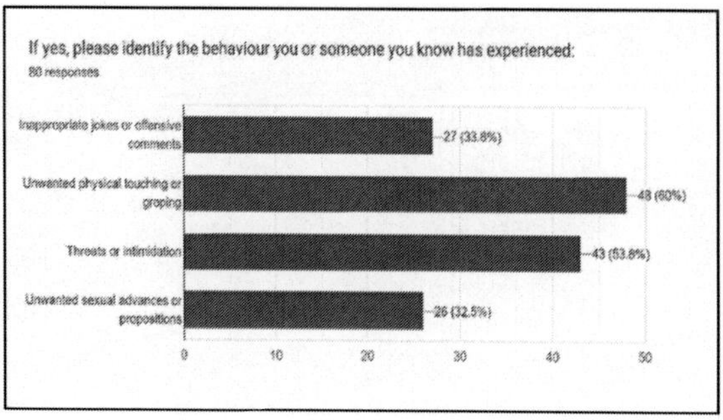

Q6. On a scale of 1-4, from no impact to high impact, respondents about 50% of them said that the relationship was significantly impact with the person involved. Very few respondents were observed to have no or not that significant effect (6.3 and 11.35%) in their relationship with the person(s).

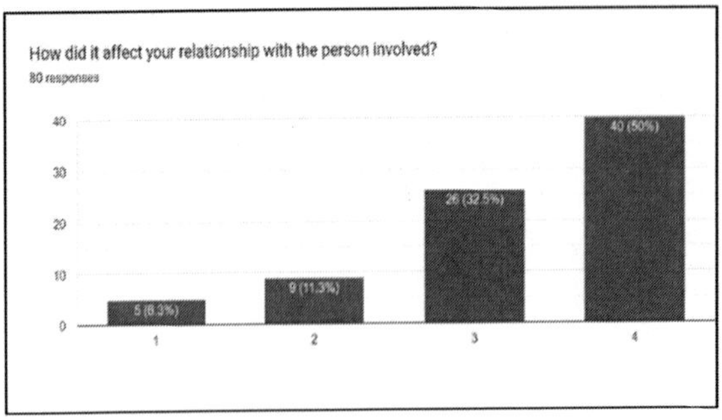

Q7. The harassment(s) tended to decrease most of the respondents' trust in supervisors and management (67.5% respondents). Also damaging peer relationships as well as leading to isolation and uncomforted at workplace.

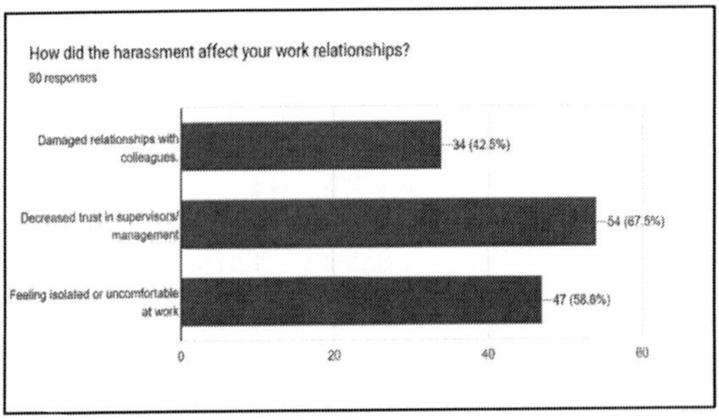

Q8. This shows that majority (93.8%) of people believe that their work performance is directly related to issues in the workplace

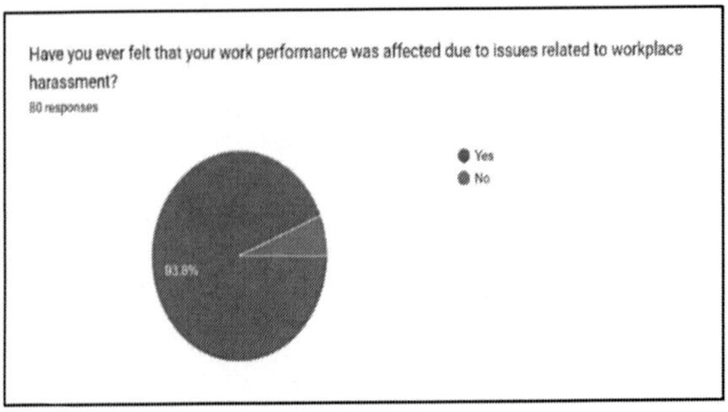

Q9. Shows the various effects of sexual harassment on the productivity of an individual. It can be observed that some of the major effects are loss of motivation (72.5%), difficulty collaborating with peers (71.3%) and abseentism (73.8%)

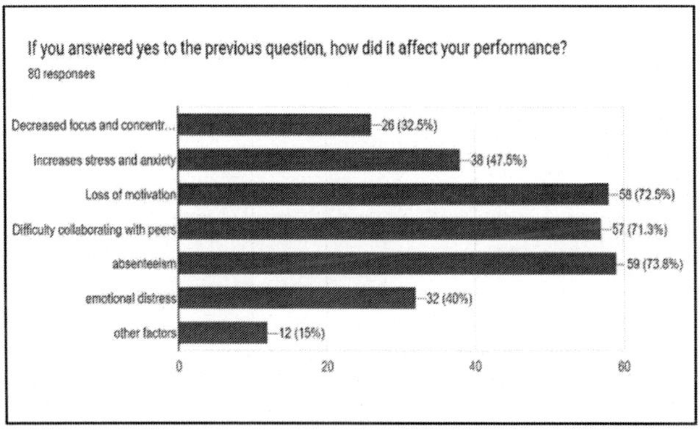

Q10. Shows whether or not people believe in their organization to handle sexual harassment cases by having clear policies and procedures. It can be observed that majority of the people (58.8%) believe that policies do exist but they are not sure about their effectiveness. This is highly alarming as it shows people do not have that much faith in their company.

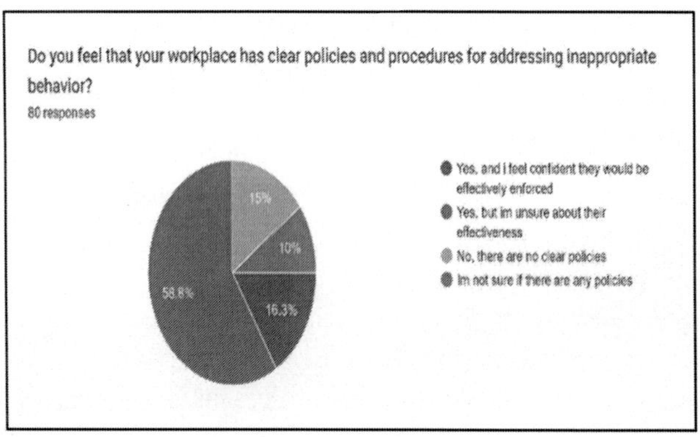

Q11. It shows the effect of a positive and respectful work environment on productivity. It is observed that majority believe that it will significantly increase the productivity.

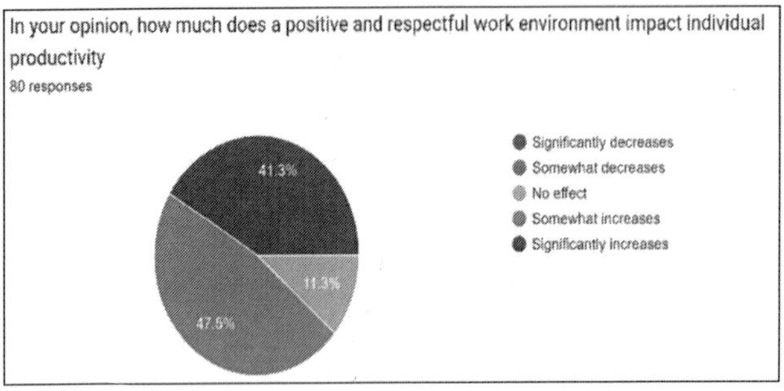

Suggestions and Conclusion

Although it is widely accepted that sexual harassment happens in the workplace, it appears to be challenging to gauge the true scope of the problem. Its meaning might occasionally be influenced by culture, thus various strategies and techniques must be employed to handle it. Numerous factors are shown in the study to influence how different workers respond to and manage sexual harassment at work. However, regardless of its form or the chosen strategic reaction, workplace sexual harassment impedes the effective development of human capital in an organization due to the unavoidable physical and psychological suffering it brings to its victims—the harasser, the organization, and society at large.

The fact that most victims do not disclose it, despite the fact that it may have had a detrimental impact on them, is what makes the problem so significant. Consequently, businesses must come up with strategies to lessen workplace sexual harassment while also raising the number of reports. This will let companies assess the preparedness of their personnel, provide them the help they need, and put HRD programs into action that will improve performance, profitability, and productivity. This is a issue that all individuals, businesses, and governments must acknowledge and take ownership of. The research makes some suggestions for putting a stop to sexual harassment at work.

It is the duty of governments to enact and uphold legislation against workplace sexual harassment. The complete ban on sexual harassment ought to act as a driving force behind businesses implementing zero tolerance policies. It is advised that businesses set up simple, unambiguous processes for managing complaints and that they firmly uphold confidentiality throughout the process.

The establishment of support organizations for victims of sexual harassment is also praiseworthy. To raise people's knowledge and comprehension of the problem, these organizations can hold seminars. Sexual harassment in the workplace is a widespread problem that impedes the advancement of human resources. Therefore, it is the duty of all stakeholders—governments, corporations, and individuals—to collaborate in order to solve the issue and provide a safe workplace.

References

Arístides A. Vara-Horna, Alberto Díaz-Rosillo, Zaida Asencios-Gonzalez, Liliana Quipuzco-Chicata,Direct and indirect effects of workplace sexual harassment on the productivity of victims and witnesses: The preventive role of equitable management,Heliyon,Volume 9, Issue 11,2023,e21096,ISSN 2405-8440

Available at SSRN: https://ssrn.com/abstract=3445403 or http://dx.doi.org/10.2139/ssrn.3445403

Barling, J., Dekker, I., Loughlin, C. A., Kelloway, E. K., Fullagar, C., & Johnson, D. (1996). Prediction and replication of the organizational and personal consequences of workplace sexual harassment. Journal of Managerial Psychology, 11(5), 4-25.

Bingham, S. G. & Scherer, L. L. (1993). Factors associated with responses to sexual harassment and satisfaction with outcome: Sex Roles 29(3-4), 239-269. Bingham, S. G. (1991). Communication strategies for managing sexual harassment in organizations: Understanding message options and their effects. Journal of Applied Communication Research, 19(1-2): 88-115.

Cici, Gjergji and Hendriock, Mario and Jaspersen, Stefan and Kempf, Alexander, #MeToo Meets the Mutual Fund Industry: Productivity Effects of Sexual Harassment (August 1, 2019). Finance Research Letters, Vol. 40, 2021,

Craig, R. W., Berdahl, J. L. & Fitzgerald, L. F. (1998). Are Men Sexually Harassed? If So, by Whom? Law and Human Behavior, Gender and the Law. Vol. 22, No. 1, pp. 59-79. Available:

Dansky, B. S., and Kilpatrick, D. G. (1997). Effects of Sexual Harassment. In O'Donohue, W. T. (ed.), Sexual Harassment: Theory, Research and Treatment, Allyn and Bacon, Boston. 152-174.

Equal Employment Opportunity Commission (2011). Sexual Harassment Charges. Enforcement and Litigation Statistics. Available: http://www.eeoc.gov/eeoc/statistics/enforcement/sexual_harassment.cfm

Fapohunda, Tinuke. (2014). Gender differences in perceptions and experiences of sexual harassment in the workplace. 1. 36-44.

https://doi.org/10.1016/j.heliyon.2023.e21096.

International Center for Research on Women – ICRW . 2018. The Costs of Sex-Based Harassment to Businesses: an In-Depth Look at the Workplace.

Merkin, R.S., Shah, M.K. The impact of sexual harassment on job satisfaction, turnover intentions, and absenteeism: findings from Pakistan compared to the United States. SpringerPlus 3, 215 (2014). https://doi.org/10.1186/2193-1801-3-215

MUELLER, C. W., DE COSTER, S., & ESTES, S. B. (2001). Sexual Harassment in the Workplace: Unanticipated Consequences of Modern Social Control in Organizations. Work and Occupations, 28(4), 411-446. https://doi.org/10.1177/0730888401028004003

Quick J.C., McFadyen M.A. Sexual harassment: have we made any progress? J. Occup. Health Psychol. 2017;22(3):286–298. doi: 10.1037/ocp0000054.

Willness C.R., Steel P., Lee K. A meta-analysis of the antecedents and consequences of workplace sexual harassment. Person. Psychol. 2007;60:127–162. doi: 10.1111/J.1744-6570.2007.00067.X.

Woodzicka, Julie & LaFrance, Marianne. (2005). The Effects of Subtle Sexual Harassment on Women's Performance in a Job Interview. Sex Roles. 53. 67-77. 10.1007/s11199-005-4279-4.

Index